# MONITORING NATURE CONSERVATION IN CULTURAL HABITATS: A PRACTICAL GUIDE AND CASE STUDIES

# Monitoring Nature Conservation in Cultural Habitats:
## A Practical Guide and Case Studies

*edited by*

CLIVE HURFORD

*Countryside Council for Wales,*
*Bangor, U.K.*

and

MICHAEL SCHNEIDER

*Västerbotten County Administration,*
*Umeå, Sweden*

 Springer

A C.I.P. Catalogue record for this book is available from the Library of Congress.


ISBN 978-1-4020-3756-6 (HB)
ISBN 978-1-4020-6565-1 (PB)
ISBN 978-1-4020-3757-3 (e-book)

Published by Springer,
P.O. Box 17, 3300 AA Dordrecht, The Netherlands.

*www.springer.com*

*Printed on acid-free paper*

# CONTENTS

# FOREWORD

The European Union Natura 2000 initiative is an unprecedented opportunity, both for nature conservation and biodiversity in general. However, the challenge facing conservationists is enormous, and we will have to be resourceful if Europe is to secure the future of its natural fauna and flora, or else risk failing to hand over the same choice to future generations.

For a 10-year period now the member countries have focused their efforts on implementing the Natura 2000 legislation and accomplishing the overarching objectives of the Birds and Habitats directives. This has certainly been a demanding exercise, and of unforeseen complexity for most of us. Now, as the implementation phase nears completion, the focus has shifted to learning to live with the long-term implications of the Natura 2000 legislation. The established network has become the single most important tool for managing biodiversity resources in European countries, while the Habitats Directive itself provides powerful conservation tools in the monitoring and reporting obligations of Articles 11 and 17.

When Sweden joined the EU in 1995, the existing members were already deeply involved in implementing the early phases of the Habitats Directive. Our understanding of the long-term implications of the Natura 2000 exercise developed gradually during this hectic phase of harmonising Swedish strategies and activities with those of the European Union. In 2001 we started to develop a national programme for monitoring the Natura network, in the context of complete integration into the national conservation networks. When we were looking around for good examples of monitoring practices, it was natural to start the search among the Life-Nature projects that were concerned with biodiversity monitoring. Here we found the then recently completed project "Integrating monitoring with management planning: A demonstration of good practice on Natura 2000 sites in Wales" run by the Countryside Council for Wales, and we were immediately attracted to the principles of using objective-driven monitoring to integrate management and conservation activities.

Subsequently, in 2002 and 2003, we ran our own national project, with the aim of establishing an integrated monitoring system to meet all Swedish nature conservation demands, while at the same time providing for the full range of monitoring activities necessary to comply with the obligations of the EU directives. After some initial analyses within the project group, we came to the fundamental conclusion that the basic principles of objectives-based monitoring could be applied to the full range of natural habitats and species constituting the natural environment of Northern Europe. I'm happy to be able to say now, on completion of the project, that we have found no reason to go back on this decision.

Our initial project work was strengthened considerably by the participation of Clive Hurford and Alan Brown, whose insights, as well as tutorial skills, produced pivotal effects during important project events, ranging from strategic broad-focus seminars to small expert group meetings on various specific topics. The fruits of this ongoing collaboration are evident in many of the contributions to this volume. Although we are still in the process of developing this approach to monitoring, I dare to say with some pride that the foundation has been laid in a work area that I trust will become a cornerstone of nature conservation practice for many years to come!

Johan Abenius

Senior Administrative Officer at the Swedish Environmental Protection Agency and Project Leader of the 2002-3 national project "Natura 2000 monitoring in Sweden".

# LIST OF CONTRIBUTORS

## EDITORS

**Clive Hurford**
Monitoring Ecologist for International Sites, Countryside Council for Wales, Bangor, UK.

**Dr Michael Schneider**
Large Carnivore Specialist and Conservation Officer, Västerbotten County Administration, Umeå, Sweden.

## CHAPTER AUTHORS

**Johan Abenius**
Principal Administrative Officer, Swedish Environmental Protection Agency, Stockholm, Sweden.

**Dr David Allen**
Environmental Monitoring Manager, Countryside Council for Wales, Bangor, UK.

**Dr Tom Brereton**
Head of Conservation Monitoring, Butterfly Conservation, Wareham, UK.

**Alan Brown**
Remote Sensing Manager, Countryside Council for Wales, Bangor, UK.

**Stephen Evans**
Botanical Society of the British Isles Recorder, Pembrokeshire, UK.

**Adrian Fowles**
Senior Invertebrate Ecologist, Countryside Council for Wales, Bangor, UK.

**Örjan Fritz**
PhD-student, Swedish University of Agricultural Sciences, Alnarp, Sweden.

**Dan Guest**
Monitoring Ecologist, Countryside Council for Wales, Bangor, UK.

**Anders Haglund**
Ecological Consultant, Ekologigruppen, Stockholm, Sweden.

**Matthias Hammer**
Coordination Officer for Bat Conservation, Erlangen University, Germany.

**Bob Haycock**
Senior Reserves Manager, Countryside Council for Wales, Pembroke, UK.

**Juha Katajisto**
Planning Officer, West Finland Regional Environmental Centre, Vaasa, Finland.

**Dr Keith Kirby**
Forestry and Woodland Officer, English Nature, Peterborough, UK.

**Dylan Lloyd**
Monitoring Ecologist, Countryside Council for Wales, Bangor, UK.

**Paul Pan**
Principal Consultant, Custom GIS, Cardiff, UK.

**Dr Terry Rowell**
Head of Environmental Change Group, Countryside Council for Wales, Bangor, UK.

**Dr Ann Salomonson**
Research Consultant, Annova forskningskonsult, Umeå, Sweden.

**Anneli Sedin**
Environmental Protection Officer, Västerbotten County Administration, Umeå, Sweden.

**Dr Graham Thackrah**
Earth Observation Data Specialist, Research Systems International, Crowthorne, UK.

.

# ACKNOWLEDGEMENTS

This book has been possible only because of the cooperation and support of very many friends and colleagues over the past ten years.

Initially our thanks must go to those who have contributed text to the book, namely (and in order of appearance); Johan Abenius, Terry Rowell, David Allen, Alan Brown, Dan Guest, Dylan Lloyd, Paul Pan, Stephen Evans, Matthias Hammer, Bob Haycock, Adrian Fowles, Tom Brereton, Keith Kirby, Anders Haglund, Örjan Fritz, Graham Thackrah, Ann Salomonson, Juha Katajisto and Anneli Sedin. Thanks also to Dan Guest and Alan Brown for preparing maps and illustrations. We are grateful to Jane Powell, Dan Guest, Keith Kirby, Graham Thackrah and Chris Millican for helping with the thankless task of copy-editing and providing invaluable comments on the content. Our gratitude is also extended to the publishing editors and technicians at Springer NL, particularly Helen Buitenkamp, Sandra Oomkes, Gerrit Oomen and Ria Kanters, for their encouragement and support throughout the writing and collation phases of the publication process.

In CCW, we would like to thank:

David Parker, Terry Rowell and Malcolm Smith for supporting the book and giving their permission to use CCW data; to the members of Life Project Team, namely Alan Brown, David Evans, Tom Hellawell, Helen Hughes, Menna Jones, Ken Perry and Terry Rowell for their collective input into the development of the habitat monitoring methods described in this book; and the more recent additions to the CCW monitoring team, that is David Allen, Dan Guest, Dylan Lloyd, Karen Wilkinson, Heather Lewis, Julie Creer, Tracey Lovering, Lesley Barton Allen and Ann Fells for building on the foundations laid by the Life Project team and for contributing to the data collection in sampling trials; the many NNR Wardens, Conservation Officers and Science Directorate Specialists who have worked with us over the past 10 years; Sam Bosanquet and Julian Woodman for their help during our preparatory work on arable weeds and neutral grasslands; Helen Jones of Training Branch for supporting the collaboration with our colleagues in Sweden; Paul Culyer and Bob Haycock for the benefit of their experience and warm hospitality during the visits by colleagues from Sweden.

In Sweden, we thank:

Johan Abenius and Björn Jonsson, of the Swedish Environmental Protection Agency and Västerbotten County Administration respectively, for ensuring that the collaboration with CCW was possible, and for their continuing support; both of the above, Anders Haglund, Ann Salomonson, Kajsa Berggren, John Jeglum, Örjan Fritz,

Sonja Almroth, Mats Jonsson and Lars Danielsson for the benefit of their time and expertise and for their warm hospitality when visitors from Wales have been in Sweden.

At University of Wales Swansea, we thank:

Charles Hipkin and Steve Wainwright for their help and support during the time that members of the CCW Life Project team were based at the university; Mike Barnsley, Graham Thackrah, Sanjeevi Shanmugam and Paul Pan, who were working in the Remote Sensing Unit at the time, for the benefit of their experience, hospitality (and equipment) during our time there, and particularly while working on the remote sensing projects at Kenfig Pool and Cors Crymlyn; the MSc Environmental Biology students that worked with us on monitoring projects for their theses, particularly Karen Wilkinson, Leila Somers, Mike Aubrey, Karl Besley, Chris Gorman, Poelo Mathobela, Adam Cooper, Emma Davies and Sandra Wilson; and all of the students who participated in sampling trials.

We would also like to thank:

Andrew Weir for supporting and co-organising a monitoring training course at the Snowdonia National Park Centre at Plas Tan y Bwlch, and the many course participants who have taken part in sampling trials; Tom Brereton, Nigel Bourne and Martin Warren of Butterfly Conservation for collaborative work, training input and general support; Reg Thorpe of the RSPB, and Ian Bullock (formerly of the RSPB) for their co-operation and hospitality during work in North Wales and on Ramsey Island; Debbie Lewis and the reserve wardens of BBOWT for engaging in monitoring trials and training sessions; to John Pinel and Tim Liddiard for encouraging, and engaging in, training for conservation workers in the States of Jersey; Rob Davies, Peter Jones, David Carrington and their colleagues at Kenfig National Nature Reserve for their assistance and generosity during the many weeks that we have spent on their site; and to Mr and Mrs Roland Lewis and David and Holly Harries for the benefit of their knowledge and experience, and for their hospitality when we were working on their land.

We wish to acknowledge the financial support of the Joint Nature Conservation Committee, Research Systems International, Västerbotten County Administration, Butterfly Conservation and Custom GIS. A colour production would not have been possible without this support.

Finally, Clive would like to thank Peter Hope Jones for being a constant source of inspiration and wisdom; Nigel Ajax Lewis, who has provided endless encouragement and support over many years; and Menna Jones and Dan Guest for being exceptionally good colleagues over the past 10 years. We would both like to thank our partners Chris and Doris, and Michael's kids Ronja and Matthis, for their endless patience, particularly during the period of collating this book.

# PART I

## AN INTRODUCTION TO CONSERVATION MONITORING

# CHAPTER 1

# MONITORING IN CULTURAL HABITATS

## An introduction

CLIVE HURFORD

*Countryside Council for Wales, Plas Penrhos, Ffordd Penrhos, Bangor, Gwynedd, LL57 2BQ*
*clive.hurford@serapias.net*

## 1.    INTRODUCTION

This book focuses primarily on the problems associated with monitoring nature conservation in cultural habitats, i.e. those habitats that are derived from human management activities; maintained by management activities; or impacted on by human activities. For practical conservation purposes, we should not differentiate between these.

We use the word 'monitoring' to mean recording the condition of habitats or species against clearly defined and measurable management aims. This is broadly in keeping with the definition of Hellawell (1991), which described monitoring as:

> Intermittent (regular or irregular) surveillance carried out in order to ascertain the extent of compliance with a predetermined standard or the degree of variation from an expected norm.

This distinguishes monitoring from other forms of ecological investigation, such as:
- *Natural history recording*, which contributes to historical archives;
- *Survey*, which is typically a 'one-off' descriptive exercise, perhaps to describe the habitats on a site or to map the distribution of a species;
- *Surveillance*, which is a repeatable survey, often used to detect trends in habitats, populations and environmental change;
- *Experimental management*, which tests the effects of different management practices;
- *Environmental impact assessment*, which assesses the likely effects of a development or incident; and
- *Research*, which is carried out to increase our knowledge about a species or habitat, perhaps through ecological modeling, population viability analysis and demographic studies.

A monitoring project should aim to incorporate the relevant information that has been generated by these exercises.

*C. Hurford & M. Schneider (eds.), Monitoring Nature Conservation in Cultural Habitats*, 3–12.
© 2007 *Springer.*

*Figure 1-1.* Globeflowers *Trollius europaeus* solar-tracking in a cattle-grazed pasture in the French Alps. This species has disappeared from many meadows and pastures in the UK as a result of agricultural improvements since the 1950s.

Much of the book centres on issues related to managing and monitoring threatened habitats on sites protected for nature conservation; sites which provide a refuge for species that are struggling to survive in the wider countryside. These species and their habitats must be able to persist in the protected areas until the wider countryside is in a suitable state to allow repopulation. The signs are not good. Rothschild & Marren (1997) found that of 182 sites in England presented to the UK Government in 1915 as potential nature reserves, more than half have been damaged or completely destroyed. Furthermore, recent monitoring results suggest that up to 80% of the habitats on Natura 2000 sites in Wales are in sub-optimal condition, and mostly declining: these are the largest and most important sites for conservation in the country. If we fail to protect the habitats and species on sites that have been designated as conservation areas, then the potential for restoring the wider countryside is slim indeed.

## 1.1 Monitoring the effects of conservation management

Monitoring is a tool of conservation management, and cannot sensibly be considered independently of it. For this reason, we have focused our attention on the role of monitoring within the context of a responsible conservation management strategy. Monitoring is particularly relevant for habitats that are derived from, or impacted by, cultural activities: where our success in maintaining the conservation value of the habitat is linked to the way that we manage it. In wilderness, or semi-wilderness situations, monitoring is less appropriate, unless we intend to take management control.

In Europe, however, most conservation management is carried out in habitats that have been shaped by human activities spanning thousands of years. These habitats include grassland, forests, heaths, sand dunes, fens, mires and many watercourses and lakes.

Since the early 1800s, the conservation value of these habitats has suffered as a result of changing agricultural practices; urban development and spread; industrialisation; commercial forestry operations; the leisure industry; and, in the less productive or inaccessible habitats, neglect. These problems have intensified over time, and we can only speculate over the scale of conservation losses during the 20th century. In the UK, for example, we know that:

- The vast majority of species-rich meadows and pastures have been ploughed-up and reseeded as permanent leys;
- Large areas of heathland have been lost or degraded through overgrazing and afforestation;
- Many fens have been drained;
- All of our raised mires have been damaged by peat-cutting, drainage or afforestation;
- Many native deciduous forests have been felled, neglected or both;
- Many of our lakes and watercourses have suffered from either acidification or eutrophication;
- Many small ponds have been lost through drainage or neglect;
- Farmland bird populations have crashed;

- Many butterfly species have suffered significant declines, both in numbers and in range, as a result of habitat loss and degradation; and
- Bat populations have suffered large declines.

This list is by no means exhaustive and is not exclusive to the UK: many western European countries have experienced similar declines. In response to these dramatic losses, many of the more important sites in Europe have now been designated for conservation, notably as part of the Natura 2000 network. The challenge we face is to secure the management of these sites so that the conservation value can be maintained or restored. As a result of these designations, habitat degradation is now a greater threat than habitat loss.

Many conservationists still believe that if we protect a site from development, natural processes will provide habitats of high conservation value. This may be true in vast expanses of wilderness, but it is a flawed concept when dealing with habitats that have been shaped, or impacted, by cultural activities. If we want to maintain or restore the conservation value associated with these habitats, we have to manage them appropriately. If we neglect a species-rich hay meadow it will become increasingly rank and species-poor before turning to scrub and eventually woodland (Grime, 2001 etc.). This general principle applies to all cultural habitats: if we want to deliver successful conservation, then first we must decide what we want the management to achieve, and then we must carry out the management. Monitoring will tell us whether or not the management is achieving its aims, and give us the opportunity to respond if the conservation value starts to decline.

## 2.      DECISION-MAKING AND ACCOUNTABILITY

The concept of deciding what to manage for, and then monitoring to see whether the management has been successful, represents a change of culture for conservation managers, and is sometimes perceived negatively as introducing accountability to conservation management.[1] It should, however, be viewed as good conservation practice: monitoring provides an early warning of potential conservation losses.

Unfortunately, managers often react indecisively when faced with the threat of conservation losses. However, if we are aware that a threatened habitat or species is declining and do not respond, we are making a management decision that will almost certainly lead to further degradation or loss. In this respect, monitoring has proved to be a catalyst for good practice in conservation management as it forces us to make difficult management decisions.

---

[1] We use the term 'conservation manager' to describe those responsible for making conservation management decisions and overseeing the conservation interest of a site.

*Photographs by Clive Hurford*

*Figure 1-2.* No sheep were present in either of these pastures at the time of the visit, yet in both cases the vegetation shows clear evidence of overgrazing in the recent past.

Other decisions focus on the types of monitoring we need on our sites. There are several options. We can do compliance monitoring to check that the land manager is adhering to a management agreement. We can monitor environmental factors, such as water table fluctuations, atmospheric deposition levels and changes in soil pH. We can monitor the 'condition' of the habitat or species; that is, to assess the condition against a predetermined standard. Also, most conservation managers would like access to surveillance data to detect trends. Ideally, perhaps, we would like to do all of these, but the resources for conservation management and monitoring are always limited. In cultural habitats, the most efficient use of those resources is to monitor the condition of the habitat or species, as this will provide direct feedback into the management of the site.

Compliance monitoring is difficult to apply effectively because we can only assess whether the appropriate management is being applied on the days that we carry out the monitoring. Even a seemingly simple task like counting the number of sheep grazing upland pastures is fraught with difficulty, and is best achieved by aerial photography. Still, these photographs will show only how many sheep were grazing the site on that particular day. At best, compliance monitoring can provide us with an indication of whether the land manager is conforming to a management agreement. Critically, it will tell us nothing about whether the management agreement is appropriate to maintaining the conservation value of the site. By contrast, the condition of the vegetation can always tell us something about how the site has been managed (see Fig. 1-2).

Monitoring environmental factors is generally an expensive process, and tends to be carried out on a countrywide scale, rather than on individual sites. Unfortunately, though probably wisely, few environmental specialists will advise on how to set limits for

*Figure 1-3.* The effects of human activities can be seen even at the more prestigious nature reserves. These Stone Pines *Pinus pinea* were widely planted in the matorral zone at Coto Doñana in the early 1700s to develop a locally important pine-nut industry.

pollution levels that will serve as an early warning of habitat degradation or species declines on individual sites. There is no doubt that surveillance of environmental factors provides valuable background information for conservation managers, but we rarely have enough site-specific information to trigger management decisions. Again, it is easier and more efficient to detect the effects of critical environmental factors by looking at the condition of the habitats. This is only possible, however, if we understand how to recognise the effects of the environmental factors.

It is not the intention of this book to undermine the importance of surveillance, as this clearly has a role to play in informing conservation management. It is a different role, however, and monitoring fills an important niche by facilitating timely and appropriate management decisions that will deliver effective conservation: there is no evidence to suggest that surveillance will do this, primarily because there is no trigger for management action.

## 3.      THE BENEFITS OF MONITORING

The most obvious benefit of monitoring is that it promotes responsible conservation management by:
- Encouraging us to consider why our sites are important;
- Prompting us to prioritise our resources for managing the most important habitats and species;

- Making us think about how we will recognise whether a habitat or species is in optimal condition;
- Leading us to consider how we will recognise whether the conservation value is under threat; and
- Providing feedback that will allow us to make an appropriate management response before the conservation value is irreparably damaged or lost.

There are also longer-term benefits. By stating why a site is important, and clearly defining what we want our management to achieve, we will overcome the problems linked to management discontinuity. There are always risks attached to a change of conservation manager, and this can have a dramatic effect on the conservation value of a site. Each individual will have their own views on why a site is important, often based on their own particular interests. For example, a botanist may feel that a site has been neglected and needs more intensive grazing, while an ornithologist may feel that the same site has insufficient cover for breeding birds and needs more scrub. Unless we clearly identify the management priority on the site, and provide guidance on where the management effort should be directed, there is a real possibility that the management effort will reflect the interest of whoever has the responsibility for managing the site.

To illustrate this point, Tables 1-1 and 1-2 show the results of a simple exercise carried out with the help of professional conservation managers. The purpose of the exercise was to demonstrate the importance of clearly defined management aims. We tested this in four different habitats, each of which had already been monitored. In each case, the conservation managers were taken to a site that was unfamiliar to them. We then asked them to walk around the site, keeping their own counsel, and consider the condition of the habitat. Later, we gave the site managers a set of clearly defined condition indicators for the habitat and asked them to repeat the exercise.

*Photographs by Clive Hurford*

*Figure 1-4.* In the absence of management to control them, invasive introductions such as Rhododendron *Rhododendron ponticum* and Japanese Knotweed *Fallopia japonica* would displace many native species.

The results show that, if there are no condition indicators to define what the management is aiming to achieve, then the chances of a new conservation manager continuing with the same management regime are no better than 50:50. By contrast, if clearly defined and unambiguous condition indicators are available, then the risk of a new recruit changing the management regime is very low. In our exercises, the conservation managers reached 100% agreement when condition indicators were available.

A further advantage of monitoring against clearly expressed condition indicators is that it identifies which sites require maintenance management and which are in need of restoration. This allows us to prioritise our resources and ensure that sites of high conservation value are secure before turning our attention to those in need of restoration.

*Table 1-1.* The results from exercises where professional conservation managers were asked to assess the condition of a habitat on their first visit to a site.

| Habitat | Number of site managers | First impression | | Monitoring result |
|---|---|---|---|---|
| | | Favourable | Unfavourable | |
| Dune grassland | 4 | 3 | 1 | Unfavourable |
| Coastal heath | 10 | 2 | 8 | Unfavourable |
| Marshy pasture | 17 | 8 | 9 | Unfavourable |
| Degraded mire | 43 | 22 | 21 | Unfavourable |
| | | | | |
| **Total** | **74** | **35 (47%)** | **39 (53%)** | |

*Table 1-2.* The results obtained from the same conservation managers when asked to reassess the condition of the same habitats against clearly defined condition indicators. This level of agreement was achieved only because the condition indicators were unambiguous.

| Habitat | Number of site managers | With condition indicators | | Monitoring result |
|---|---|---|---|---|
| | | Favourable | Unfavourable | |
| Dune grassland | 4 | 0 | 4 | Unfavourable |
| Coastal heath | 10 | 0 | 10 | Unfavourable |
| Marshy pasture | 17 | 0 | 17 | Unfavourable |
| Degraded mire | 43 | 0 | 43 | Unfavourable |
| | | | | |
| **Total** | **74** | **0** | **74 (100%)** | |

From a management and monitoring perspective, the importance of unambiguous condition indicators (which are a form of 'ecological shorthand' for describing a habitat in a state of high conservation value) cannot be overstated, because they:

- Describe, in terms that are accessible to a non-specialist, what we want our management to achieve;
- Enable us to design efficient and reliable monitoring projects; and
- Allow us to prioritise repeat monitoring for sites where we are uncertain of the condition.

The first point is vital. There is no point in asking farmers or foresters to maintain and enhance the biodiversity on their land: this will make no more sense to them than it does to anybody else. It will be equally meaningless if we express our management aims in technical terms, for example, by referring to the extent of the various plant communities present. We must express our management aims in terms that can be accessed by the conservation manager, the monitoring ecologists and the land managers. This is possible, even for the more complex habitats and difficult species, if we make good use of the knowledge that is already available from research and survey. Perhaps the major benefit of monitoring, however, comes from having reference points on sites that will give the conservation managers the confidence to make appropriate management decisions. This confidence can be derived only from reliable, evidence-based, monitoring projects.

## 4.     THE CASE STUDIES

The case studies in this book were developed to answer very specific questions in relation to the conservation value of different sites. The monitoring methods are efficient and typically require a low level of recorder expertise. The obvious benefit of this is that, with a minimum of training, relatively inexperienced field surveyors, conservation volunteers and land managers can carry out the monitoring. This type of monitoring places a high demand for expertise on the developmental phase of the project in order to increase the reliability of the monitoring result and reduce the skill level needed for field recording, which is the repeatable phase of the project.

The primary aim when developing the monitoring methods was to reliably assess the condition of the habitats and species as efficiently as possible. This involved:

- Focusing only on what we needed to know;
- Using existing knowledge to identify the key attributes;
- Minimising differences between observers, by using measurable attributes; and
- Using survey information combined with logic to identify the most appropriate locations to monitor.

Consequently, many of the methods applied in the case studies are underpinned by logical inference drawn from our existing knowledge of the sites and the habitat or species: they are not statistical methods. These monitoring methods are often the most appropriate means of answering practical questions about conservation management in cultural habitats. We are not recommending that they are adapted for research purposes, though some of the issues that we raise in relation to measurability are relevant to any ecological sampling exercise. We deal with problems of measurability in some detail in Part III of the book.

We have deliberately kept the focus on a small, but difficult, subject area, and have preferred to cross-reference to other texts than repeat work in other publications. Although the book is not intended to be an inventory of field methods, we draw attention to the strengths and weaknesses of the more widely used forms of data collection. A more exhaustive account of field methods can be found in other publications, e.g. Bonham

(1989) and Kent and Coker (1992). Similarly, there are existing texts that deal adequately with issues related to sampling and statistics, e.g. Krebs (1989). On the occasions that we introduce new concepts or methods, these are described and discussed in detail.

The case studies demonstrate how monitoring can be applied to a variety of different habitats and species, and on sites ranging in size from less than a hundred hectares up to several thousand hectares.

## 5. REFERENCES

Bonham, C. D. (1989). *Measurements for terrestrial vegetation.* New York: John Wiley

Grime. J.P. (2001). *Plant Strategies, Vegetation Processes and Ecosystem Properties (2nd Ed.).* John Wiley & Sons. Chichester.

Hellawell, J. M. (1991). Development of a rationale for monitoring. In B. Goldsmith (Ed.), *Monitoring for Conservation and Ecology* (pp. 1-14). London: Chapman and Hall.

Kent, M., & Coker, P. (1992). *Vegetation description and analysis - a practical approach.* London: Belhaven Press.

Krebs, C. J. (1989). *Ecological methodology.* New York: Harpers Collins.

Rothschild, M. & Marren, P. (1997). *Rothschild's Reserves: Time and fragile nature.* Harley Books. Colchester.

# CHAPTER 2

# THE RELATIONSHIP BETWEEN MONITORING AND MANAGEMENT

Terry Rowell

*Countryside Council for Wales, Plas Penrhos, Ffordd Penrhos, Bangor, Gwynedd, LL57 2BQ*
*T.Rowell@ccw.gov.uk*

## 1.    INTRODUCTION

Most conservation managers are faced with the challenge of linking monitoring to management and reporting their success in managing sites, habitats or species. Both the reports they write and the management-related decisions they make must be able to cope consistently with a wide variety of different sorts of habitats and species. This chapter presents a model for the role of monitoring in conservation management which is the basis for the case studies in this book.

## 2.    TRENDS ARE DIFFICULT TO INTERPRET INTO MANAGEMENT ACTIONS

Much of the regular observation made for conservation purposes consists of some kind of *surveillance* – repeated observations made with the intention of discerning trends (see Chapter 4). These trends are usually interpreted as "increasing" (or "improving"), "stable", "declining", or similar categories. While these descriptions can easily be accumulated into totals for reporting disparate habitats and species, they are not easy to interpret. Is "stable" at a low level the same thing as "stable" at a high level? Is "decline" from a high level the same thing as "decline" from a low level? Such questions are critical to both local management decisions and broader-scale strategic planning. One possible answer is to introduce so-called *alert limits* of the sort used by the British Trust for Ornithology (Crick *et al.*, 1998). Surveillance data are scanned regularly for changes of more than 25% (a low-level alert) or 50% (a high-level alert) over an agreed period. Exceeding these limits warns the conservation community of the need to pay additional attention to the population in question.

*C. Hurford & M. Schneider (eds.), Monitoring Nature Conservation in Cultural Habitats, 13–22.*

# 3.    A CONCEPTUAL MODEL LINKING CONDITION TO MANAGEMENT

Alert limits refer only to previous values of the population, not to the levels we are trying to achieve in our conservation management. We can keep the idea of alerts or triggers, but use a generalised conceptual model to link the condition of habitats and populations to human activities by comparing the actual condition to the desired condition. This sort of model can provide a complete set of categories to describe the link between monitoring, reporting, and management actions. The condition of the habitat or population is linked to events through decision processes rather than abstract descriptions of trends.

The model is drawn on a traditional two-axis graph, with time advancing on the *x*-axis, and condition on the *y*-axis (Fig. 2-1). According to the model, activities - including management - can be either positive (restoration or maintenance management) or negative (e.g. damaging activities), triggering *recovery* or *decline* in condition. Alternatively, the habitat or population can be *maintained* in a good condition or at least suffer *no change*.

To illustrate this, suppose the habitat or population is in the condition it should be if it is well conserved and managed. This is referred to as *optimal* condition (Rowell, 1993) or *favourable* condition (Joint Nature Conservation Committee, 1998). With appropriate management, this is the usual state of the feature. If a damaging activity happens, the habitat or species may be affected insignificantly, mildly, or anything through to total, virtually instantaneous, destruction. In the model, we can show these possible degrees of decline as a family of curves of varying steepness, up to the vertical drop signifying destruction. If there is no management intervention, the decline represented by any of these curves, may eventually result in loss of the habitat or population, or may stabilise at a lower, less desirable level.

Assuming the worst has not occurred, the good conservation manager curbs the damaging activity and instigates some form of restoration management. If effective, this sets the habitat or population on the road to recovery, which can be shown as one of many possible upward trajectories. Of course, the management may only stabilise the habitat or species at a lower, less desirable level. At worst, the management doesn't work, and the habitat or population continues on the downward path until it is completely lost.

If the management works as hoped, the habitat or population eventually recovers to the optimal (favourable) condition. At this point, we will often need a change in management to ensure maintenance of the habitat or population into the future. A subtlety of the model, not shown in Fig. 2-1, is that a recovering habitat or population should not be regarded as reaching favourable condition when it simply passes the lower limit of acceptable condition during maintenance management. This is because of the necessary link to management. A switch back to maintenance management at this point might well result in the habitat or population stabilising at too low a point, a point where normal fluctuations continually take it below the lower limit. The trick is to set a recovery target within the limits, and to continue with restoration management until that point is reached.

It is easy to envisage the need for a restoration target with a population, perhaps less easy with a habitat (see Example 1). Considering the extent of a habitat with an advancing front of, say, scrub, a good manager would not cut back only to the lower limit, but would usually take it much further back to give some management breathing space. Where we are concerned with managing habitat quality, we have only to remind ourselves that most habitats are collections of populations for it to be easy to see the relevance of the recovery target.

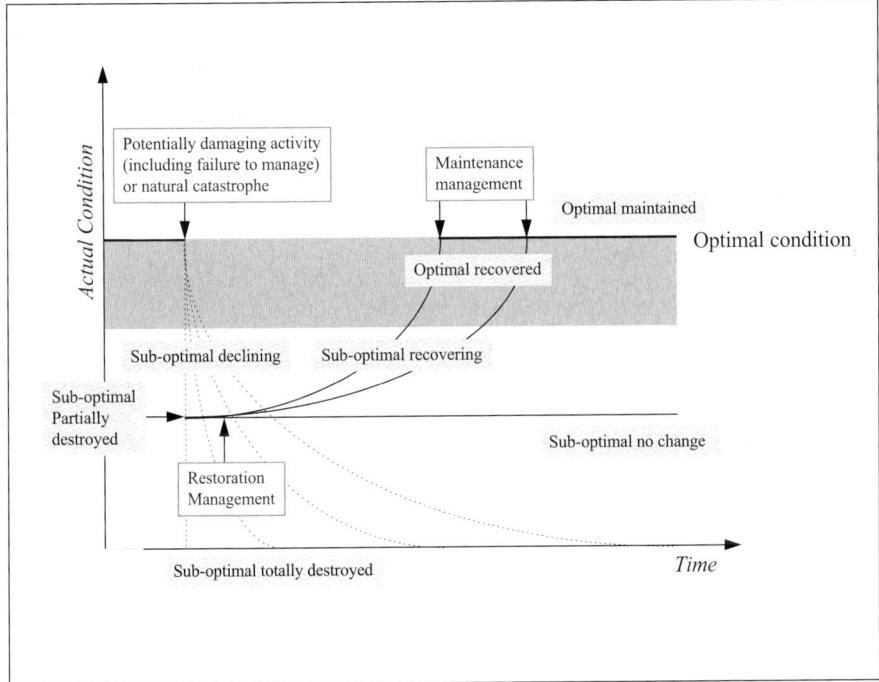

*Figure 2-1.* A simple conceptual model linking habitat or population condition to management. Full explanation in the text.

## 4. EXCEEDING UPPER LIMITS

For clarity, the model shown in Figure 1 only illustrates decline and recovery below a lower limit. Some features might become unfavourable because they exceed an upper limit. Examples might be where a predatory population grows to a size where it is beginning to push a conserved prey species beyond its own lower limit. Similarly, a habitat might expand above its upper limit to the detriment of an adjacent conserved habitat. Some aspects of habitat quality may be subject to upper limits, especially when one remembers that the vegetation part of a habitat is composed of a set of populations. It may be difficult to envisage setting restrictions on species diversity on some habitats such as chalk grassland, but even this may be necessary. For example, the extent of species-rich habitat may need to be restricted where species such as Lulworth Skipper (*Thymelicus acteon*) require ranker grassland for breeding.

---

**Example 1 – Sycamore *Acer pseudoplatanus* in UK Tilio-Acerion woodlands**

Although Sycamore *Acer pseudoplatanus* is a native component of Tilio-Acerion woodland in continental Europe, the species is not native in the UK. As such, we might reasonably want to set an upper limit for Sycamore in UK woodlands.

If we decide, for example, that we are prepared to tolerate Sycamore forming 20% of the forest canopy before we become concerned about it: as soon the species exceeds that we are committed to a management response. Without a recovery target this could be a minor exercise, perhaps involving the removal of one or two trees.

This is unlikely to solve the problem, however, as mature Sycamores will now perennially form around 20% of the canopy, and produce vast numbers of seeds and saplings. This type of response stores up problems for the future, and commits us to regular monitoring and management. Furthermore, we would now be attempting to maintain Sycamore at the level that we originally identified as a cause for concern.

If, after the lower limit was breached, we set a recovery target for <5% Sycamore in the canopy, our management response would be to remove a large number of mature Sycamores over a short period of time, and then to check periodically to ensure replacement by native species (by removing viable Sycamore saplings).

---

## 5.  CONDITION CATEGORIES LEAD TO MANAGEMENT ACTION

According to the model, then, a habitat or population can be one in of several states at any particular point in time:

- It is in *optimal condition*, and comparison with a previous observation shows that it has been *maintained*.
- It has *recovered* from a previously *sub-optimal* condition and is now in *optimal condition*.
- It is in *sub-optimal condition*, and comparison with a previous observation shows that it is *declining*.

- It is in *sub-optimal condition*, and comparison with a previous observation shows that it is *recovering*.
- It is in *sub-optimal condition*, and comparison with a previous observation shows that it is has *not changed*.
- It has been *partially destroyed*.
- It has been *totally destroyed*.

These seven states can be used to transmit simple information to site managers about the state of the important features of their sites. The aim here is to fuel decision-making directly from monitoring. For instance:

- *Optimal-maintained* indicates successful maintenance management that should be continued.
- *Optimal-recovered* indicates a need to switch from recovery management to maintenance management.
- *Sub-optimal-declining* indicates a need for recovery management, or that current management is simply not working.

## 6. CONDITION CATEGORIES PROVIDE REPORTING CATEGORIES

Once we have determined the condition of a feature at a site, we report it to the conservation manager who will take some management action. Collating the condition categories for a particular feature across a region or country provides a higher-level report on condition. You will recall that we made a similar point about trend categories, suggesting they could be collated but did not always mean the same thing (we posed the question "is decline from a high position the same as decline from a low position?"). The position is different with condition categories; the meaning is always identical in that optimal condition represents the management aim at a particular site. Collating the data can tell us whether management aims are being met across a region or country. This is a useful basis for policy-level decisions. Comparing two or more features in this way is a potential aid to prioritisation for resources and action.

## 7. WHAT IS OPTIMAL CONDITION

If we are to base our monitoring on this model we have some crucial decisions to make *before* we do any monitoring. Most critically, we must define optimal condition for this habitat or species on this site.

In the UK conservation agencies, suggestions for optimal condition have included the condition of a habitat or population at the time it was first identified as important – perhaps when a site was designated. Unfortunately, this can be seen as rather arbitrary; and in practice we rarely have suitable information available. Another suggestion (for vegetation) has been to use the descriptive tables of the UK's National Vegetation

Classification (e.g. Rodwell 1991). Once again, what might seem obvious definitions are impractical because they were never intended to describe condition in this way (Rodwell 1996, Brown 2000), and are mathematical constructs rather than descriptions of real vegetation.

The only practical statement of optimal condition is (or should be) our management aim for the habitat or population in question. This does mean that our aims must be well drafted, avoiding imprecise appeals to "maintain or enhance" or, worse still, setting out a standard management prescription for a habitat. How to turn management aims into performance indicators that can be reliably monitored is explained in Part III of this book, and is illustrated by the examples in the case studies.

Accepting what we are aiming for as the statement of optimal condition makes good sense, allowing us to accommodate local variations in habitats that we want to conserve. It also makes explicit the link between this form of monitoring and conservation management. Recording our aims in a management plan, along with the performance indicators and details of the monitoring method, gives a complete audit trail for the monitoring result, providing all the reasoning is clearly set out. Anyone should be able to track back the meaning behind a statement about the condition of a particular habitat, to its root.

## 8.    TURNING A MANAGEMENT MODEL INTO A MONITORING MODEL

The model, as presented so far, is only conceptual; while it may represent how we think about change, this is not how we actually see it on the ground. Our monitoring, rather than giving us relatively smooth curves composed of apparently precise data, is simply a series of snapshots of relatively imprecise data. Where we have many sites to look after, there may be several years between each set of observations. However frequently we visit, we're unlikely to witness the damaging event that caused the fall into sub-optimal condition.

This was illustrated by Brown (2000) in a way that begins to show how the model can help simplify our monitoring (Fig. 2-2). This version of the model also clarifies how the recovery target can be used. Working from left to right:

- The first visit draws a conclusion of optimal condition based on data falling anywhere between the upper and lower limit. The management conclusion should indicate maintenance management. *Note that we can draw a conclusion, and act on it in respect of management, having collected only one dataset. This is an important consequence of applying the model (and making decisions about our management aims and performance indicators before monitoring) – we don't have to make several visits followed by trend identification and interpretation.*
- The second visit concludes sub-optimal condition based on data falling anywhere below the lower limit. There is, at this point, the possibility of partial or complete destruction of a habitat or a population. In the case of a habitat, and for some species, this will be obvious.  For other species it will not, and they are likely to be recorded

simply as sub-optimal. This is unavoidable unless the population is kept under more or less constant surveillance. At this point, restoration management should be instigated.

- The third visit still concludes sub-optimal condition based on data falling anywhere below the recovery target, even though it may actually exceed the lower limit.

- The fourth visit concludes optimal-recovered condition based on data falling above the recovery target and below the upper limit. The management should be changed to a maintenance regime at this point.

- The fifth visit concludes optimal-maintained condition based on data falling anywhere between the upper and lower limits.

All this has been achieved with simpler or fewer data compared to those we might have to collect to demonstrate a trend over a short enough period of time to be able to respond with management changes. The simple expedient of deciding on our management aims before monitoring allows us to make a decision after every monitoring visit. The alternative is to wait for sufficient data to accumulate to demonstrate a trend; and we may still be unclear on how to react to it.

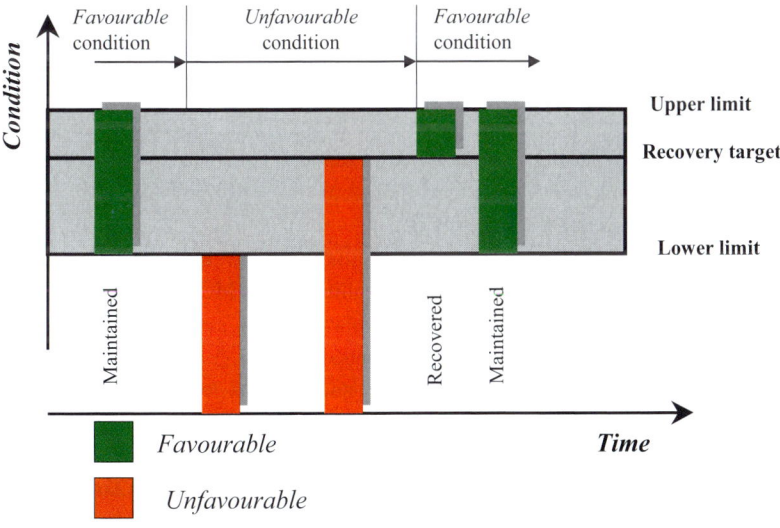

*Figure 2-2.* A statistical version of the model (from Brown 2000) showing the possible uncertainty in the observations for drawing any particular conclusion about condition; a full explanation is in the text.

Unfortunately, the sub-divisions of the sub-optimal category require trend data and the more intensive sampling this usually implies. It is worth considering how we might avoid more intensive data collection. One possibility is to make use of our knowledge of management, and the confidence we have in our management methods. If we are applying proven management methods, then perhaps we can assume that the recovery process is underway. If we're not, then we should assume decline, unless we know that spontaneous recovery is common under similar circumstances. But are these assumptions reasonable? If we are applying management that we have little confidence in, then we should be asking more complex questions than can be answered by this simple monitoring, or by surveillance. In cases like this, we need to find out more about management methods, through systematic literature review (Pullin & Knight, 2003), small-scale management experiments, or by taking a larger-scale adaptive management approach (see below).

Now that we have a monitoring version of our model of change, it is possible to begin to develop a statistical version. Brown (2000) began this process by showing that the model operates under three separate sets of hypotheses, depending on the situation:

- When the feature is, according to previous assessments or received opinion, in favourable condition and there is an upper limit, we test the null hypothesis of favourable condition against two alternative hypotheses of unfavourable condition (one above the lower limit, and one below).
- As above, but with no upper limit, we test the null hypothesis of favourable condition against the alternative hypothesis of unfavourable condition below the lower limit.
- When the feature is believed to be in unfavourable condition, we test the null hypothesis of favourable condition *between the upper limit and the recovery target, effectively favourable-recovered*, against the alternative hypothesis of unfavourable condition below the recovery target.

It is not the aim of this chapter to develop the statistical aspects of the model any further. Readers can consult Brown (2000) who made a statistical power analysis of the model, and provided practical calculations for sample sizes and power analysis in respect of normalised cover data and frequency data.

It's worth, however, drawing out the similarities and differences between this approach and that taken by adaptive management (see Meffe *et al.*, 1997). Both approaches use hypothesis testing. Adaptive management assumes that management methods are always uncertain in their effectiveness, and that constant experimentation is therefore necessary; the design, data collection and analysis requirements appear quite onerous (Sit & Taylor, 1998). The model used in this book can use minimal data to drive decisions about management changes that we already understand. In its simplest form, it can detect if maintenance management is failing, but needs to use surveillance techniques to detect failure or success of recovery methods in reversing declines below the restoration target.

## 8.1    Simplification

We have already seen how using what we already know, applied through this model, can simplify our monitoring. There are further steps in simplification we can take, and many of these will be illustrated in the case studies that follow. A key aspect of the model is that it uses thresholds, and the efficiencies that can be gained as a result are considerable. If our criteria for optimal condition include, for instance, the requirement that there should be no patches of bare ground above, say, 0.5 m$^2$ then we will know that we have sub-optimal condition as soon as we see the first patch of this size. We don't need to do any further data collection to establish the condition of the habitat.

This sort of approach will be developed further elsewhere in the text; it has considerable potential for helping us make early decisions with minimal observations. There will, of course, be times when we want to collect more complete datasets, perhaps to provide more information to aid management, or simply to increase our confidence in our minimalist approach.

## 9. LIMITATIONS OF THE MODEL

While monitoring based on the model can provide powerful information that can be used for decision-making, we should recognise that there are limitations to the conclusions we can draw from monitoring. Sometimes, site managers are disappointed when they learn that it will not tell them *how* to change their management, only that it needs to be changed. The model provides a basis for, probably, the simplest monitoring that could be done in the field of nature conservation. It provides an instant result that could, and should, be followed by an instant decision, and action on the ground. It is worrying to see site managers leaning towards anything more complex and less informative. Yet we know that much of the "monitoring" instigated on our nature reserves does not have the statistical power, or the rigour, to provide us with useful information over even quite long periods of time.

Where managers are uncertain about the effectiveness of the management methods they employ, they should try to boost their confidence either by learning more, or by demanding suitably stringent tests of management methods by competent field experimenters.

## 10. IN SUMMARY

This simple model of change provides us with a logical set of categories for informing basic conservation management decisions. We can base then our monitoring on tests of whether our management aims are met or not. This is much simpler, and potentially more efficient, than other forms of monitoring that try to distinguish trends or use experimental designs.

Elsewhere in this book are many examples of this approach being used across Europe for monitoring both habitats and species. While the details of implementation may differ considerably, the underlying model is always the same.

# 11. REFERENCES

Brown, A. (2000). *Habitat Monitoring for Conservation Management and Reporting. 3: Technical Guide.* Life-Nature Project no LIFE95 NAT/UK/000821. Integrating monitoring with management planning: a demonstration of good practice in Wales. Countryside Council for Wales, Bangor.

Crick, H.Q.P., Baillie, S.R., Balmer, D.E., Bashford, R.I., Beaven, L.P., Dudley, C., Glue, D.E., Gregory, R.D., Marchant, J.H., Peach, W.J. & Wilson, A.M. (1998). *Breeding Birds in the Wider Countryside: their conservation status (1972-1996).* BTO Research Report No. 198. BTO, Thetford.

Hellawell, J.M. (1978). *Biological Surveillance of Rivers. A Biological Monitoring Handbook.* Water Research Centre, Medmenham & Stevenage.

Joint Nature Conservation Committee (1998). *A Statement on Common Standards Monitoring.* Peterborough.

Meffe, G.K., Carroll, C.R. & Contributors (1997). *Principles of Conservation Biology, Second Edition.* Sinauer Associates, Sunderland, Mass.

Pullin, A.S. & Knight, T.M. (2003). Support for decision making in conservation practice: an evidence-based approach. *Journal for Nature Conservation* **11**, 83-90.

Rodwell, J.S. (1996). The National Vegetation Classification and Monitoring. Report to the Countryside Council for Wales. Unit of Vegetation Science, Lancaster University, Lancaster.

Rodwell, J.S. (Ed.) (1991). *British Plant Communities. Volume I. Woodlands and Scrub.* Cambridge University Press, Cambridge.

Rowell, T.A. (1993). Common Standards for Monitoring SSSIs. Joint Nature Conservation Committee, Peterborough.

Sit, V. & Taylor B. (Eds) (1998). *Statistical Methods for Adaptive Management Studies. Land Management Handbook No 42.* Ministry of Forests, British Columbia, Victoria, BC.

# PART II

## TRADITIONAL APPROACHES TO DATA COLLECTION

# CHAPTER 3

# THE ROLES OF SURVEY

CLIVE HURFORD

*Countryside Council for Wales, Plas Penrhos, Ffordd Penrhos, Bangor, Gwynedd, LL57 2BQ*
*clive.hurford@serapias.net*

## 1.    INTRODUCTION

Many of the data collection activities undertaken by conservation bodies can be collated under the general heading of 'survey'. Hellawell (1991) defines survey as:

An exercise in which a set of qualitative or quantitative observations are made, usually by means of a standardised procedure and within a restricted period of time, but without any preconception of what the findings ought to be.

Survey has several important roles to play in conservation management and monitoring. Initially, we use it to identify sites of conservation value, and then we use it to assess the sites in terms of their international, national and regional importance. We also use survey information to identify the conservation potential of a site. In this context, historic survey data are particularly valuable, as they can provide evidence of conservation losses and give conservation managers the confidence to make a positive management response.

For management and monitoring purposes, we would like the following survey information to be available:

- The approximate extent of the key habitat;
- The approximate location of the key habitat;
- The potential distribution and extent of the key habitat;
- The diversity of species associated with the key habitat;
- The distribution of important species on the site; and
- The population size of important species on the site.

*C. Hurford & M. Schneider (eds.), Monitoring Nature Conservation in Cultural Habitats*, 25–34.
© 2007 *Springer.*

In reality, conservation managers rarely, if ever, have access to all of this information, which means that management decisions have to be founded on incomplete survey data. This chapter looks at how existing survey information can be incorporated to best effect in management strategies and monitoring projects.

## 2.      USING EXISTING SURVEY DATA TO INFORM MANAGEMENT DECISIONS

The most valuable survey data describe a habitat or species population when it was perceived to be in optimal condition, and pre-date known periods of decline. Unfortunately, very few data sets can be claimed, with certainty, to do this; though the historic distribution of forests in England is unusually well documented, e.g. Rackham (1995). Often, the best that we can hope for are data sets extending back 50 years or so that offer an insight to the historic distribution of a habitat or species. Most surveys, however, have been carried out more recently, typically since the 1970s. These still offer a useful comparison with the current situation, but are less likely to provide evidence of the true potential for restoration.

Given the rate of conservation losses since the early 1800s, and particularly since the 1950s, it would be irresponsible to accept the current condition of any threatened habitat or species as being optimal without at least exploring the potential for restoration. The challenge for conservation managers is to identify how their sites can best contribute to the restoration of threatened habitats and species on a broader scale. Incorporating information from existing data sets is a good place to start. Survey data commonly exist as:

* Habitat maps;
* Quadrat data;
* Species lists; and
* Species distribution maps.

The following sections highlight how existing survey data can contribute to well-informed conservation management decisions.

## 2.1     Habitat maps

Habitat maps are a vital source of information for conservation managers, because often the only practical response to degradation or loss is habitat restoration. Therefore, we need to know the approximate extent and distribution of the key habitats on our site before deciding where to prioritise our management effort. We know that some habitats are intrinsically of higher conservation interest than others, e.g. native broad-leaved woodlands as opposed to conifer plantations, but even these habitats will vary in their conservation value as a result of their management history. Consequently, we are unlikely to obtain reliable information on the true conservation value of the vegetation from a habitat map unless the survey was designed specifically to do this (see Chapter 9).

In the context of developing a monitoring project, the main purpose of a habitat map is to prompt and inform management decisions. Any habitat map, new or old, will provide a framework for deciding what we want our management to achieve and where we want to achieve it. For monitoring purposes at least, the habitat map has served its purpose when we have made these decisions.

We should resist any temptation to use a habitat map as a baseline for monitoring, because the combined effects of imprecise habitat definitions and imprecise field mapping methods will undermine the validity of the exercise. These sources of inaccuracy are discussed briefly below.

## 2.1.1    Imprecise habitat definitions

The two longest rivers in the world are the Amazon flowing into the South Atlantic and the Nile flowing into the Mediterranean. Which is the longer is a matter of definition rather than measurement (The Guinness Book of Records).

Conservation managers, researchers and bureaucrats all tend to have unrealistic expectations of habitat maps and can rarely refrain from deriving estimates of habitat area from them. Most habitat maps, however, are not accurate enough to warrant this level of attention.

The major problem is one of definition. The effect that imprecise definitions can have on area estimates is demonstrated in Table 3-1. This table lists area estimates for Great Britain that have appeared in various 'well respected' publications since 1950. The range of variation in these estimates is believed to arise from different interpretations of what constitutes Great Britain, e.g. whether one or more of inland waters, the Isle of Man, Northern Ireland and the Channel Islands were included in the area measurements (Maling, 1989).

Area estimates derived from site habitat maps will suffer from an even greater scale of variation (relatively of course), for the same basic reason, i.e. that we have not provided unambiguous definitions for the habitats that we want to map (Cherrill & McClean, 1999). Before we can obtain accurate area estimates from maps we have to address the question of where one habitat ends and the next one starts, e.g. at what point does heathland stop being heathland and become either grassland or woodland? No habitat classification system deals adequately with this issue. In practice, this means that the boundaries between habitat types are open to observer interpretation.

Consequently, most habitat maps show clearly demarcated habitat patches adjoining each other, and pay no attention to boundary vegetation. This is a problem for conservation managers, because the area of habitat least well defined is often the area where loss or degradation is most likely to occur and, conversely, where restoration is most likely to succeed.

*Table 3-1.* The area of Great Britain (in km$^2$) according to different, apparently reliable, authorities over the period 1945-1985 (adapted from Maling, 1989).

| Source | Area - km$^2$ |
|---|---|
| | |
| Chambers' Encyclopaedia, 1950 | 230, 614 |
| Columbia Lippincott Gazetteer and Geographical Dictionary, 1952 | 299, 848 |
| Encyclopaedia Britannica, 1959 | 243, 187 |
| The Statesman's Yearbook World Gazetteer, 1975 | 229, 868 |
| Reader's Digest Great World Atlas, 1962 | 243, 995 |
| The Statesman's Yearbook, 1980-81 | 230, 609 |
| The Europa Yearbook, 1985 | 244, 103 |
| | |

## 2.1.2     Imprecise mapping methods

As there is no requirement for habitat surveyors to have a background in cartography or cartometry, we should not expect habitat maps to be either accurate or precise.    The most frequently used field mapping methods are based on either:

- Visual homogeneity, as identified on aerial photographs (e.g. Avery, 1962 and Graham & Read, 1986) or remote sensing images; or
- Vegetation homogeneity as recognised by the surveyor in the field, with the location of the homogeneous stands marked directly onto base maps of the site.

In both cases, the surveyors use individual experience and intuition to decide what is homogeneous and exactly where they are on a site.  So even if the surveyors were given unambiguous habitat definitions, neither of these methods would give rise to accurate (or precise) survey maps, because different surveyors would draw the habitat boundaries in different places (excepting boundaries clearly demarcated by semi-permanent features such as fence-lines and walls).  Furthermore, there is a real risk of areas of habitat being overlooked, either through misinterpretation of the remote images or observer inexperience.

Unless we are prepared to invest more time, effort and resources into the production of habitat maps, then we must learn to have realistic expectations of the final product.  The best that we can hope for is a rough approximation of the extent and distribution the habitats on the sites.  In reality, we do not need to know the precise area of a habitat: we simply need to know whether we have enough of it and whether it is in a state of high conservation value.  We can obtain this information from a well-designed monitoring project.

## 2.1.3     Working with appropriate classification systems

Several habitat classifications are operational in Europe; these can be divided into two groups. At the European scale there is the CORINE Palaearctic classification (European Communities – Commission, 1991), from which the Annex I habitat classification was derived, and the EUNIS classification that (in 2005) is nearing completion. Each of these works, or can work, at the broad habitat level. Alongside these, there are several regional classification systems, such as those for the Nordic Countries, Central Europe and the UK, which operate at plant community level.

Most of these systems are well aligned, notably the CORINE, Annex I Habitats, Vegetation Types of the Nordic Countries and EUNIS classifications: the others are less well aligned, but have been interpreted to allow integration with the Annex I habitat classification.

Although maps based on any of these classifications can be used to facilitate management decisions, we have found that maps based on broad habitat classifications are better suited to the process for two reasons, a) because habitat management is mostly applied at the broad habitat level, e.g. heaths, hay meadows and pastures, and b) because land managers (who are less likely to be familiar with the terminology of plant community classifications) can relate to them. An advantage of the Annex I habitat classification is that it operates mostly at a level that is relevant both to conservation managers and land managers, making communication relatively painless.

The practical advantages of working with broad habitat classifications revolve around efficiency. For example, they tend to be carried out at a relatively coarse scale, typically on 1:10 000 maps, making data collection less time consuming. Also, they are better adapted for aerial photo interpretation, as the boundaries visible on aerial photographs typically represent the boundaries between broad habitat types (or dominant species at least). Furthermore, analysis of survey maps has indicated that assessments of the extent of the broad habitat are likely to be more reliable than assessments of extent made at finer scales of classification (Stevens *et al.*, 2004). Therefore, if aerial photographs are available, the results from surveys focused on assessing the extent of the broad habitat types are likely to be more reliable than the results from surveys that focus on assessing the extent of the component plant communities.

When we were developing a monitoring project at Cors Crymlyn (see Chapter 35), for example, the only available habitat map dated from the mid-1970s (Headley, 1990). This map showed an extensive area of transitional fen vegetation along the western edge of the site. In the late 1990s, this area was dominated by *Phragmites australis*. We were confident that this represented a real change because:

- The original surveyor was unlikely to have overlooked dense stands of *Phragmites* in this area, because it is the most accessible edge of the fen; and
- Cattle grazing had ceased along this edge of the fen, and invasion by *Phragmites* was a predictable effect of this.

The critical point here is that, an old map showing the distribution of the broad habitats on the site can be more relevant for management purposes than data gathered during a more recent survey.

## 2.1.4    Distilling information from NVC maps in the UK

The commonest form of habitat map on sites of conservation interest in the UK will be based on the National Vegetation Classification (NVC). In terms of the inaccuracies associated with unambiguous definitions and imprecise mapping methods, these maps will be no better or worse than any other habitat map. Similarly, although we can broadly associate different communities (or sub-communities) with being of generally higher or lower conservation value, there is no absolute link: most plant communities can exist as either species-rich or species-poor examples of the type, and may or may not have important species associated with them. Therefore we cannot reliably obtain habitat quality information from an NVC map. For example, the M25 *Molinia caerulea-Potentilla erecta* mire community can be relatively species-rich and support a large population of the Marsh Fritillary *Eurodryas aurinia* butterfly, or it can be a virtual monoculture of rank *Molinia caerulea*, with no associated species of note. An NVC map will not differentiate between these: it will simply show a stand of M25 vegetation.

There are, however, advantages to having an NVC map of your site. Firstly, it will allow ecologists who are familiar with the way that the communities interact to identify the potential for increasing the extent of the various plant communities on a site: this information is less accessible to non-specialists. Secondly, it gives you access to the relevant 'Succession and Zonation' sections in the British Plant Community volumes (Rodwell, 1991 *et seq.*). These sections often draw attention to the expected direction of change as a result of the likely management impacts, and may even suggest appropriate indicator species. This is valuable information that can inform both management and monitoring decisions.

A disadvantage of the NVC is that relatively few land managers have a working knowledge of the classification. In practice, this means that the information locked up in NVC maps is accessible to only a relatively small number of highly specialised individuals, few of whom are directly involved in practical conservation management. Therefore, if the NVC is going to make a positive contribution to the restoration of habitats in the UK, the relevant specialists will have to learn to distil the critical information from the survey maps and make a conscious effort to avoid communicating it in the coded jargon that has become the currency of the classification. In the right hands, the NVC can be a useful management tool, while in the wrong hands; it can be a serious barrier to effective communication.

## 2.2    Site descriptions and quadrat data

Many of our older data sets began their existence as baseline data for surveillance projects that were subsequently abandoned, typically because the original transects or plots could not be relocated. These data can still provide conservation managers with the confidence to make positive management decisions.

For example, Godwin (1936) described large areas of the raised bog surface at Cors Caron in Mid Wales as a regeneration complex (now commonly referred to as a 'hummock and hollow' complex) typified by a high cover of *Sphagnum pulchrum*. When we visited the site in 1996, however, it was clear that the raised bog surface had undergone a dramatic change, both in structure and species composition. *Molinia caerulea* was dominant over much of the bog surface, and vegetation resembling the regeneration complex described by Godwin was generally scarce and only locally distributed (Hurford & Perry, 2000). This change had occurred over 60 years: a relatively short period in the life of a raised bog. The shift from regeneration complex to dominant *Molinia* is most likely to have occurred as a consequence of drying out at the bog surface, though atmospheric depositions of nitrous oxides is also known to benefit *Molinia* (Tomassen *et al.*, 2003) and could have accelerated the process. Either way, the description of the site in the 1930s made it possible to be certain that the bog surface had undergone considerable change over a relatively short space of time, and gave the conservation manager the confidence to initiate a programme of restoration management.

Similarly, at Whiteford Burrows in south-west Wales, baseline surveillance was carried out in the 1970s to study trends in the development of dune slack vegetation (Hughes, 1981). The sampling areas were marked with wooden posts and then recorded on a site map. Despite this, the first attempt at repeat sampling, in 1994, failed because we could not relocate the wooden posts: this is a recurring theme for habitat surveillance projects. This ruled out the opportunity of collecting matched pairs of data for direct comparison. However, by relocating the general sampling areas (from the map) and then referring back to the original data, we could conclude, beyond any reasonable doubt, that a) there had been a decline in species diversity, and b) the vegetation was no longer capable of supporting Fen Orchids *Liparis loeselii*, an Annex II species on the EC Habitats Directive. This information, combined with evidence from aerial photographs showing that no new dune slacks had formed at Whiteford in the interim period, indicated that the conservation value of the humid dune slack habitat on the site was declining.

On a more general note, unless you know that the vegetation samples were distributed randomly across the site, or have a map showing where the samples were taken, it is probably safe to assume that quadrat data were collected from the best, or most important, habitat patches on the site. There is no way of being certain of this, but given the choice of collecting quadrat data from the more interesting and species-rich stands of vegetation or from species-poor stands, most botanists would opt for the former. This is particularly true if the data were collected to demonstrate the conservation value of the site.

If we are prepared to make this assumption, we can collect new data from the best patches of habitat and compare the two data sets, focusing on changes in species composition and co-existence. If successional processes have prevailed, a number of species will have either disappeared or become scarcer in the samples: these species could be useful site-specific indicators of habitat quality (Chapter 11).

## 2.3    Site-based species lists

Species lists are probably the most widely available form of survey data. Many sites will have a list of the birds, higher plants and butterflies recorded there, reflecting the more popular recreational interests of naturalists. Comprehensive lists of other species groups are less common, except at the better-known sites for them. In many cases, casual observations by local enthusiasts or visiting naturalists will form the bulk of the survey data.

*Figure 3-1.* Purple Moor grass *Molinia caerulea*, the pale grass dominating the cut-over raised bog habitat in this picture, covers large areas of the bog surface at Cors Caron, including much of the primary bog surface. Descriptive survey information from the 1930s suggests that this has happened relatively recently, and is indicative of drying out.

When reviewing species lists, we must take into account how the data were gathered. If the list comprises casual observations collected over years or decades, we cannot assume that the species occupied the site at the same time, or indeed could occupy the site at the same time. We will simply have a record of species that have been able to occupy the site, perhaps very temporarily, at various stages in its development. Lists generated through this process are probably the least useful form of site-specific survey information. We could compile an impressive species list for any semi-natural grassland site if we removed the management and simply recorded all of the species that came and went during the various phases of succession.

By contrast, if a species list was compiled over a short period, perhaps one or two summers, it is much more likely to comprise species that co-existed on the site. This will allow comparison with the current situation, and could be used to inform the selection of a site-specific indicator assemblage. This will depend on whether or not the aim of the management is to restore the site to its condition at the time of the original survey.

## 2.4    Species distribution maps

For many species, survey information exists only in the form of a distribution map, often plotted as dots on a grid. Distribution maps vary in scale, depending on the area covered by the map. For example, maps showing the distribution of breeding birds in a county may be mapped on 2 x 2 km grid, e.g. Hurford & Lansdown (1995), whereas national plant distributions are more likely to be mapped on a 10 x 10 km grid, e.g. Preston *et al.* (2002) and European-scale maps will use a 50 x 50-km grid, e.g. Mitchell-Jones *et al.* (1999). The most useful distribution maps will also provide some indication of abundance, typically by using different sized dots to represent species density, e.g. <5 pairs, 5-50 pairs, and >50 pairs. Unfortunately, maps with abundance data are relatively scarce, and tend to be restricted to bird surveys.

Even in the absence of information on abundance, these maps offer us a basic level of information that we can incorporate into a conservation strategy: by identifying either the current or potential range of a species. When we combine this information with the knowledge available through research, i.e. the requirements of the species and the limiting factors, we can identify the most appropriate areas for range expansion. We can also use distribution maps to predict which species could or should be present on sites with suitable habitat.

It is difficult to assess the error associated with species distribution maps, but there are a few general points that we should consider. Firstly, the accuracy of distribution maps is likely to decrease with the size of the area of search, primarily because it is more difficult to be certain that a species is absent from a 50 x 50 km square, than it is from a 2 x 2-km square. This is particularly the case for species that are shy or thinly distributed (or both). Furthermore, the organisers of a survey will often check records of a rare species before they are published, so rare species are always more likely to be under-recorded on distribution maps than over-recorded. Finally, conscientious recorders seldom say something is present when it is not: they are far more likely to overlook a species. Taken together, this suggests that distribution maps are generally more likely to give a conservative estimate of the distribution of a rare species than provide an overly optimistic picture of its distribution.

The danger of distribution maps, of course, is that they can give a false impression of abundance and create an air of complacency with respect to species that have suffered steep declines in numbers while maintaining their overall distribution. For this reason, it is good practice to try and incorporate some estimates of abundance to underpin a distribution map: the map alone will not provide an early warning of decline.

## 3.    IN SUMMARY

Most sites of high conservation value will already have survey data available in one form or another. Often, enough site-specific information will be available to make responsible management decisions. Therefore, before committing resources to collecting additional survey information, we should decide whether new information would make our management decisions any easier. We should also consider whether the new survey information would be any more reliable than our existing data. If, ultimately, we do decide to collect new survey data, the new survey should not attempt to be 'all encompassing', but should focus on filling the critical gaps in our knowledge. Chapters 9, 10 and 11 suggest ways of doing this.


# 4.    REFERENCES

Avery, T. E. (1962). Interpretation of aerial photographs: an introductory college textbook and self-instruction manual. Minneapolis: Burgess Publishing Co.

Cherrill, A. & McClean, C. (1999). Between-observer variation in the application of a standard method of habitat mapping by environmental consultants in the UK. Journal of Applied Ecology **36**, 989-1008.

European Communities - Commission. (1991). EUR 12587 - CORINE biotopes manual - a method to identify and describe consistently sites of major for nature conservation. Luxembourg: Office for Official Publications of the European Communities.

Godwin, H. C. V. M. (1939). The ecology of a raised bog near Tregaron, Cardiganshire. Journal of Ecology 27, 313-363.

Graham, R., & Read, R. E. (1986). *Manual of aerial photography*. London: Focal Press.

Headley, A.D. (1990). Preliminary Report on Management Proposals for Crymlyn Bog. Unpublished report to CCW.

Hellawell, J. M. (1991). Development of a rationale for monitoring. In B. Goldsmith (Ed.), *Monitoring for Conservation and Ecology* (pp. 1-14). London: Chapman and Hall.

Hughes, M.R. (1981). Whiteford National Nature Reserve; Description and classification of dune slacks and their vegetation. Nature Conservancy Council report. South Wales Region.

Hurford, C. & Lansdown P.G. (1995). Birds of Glamorgan. D. Brown & Sons. Bridgend.

Hurford, C. & Perry, K (2000). *Habitat Monitoring for Conservation Management and Reporting. 1: Case studies*. Life-Nature Project no LIFE95 NAT/UK/000821. Integrating monitoring with management planning: a demonstration of good practice in Wales. Countryside Council for Wales, Bangor.

Maling, D.H (1989). *Measurements From Maps: Principles and methods of cartometry*. Pergamon Press. Oxford.

Mitchell-Jones, A.J., Amori, G. Bogdanowicz, W. Kryštufek, B., Reijnders, P.J.H., Spitzenberger, F., Stubbe, M., Thissen, J.B.M., Vohralík, V. & Zima, J. (1999). The Atlas of European Mammals. T. & A.D. Poyser. London.

Preston, C.D., Pearson. D.A. & Dines, T.D. (2002). *New Atlas of the British and Irish Flora*. Oxford University Press. Oxford.

Rackham. O. (1995). Trees and Woodland in the British Landscape. Weidenfield & Nicholson, London.

Rodwell, J.S. (Ed.) (1991). *British Plant Communities. Volume I. Woodlands and Scrub*. Cambridge University Press, Cambridge.

Stevens, J.P., Blackstock, T.H., Howe, E.A. & Stevens, D.P. (1999). Repeatability of Phase I habitat survey. *Journal of Environmental Management*, **73**, 53-59.

Tomassen, H.B.M, Smolders, A.J.P., Lamers, L.P.M. & Roelofs, J.G.M. (2003). Stimulated growth of *Betula pubescens* and *Molinia caerulea* on ombrotrophic bogs: role of high levels of atmospheric nitrogen deposition. Journal of Ecology: 91, Issue 3, p.357.

# CHAPTER 4

# SURVEILLANCE

### DAVID ALLEN

*Countryside Council for Wales, Plas Penrhos, Ffordd Penrhos, Bangor, Gwynedd, LL57 2BQ*
*D.Allen@ccw.gov.uk*

## 1.    INTRODUCTION

Think of some of the most celebrated ecological or conservation monitoring projects and what comes to mind?  The Rothamsted Park Grass plots (Williams, 1978) perhaps, or maybe one of several national bird monitoring programmes, e.g. the UK's Common Bird Census[1] (Marchant *et al.*, 1990) or the North American Breeding Bird Survey, e.g. Cooke ( 2003). A search through the scientific literature will reveal many others.

Were we to conduct a thorough search we would find a diverse mix of such projects, conducted at a range of spatial and temporal scales, using different techniques and designed to address different objectives.   But all would be loosely described as 'monitoring'.  Looked at more closely, however, they would actually fall into several categories of different if related activities - survey, monitoring, surveillance, etc.  The type of activity we wish to engage in has important implications for how we go about designing and implementing our project, and the clearer we are from the outset about what we wish to achieve, the more likely we are to be successful.

Much nature conservation activity that has been described as monitoring in fact comes close to the definition of surveillance by Hellawell (1991):

> An extended programme of surveys, undertaken in order to provide a time series, to ascertain the variability and/or range of states or values which might be encountered over time (but without preconceptions of what these might be).

It is this type of study that is the focus of this chapter: in particular the relationship between surveillance and monitoring – and how the two types of study can complement one another.

---

[1] Since 1994, replaced by the Breeding Bird Survey (Marchant, 1994).

*C. Hurford & M. Schneider (eds.), Monitoring Nature Conservation in Cultural Habitats, 35–42.*
© 2007 *Springer.*

## 2.    SURVEILLANCE OR MONITORING:  DEFINING OUR TERMS

We need to be clear about what it is that distinguishes surveillance from the monitoring studies that form the backbone of this guide.  Hellawell's (1991) definition of monitoring provides some clues:

> Intermittent surveillance carried out in order to ascertain the extent of compliance with a pre-determined standard or the degree of variation from an expected norm.

From this definition, it emerges that monitoring is, in fact, a particular type of surveillance, where the emphasis is on assessing the recorded level of some measure against a form of benchmark; it is the benchmark that is the distinguishing feature of monitoring.  It is clear from this that monitoring is an example of hypothesis-testing - in monitoring we establish, in advance, a view of what we are aiming to achieve, and then we collect data to determine whether or not we have achieved it.

The model described in Chapter 2 goes a step further by making this benchmark the pivotal point at which decisions are made about changing management; the monitoring result is a trigger for action, prompting a response designed to deliver management objectives.  In this form of monitoring, this benchmark is an expression of what we are aiming to achieve through our conservation management – the target condition for the habitat or species of interest.  And it is both the existence of this benchmark – defined in advance of the activity – and its role as a trigger for action that marks out this monitoring as different from surveillance.

Surveillance, by contrast, is concerned with the detection of change rather than with whether a benchmark has been met.  It is the tracking of some entity – for example, a species population, habitat composition, or some physical or chemical variable in the environment - through time that is its characteristic feature, usually without any predetermined view as to what sort of change is expected.  In surveillance there is usually – at the outset of a project at least – no explicit hypothesis that is being tested.

A consequence of this is that surveillance tends to collect data on a wider range of variables than monitoring, which targets only those required to test a hypothesis about the condition of a habitat or species.  It follows from this, that for a given amount of resource, monitoring is able to collect data from a larger number of locations than surveillance.  In monitoring we tend to collect a few data from many locations; in surveillance, a lot of data from comparatively few locations.

## 3.    A CASE STUDY – MONITORING OR SURVEILLANCE?

From the late 1980s, the UK Ministry of Agriculture, Fisheries and Food (MAFF, now Defra) instigated a programme of monitoring on its Environmentally Sensitive Areas (ESA) agri-environment scheme.  This scheme operated in distinct parts of the UK, and provided payments to farmers in return for their adopting sympathetic management prescriptions e.g. stocking within certain limits. The aims of this scheme were reflected in a hierarchical structure of environmental aims, environmental objectives and performance indicators (Box 1), and monitoring was carried out to determine whether these were being met.

---

**Box 1. The hierarchical structures of aims, objectives and performance indicators associated with ESAs.**

Environmental Aim (Common to all ESAs): To maintain and enhance the landscape, wildlife and historic value of the area by encouraging beneficial agricultural practices.

Environmental Objective (an example): To maintain and enhance the wildlife conservation value and landscape quality of open moorland.

Performance Indicator (an example): Vegetation that is characteristic of less agriculturally improved meadows, pastures and rough grazing does not deteriorate on land under agreement.

---

Grassland monitoring was an important part of this work and typically addressed a performance indicator of the form shown in Box 1. Similar performance indicators were defined for other habitat types (e.g. heather moorland), for breeding and wintering birds, and for a range of landscape features.

From 1993 onwards, grassland was monitored using a method based on fixed plots of 32 nested quadrats, and the occurrence of all species was recorded (Critchley & Poulton, 1998). Repeat recording of the plots took place a few years after initial recording and the data were analysed for evidence of change. The analysis was based on a measure known as a *suited-species score*, a composite score based upon the traits and habitat preferences of individual species, and reflecting the extent to which vegetation contained species suited to the conditions that the ESA scheme was designed to encourage (Critchley *et al.*, 1996). Conclusions were drawn about whether, for example, species suited to grazed conditions or high soil moisture content (in line with ESA management prescriptions) had changed.

So here we have a programme of repeat vegetation recording, with a benchmark (the performance indicator) against which the success or otherwise of the ESA scheme was to be evaluated. Also present in this example are hypotheses about the type - though not at this stage the magnitude - of change expected under the management prescriptions (e.g. lower grazing is expected to result in a decrease in species suited to grazed conditions; cessation of fertiliser applications in an increase in species suited to low nutrient availability; and so on). In short, the work displays several of the characteristics of the sort of monitoring project featured in this book.

Where this work differs from monitoring is primarily in that the benchmark against which success is evaluated is not fully quantitative. This leads to difficulties in concluding whether any changes detected have been large enough to meet expectations – we may be able to conclude that quality has been maintained but drawing conclusions about enhancement are more difficult – do we conclude that any increase represents an ecologically significant enhancement? And there is a potential knock-on effect of this uncertainty – are we able to decide whether we need to change management based on the results of the monitoring? Another difference is that the data collection is more comprehensive – observers set out to record all species – than in the more focused activity of monitoring.

Later in this programme of work, an approach was developed that lent itself to the establishment of increasingly quantitative targets. This was based on calculating reference values from stands of vegetation of recognised quality, as a benchmark against which the recorded measures for stands of vegetation under ESA management could be compared (Critchley *et al.*, 1999; Critchley, 2000; Fowbert & Critchley, 2000).

## 4.    THE ROLE OF SURVEILLANCE

### 4.1    Surveillance as a starting point

The above case study is an example of how a surveillance project can be modified in the direction of a monitoring project through the introduction of a benchmark. (A different approach is taken in the 'raising of alarms' about bird populations; here an analysis of patterns of change in time series data is used to determine whether additional conservation action is required (Greenwood *et al.*, 1995)).

There may sometimes be advantages in adopting such an approach, particularly where we are not confident in our understanding of the habitat or species population we are interested in, and the factors acting upon it. In such cases, surveillance can be used to increase our understanding of the feature we wish to monitor[2], allowing us to gain two important pieces of knowledge:

- An understanding of the relationship between the feature and factors acting on it (to allow us to focus on a small set of key attributes that are responsive to those factors), and;
- An understanding of typical fluctuations (to allow us to set appropriate target levels).

In some cases, where a historical archive exists, it may be possible to look back in to the past to do this – perhaps where we have an archive of aerial photographs, or where a long run of population data has been collected. For example, it is clear from the aerial

---

[2] In those cases where our understanding of the feature is poor we may need to carry out additional investigative research to identify relationships between the feature and its environment.

photograph archive that many coastal dune systems in Wales, such as those at Kenfig (Chapter 32), have been stabilising since the 1940s. Even where we have no data for a particular site under study, it may be that we can draw on existing data from other similar or nearby sites.

There is also often a very practical advantage in undertaking some surveillance in advance of monitoring, especially of habitats – the intensive observation of the feature and the collection of comprehensive data act to increase the observer's familiarity with the subject under study. This can help to minimise problems resulting from species misidentification. If more than one observer is involved in recording, setting up surveillance plots provides an early opportunity to standardise approaches to measurement, and to harmonise measures such as estimates of species covers or species counts, which otherwise can differ greatly between observers.

## 4.2　Surveillance as a means of reducing risk

Condition monitoring is, almost by definition, a minimal activity – the focus is on collecting data on as few measures as are necessary to draw conclusions about condition, and to determine whether any change in management is required. But by focusing on only a few key attributes of a habitat or species, chosen to reflect those factors we believe most likely to influence it, there is a risk that we might fail to detect the action of some factor that we had considered unimportant or of which we were unaware.

We can use surveillance with its more comprehensive data collection as a check on our more streamlined monitoring projects. It acts to validate the decisions we made early on about which attributes of the features we would record. A series of surveillance plots can provide the confidence that our focus on just a few attributes in our monitoring is justified and that we are not missing other changes.

Another risk we take in monitoring is that of using the wrong benchmark. A time series of surveillance data provides a basis for reviewing and, where appropriate, amending such targets in light of increased knowledge of temporal and spatial variation, and of resilience (the ability to recover after a decline).

## 4.3　Surveillance as context

Surveillance projects are likely to be most powerful when they have been set up across a series of sites. On any single site, because of the intensive nature of the recording, it is usually only possible to record a small number of samples; consequently, the ability to detect anything but large change is small. By pooling data across a series of sites (assuming we have a valid sampling strategy), our ability to detect change increases.

There are many examples of large-scale surveillance projects, e.g. the EMAP project in the US (Stevens, 1994), Countryside Survey in the UK (Haines-Young *et al.*, 2000). Generally, these projects use data collected from a large number of sample locations to produce a picture of change in selected variables at regional and larger scales. With the exception of the sampled locations, however, they do not provide the basis for describing trends at any particular individual location. Instead, such projects provide a general picture, describing wider trends in the environment, which we can compare with our observations on any particular site. This is important contextual information that we should consider in reviewing the results of our monitoring – perhaps we are failing to meet our objectives for a site not because of a failure with our management but because of larger-scale trends (e.g. climate change) or off-site influences (e.g. impacts on a migratory bird species at its wintering grounds rather than at a breeding site we are managing for it).

## 4.4    Surveillance as the link to other data sets

The streamlined nature of monitoring can make for a specialised form of data collection; for example, in the use of composite attributes (see Chapter 11). This can act as a constraint on any wider use of the data; because of the site-specific nature of attributes used in condition monitoring, it will often not be possible to pool the data from separate projects.

Surveillance projects, by contrast, tend to collect data in a less specialised form, and there is greater potential for combining data across sites, as well as for using the data for other purposes, and for combination with other types of data (e.g. remotely sensed data). An extension of this is the collection of potential explanatory variables at sample locations, which can be used in correlative analysis with the surveillance data to suggest avenues of further investigation when attempting to diagnose causes of change.

## 5.       SETTING UP A SURVEILLANCE PROJECT

That surveillance is a less focused activity than monitoring does not mean that there are no decisions to be made in setting up a surveillance project. As with all types of investigation, a clear set of objectives is a prerequisite for a properly designed project. And while we may begin a surveillance project with no explicitly defined hypotheses about the type, direction and magnitude of change we are interested in[3], we do at least need to consider the following questions:

- What is it that we wish to be able to detect change in (area of habitat, quality, breeding population of a particular species)?
- What are the defining boundaries of the entity about which we want to draw conclusions? For example, do we want our conclusions to apply only to one particular stand of vegetation on a site or to all similar stands on the site?
- Having defined the boundaries of the entity under study, can we collect data from all parts of it (a census) or from only a limited sample? And if the latter, how do we decide where to take our samples? And what is the basis for our considering these sample data to be representative of the entity as a whole?

---

[3] There may be cases where we do have a clearly stated hypothesis which we wish to test, for example 'detect a decline of 50% over a 25-year period (-2.7% per year) with a probability of 90% (Hatch 2003). In such cases, we can use statistical power analysis to help in the study design (e.g. Thomas & Krebs 1997).

- And once we've decided upon sample locations, exactly what are we going to record, and at what time of the year?

Some of the more difficult questions here relate to sampling, which is covered elsewhere in this book (see Chapter 5), and which is an issue of equal relevance to both monitoring and surveillance. Also as relevant to surveillance as to monitoring are the issues of measurability and repeatability (see Chapters 10 and 11). The reader is referred to those chapters for more specific advice on these issues.

All surveillance is dependent on some form of repeated measure and this requires us to have identified some fixed unit within which to repeat our recording. This does not, however, mean that surveillance requires repeated recording of fixed quadrats or particular individuals within a population. This is a valid method but only one of several options available to us. More fundamentally, we need to fix the unit at the level at which we want to draw conclusions and this will often be at a larger scale than individual quadrats.

For example, if we wish to describe trends in the composition of a field of neutral grassland it is the field that is our fixed unit; with a species, it might be the population within a defined geographical area. Once these units have been defined we then have the option of recording from the whole of the unit (a census) or from a sample of sub-units within it; and these sub-units may themselves be fixed or not. The advantage of fixed sub-units - a repeated measures design - is that, within the limits of measurement accuracy, any recorded change is a measure of real change (Green, 1993) – *Lolium perenne* has increased in cover within the quadrat, sward height has decreased or the number of nesting chough has stayed the same. The disadvantage of fixed sub-units are that they may, over time, become less representative of the whole unit.

With non-fixed sub-units, a separate sample is chosen at each recording time – a re-randomization design – but while this approach ensures that a representative estimate of change is maintained over time, this is at the expense of precision. A compromise would be to replace a small proportion of the sub-units each time a measurement is made, i.e. introducing turnover to the repeated measures design. This latter design also has the benefit of reducing the impact of repeat visits and measurement on the subject; although the more sensitive the subject is to disturbance, the stronger is the argument for using a complete re-randomisation design.

Ending on a practical note, the more comprehensive recording required in surveillance will often require a greater level of expertise than condition monitoring, where it is often possible to train relatively inexperienced observers to measure the few attributes that will be used. It is important that observers on surveillance projects are fully trained, and that quality assurance measures are in place.

# 6.   REFERENCES

Cooke, F. (2003). Ornithology and bird conservation in North America – a Canadian perspective. Bird Study 50: 211-222.

Critchley, C.N.R. (2000) Ecological assessment of plant communities by reference to species traits and habitat preferences. Biodiversity and Conservation 9: 87-105.

Critchley, C.N.R. & Poulton, S.M.C. (1998) A method to optimize precision and scale in grassland monitoring. Journal of Vegetation Science 9: 837-846.

Critchley, C.N.R., Burke, M.J. & Fowbert, J.A. (1999) Sensitivity and calibration of plant community variables for grassland monitoring in English ESAs. ADAS report to MAFF.

Critchley, C.N.R., Smart, S.M., Poulton, S.M.C. & Myers, G.M. (1996). Monitoring the consequences of vegetation management in Environmentally Sensitive Areas. Aspects of Applied Biology 44: 193-201.

Fowbert, J.A. & Critchley, C.N.R. (2000) Calibration of plant community variables for mires, wet grasslands and upland communities. ADAS report to MAFF.

Green, R.H. (1993) Application of repeated-measures designs in environmental impact and monitoring studies. Australian Journal of Ecology 18: 81-98.

Greenwood, J.J.D., Baillie, S.R., Gregory, R.D., Peach, W.J. & Fuller, R.J. (1995). Some new approaches to conservation monitoring of British breeding birds. Ibis 137: S16-S28.

Haines-Young, R.H., Barr, C.J., Black, H.I.J., Briggs, D.J., Bunce, R.G.H., Clarke, R.T., Cooper, A., Dawson, F.H., Firbank, L.G., Fuller, R.M., Furse, M.T., Gillespie, M.K., Hill, R., Hornung, M., Howard, D.C., McCann, T., Morecroft, M.D., Petit, S., Sier, A.R.J., Smart, S.M., Smith, G.M., Stott, A.P., Stuart, R.C. and Watkins, J.W. (2000) Accounting for nature: assessing habitats in the UK countryside. DETR, London.

Hatch, S.A. (2003) Statistical power for detecting trends with applications to seabird monitoring. Biological Conservation 111: 317-329.

Hellawell, J.M. (1991) Development of a rationale for monitoring. In Monitoring for Conservation and Ecology (ed. B. Goldsmith) Chapman & Hall, London.

Marchant, J.H. (1994) The new Breeding Bird Survey. British Birds 87:26-28.

Marchant, J.H., Hudson, R., Carter, S.P. & Whittington, P. (1990) Population Trends in British Breeding Birds. British Trust for Ornithology, Tring.

Stevens Jr., D.L. (1994) Implementation of a National Monitoring Program. Journal of Environmental Management 42: 1-29.

Thomas, L. & Krebs, C.J. (1997) A review of statistical power analysis software. Bulletin of the Ecological Society of America 78: 126-139.

Williams, E.D. (1978) Botanical composition of the Park Grass plots at Rothamsted: 1856-1976. RES, Harpenden.

# CHAPTER 5

# STRATEGIC SAMPLING

ALAN BROWN

*Countryside Council for Wales, Plas Penrhos, Ffordd Penrhos, Bangor, Gwynedd, LL57 2BQ*
*A.Brown@ccw.gov.uk*

## 1.    INTRODUCTION

### 1.1    The need for sampling

Sampling is one of the most important aspects of any practical monitoring project. Only rarely can we make a complete record of a habitat or species population at a protected site. Unless it is conspicuous, small and confined within a small area, we must draw some conclusion about the condition of the whole feature from measurements made in a carefully chosen sample of habitat units or individuals. Even if complete recording is possible, sampling is usually quicker, cheaper, less damaging, and more accurate than a nearly completed census, because it is unbiased. When we find that sampling is impractical, we may have proven that, in this case, management cannot rely on monitoring to detect a significant decline (see Taylor and Gerodette, 1993).

### 1.2    Strategic sampling

The aim of strategic sampling is to minimise the effort in collecting new sample data in a monitoring project, that is, a project is undertaken to conclude whether or not a habitat or species feature is in a favourable condition.

If we are using statistical methods to estimate parameters, effort is minimised within the constraints of a specified degree of accuracy (freedom from bias) and repeatability or precision, expressed as confidence or credibility limits. If testing a hypothesis, this will be tested within acceptable limits of type 1 and type 2 statistical errors.

*C. Hurford & M. Schneider (eds.), Monitoring Nature Conservation in Cultural Habitats, 43–54.*
© 2007 *Springer.*

Strategic sampling can be considered before or after data collection. We can see how best to use an existing set of sample measurements to give the most accurate, most repeatable conclusion about the whole feature. Alternatively, working backwards from what would be a good enough conclusion, we can calculate the minimum amount of fieldwork required.

## 1.3    The layout of the chapter

Textbooks on classical sampling typically need several hundred pages to describe the choice of practical sampling schemes – and can be daunting to the non-specialist. This chapter aims to condense and review these methods in a non-mathematical way to give just enough insight into what choices there are and suggest where to look for practical guidance. We will cover both classical probability-based methods and other methods more commonly used in habitat monitoring in the UK.

The chapter uses examples of habitat monitoring because they can be illustrated more graphically than projects based on species. The first part of the chapter sets out a philosophy of approach and provides some definitions and descriptions of data types. This is followed by a short description of simple random sampling. Strategies for improving the efficiency of probability-based sampling and fieldwork are then presented as modifications to this basic method, followed by a section on using non-probability methods. The final section gives recommendations for further reading and developing practical solutions.

## 1.4    Data categories

We will distinguish between new habitat data collected in a monitoring project at a series of sample points and the contextual information used to help plan data collection and interpret the results. We assume that strategic sampling methods are required to interpret existing sample data or – more powerfully – plan new sample data collection.

### 1.4.1    New sample data

Habitat sample data is typically collected by making observations or measurements at some pre-determined pattern of points in the field called a sample layout, organised within a sampling frame. Landscape ecologists call this geostatistical data, contrasting it with point data where the location is the primary information (for example, the location of isolated ancient oak trees in a parkland). For brevity, we will refer to geostatistical data simply as sample data.

### 1.4.2    Contextual information

Useful contextual information is usually not confined to sample points. A survey may produce a thematic map in the form of polygons, each of which is classified into one or more categories of vegetation, or condition if it is a condition survey. In a geographical information system (GIS) these are stored as vectors. Imaging remote sensing instruments collect information in the form of a raster grid of similar-sized units (referred to as pixels because of the way they appear on a computer screen). Pixels may be analysed as individual values or grouped into objects and classified into a thematic map. Both the vector polygons and raster grid are examples of lattice data. Data values for polygons or pixels may be continuous (on an interval or ratio scale, depending on calibration) or thematic (also called a nominal scale).

In other types of analysis (notably analysis of remote sensing images) the lattice data is of primary interest; but here it will be treated as contextual information.

Contextual information may be more subjective. The manager of a small site may know which is the area of habitat in best condition without drawing a map or making systematic measurements. It may be obvious to them that part of a field next to a road with tall hedges is trampled and overgrazed where stock habitually shelter from the rain or wait for supplementary winter feed. These observations will be referred to as *gradients* because they describe relative changes in condition with direction or location, including how one part of the site is 'better than' another.

Knowledge can also be expressed as subjective prior probabilities in a Bayesian statistical analysis. For example, a site manager may be able to use a probability value between 0 (never happens) and 1 (always happens) to express their belief about the likelihood of different parts of a heathland being subject to an accidental burn in the next five years, based on a subjective assessment of the amount of woody biomass.

## 1.5    Using contextual information

The classical approach to science based on experimental design and the analysis of variance excludes most types of contextual information. We are taught that combining subjective results into the analysis will introduce bias, and that bias threatens the universal application of the conclusions. This belief may not be helpful in our case, where we are not concerned with drawing universal conclusions but making the best decisions about the management of a particular feature on a particular site, over a particular time interval and where external influences are unavoidable. Moreover, Hurlbert (1984) and subsequent studies of BACI (before-after-control-impact) designs show that there is no complete equivalent of experimental work in field studies that can eliminate the effects of confounding external variables.

Instead, we will regard contextual and sample data as related through a process of updating our 'state of knowledge' – more specifically the state of knowledge of the site manager in relation to habitat management. Though to some degree this knowledge is personal and subjective, if it is made transparent by expressing attributes and targets (shown by the case studies in this volume), it can be shared and made more or less consistent between individuals and organisations. In statutory nature conservation, our state of knowledge is built up through alternating phases of data collection and decisions, including survey work, deciding the aims of management and monitoring.

## 2.     SAMPLING STRATEGIES

### 2.1     Simple Random Sampling

Simple random sampling without replacement is the simplest form of non-strategic sampling – here 'without replacement' means that we avoid visiting and recording the same point twice. It uses no contextual information, except to define the area of study (AOS) as the whole feature. A simple method is to place a grid of co-ordinates (the sampling frame) across a map of the AOS, and take random numbers to determine the x and y co-ordinates of sample points, rejecting any duplicates or points that fall outside the habitat. Measurements and observations are made at the corresponding points in the field.

This method gives unbiased estimates of two statistical parameters, the population mean and variance of the measured variable, derived from the sample mean and sample variance. (The confusing term 'population' refers in this case to the complete set of sample points that might be measured, if we had enough time; so population statistics or parameters are the true, completely repeatable values we would get if we measured every point). The unbiased estimates are a consequence of the design of the method, here guaranteed by equal probability sampling, in which every part of the AOS has an equal chance of being sampled.

### 2.2     Parameter estimation from simple random sampling

Let us assume the measurement at each point is a continuous variable such as vegetative grass height in a field, measured using a standard weighted disc and ruler (illustrated in Fig. 1A). The sample mean is calculated by dividing the sum of height measurements by the number of sample points. This is the unbiased estimator of the population mean. The sample variance is the sum of the 'squared deviations' from the sample mean. This is calculated by taking the value at each point minus the sample mean; squaring the result (this makes the values positive, otherwise they cancel out); summing the squares for every point and dividing by the number of points. With slight inflation to compensate for the finite sample size, this is the best estimator of the population variance.

As the sample size increases, these sample estimates will move, on average, closer to the true values for the statistical population. Any given degree of statistical confidence – usually shown as confidence limits – can be bought with enough sampling effort, provided of course that consistent measurements can be made in the field: see comments on minimising observer error in Chapter 10. However, we cannot predict how many sample points are needed to give the required degree of confidence until we have started sampling, because we also need an estimate of variation in grass height, summarised by the sample variance. If the field is very heterogeneous, variance may be high and more samples will be needed for a given degree of confidence. The increase is not linear but proportional to the square: as a rule of thumb, to halve the confidence limits the sample size has to increase by a factor of four.

## 2.3 Minimising variance

In simple random sampling, the population mean and variance are only parameters in the Normal or Gaussian statistical model. In this application, the variance is a secondary parameter that is necessary for estimating mean grass height. However, for our purposes, any statistical model will do providing it is part of a sampling strategy giving unbiased estimates. If we can find a sampling method and model that happens to give a smaller sample variance, we can reduce the sample size. This is possible if we incorporate some of the contextual information ignored by simple random sampling.

## 2.4 Stratified random sampling

High variance is the result of having many sample values much higher or lower than the sample mean. A field with a gradient of grazing might have areas of very short grass and others with tall grass. Here the mean height for the whole field is in between the extremes and close to neither – with the consequence that the variance will be very high.

Using our contextual knowledge of the gradient, we could divide the field into areas of short, medium and tall grass (shown for clarity as separate fields in Fig. 1B). We can then calculate local means for each division (or stratum) and calculate corresponding local variances. Since the local means are closer to the range of actual grass heights in each of the three areas, the variance in each will be much lower and the precision of our estimate of the mean value much greater – despite the smaller sample sizes. By incorporating contextual knowledge of a gradient, we have bought extra precision for little additional work.

Stratified random sampling can be considered at the planning stage (pre-stratification) or for already existing data (post-stratification) and can use a gradient, a thematic map of condition categories or knowledge of grazing history. For planning stratified random sampling, textbooks commonly give formulae for calculating the optimal number of strata based on initial estimates of variance in a type of two-stage sampling. The advantage of this approach is that it allows a different density of samples in each stratum according to the variance – so collecting more information where there is more variation. Though this is equal probability sampling within each stratum, different parts of the habitat now have unequal probabilities of being included in the sample – but nevertheless still gives unbiased estimates. The most powerful methods incorporate contextual information at the design stage and depart from the simplest method of paying equal attention to all parts of the AOS.

## 2.5 Ratio and regression estimators

Stratified random sampling uses thematic contextual information. Ratio and regression estimators extend this idea to take advantage of any continuous supplementary variables that might be used to predict the measured values at any point. Staying with the example of the grazed field, supposing we had some lattice data estimating productivity or biomass at any sampling point. At each point we then have two variables: a real measure of height

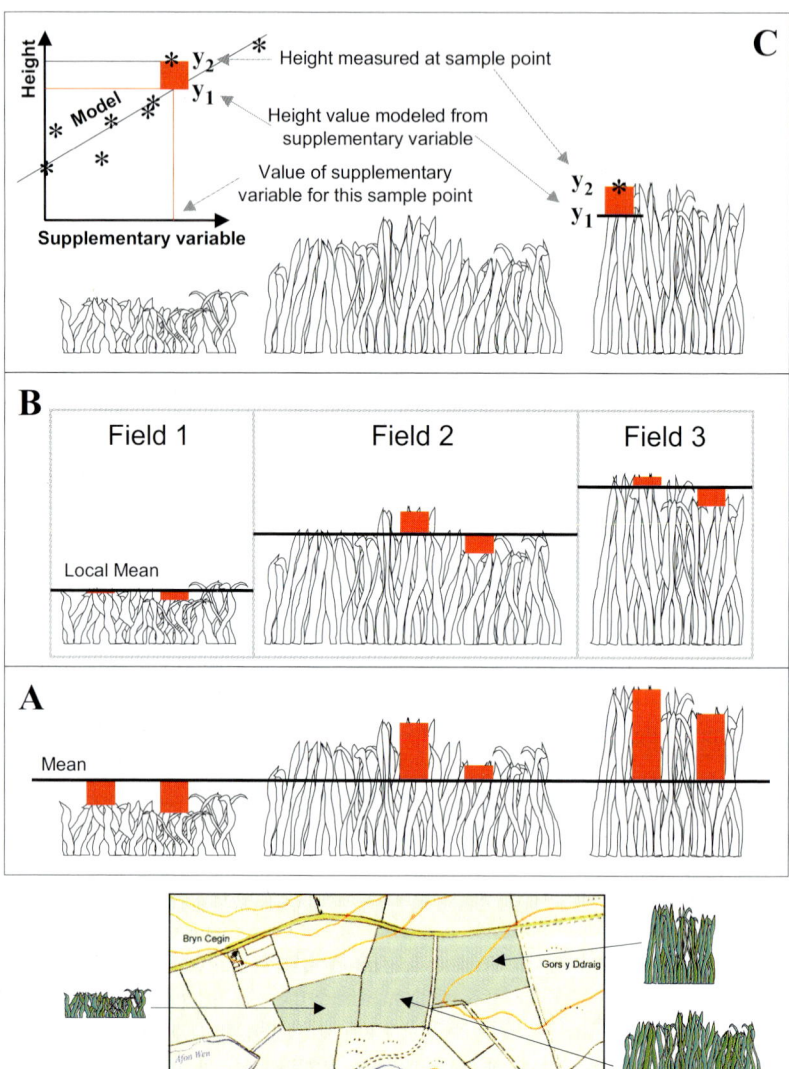

*Figure 5-1*: Random sampling (Fig. 5-1A), Stratified random sampling (Fig. 5-1B) and Regression estimators (Fig. 5-1C). In each case, the differences between the actual value and mean value are shown in red. These differences will be squared when variance is calculated – so the smaller amount of red in 1B compared with 1A corresponds with a much lower variance. The model in 1C is a regression line, with sample points shown as asterisks. Here the red bar represents the difference between the actual value measured in the field and a value calculated using the regression model from the value of a supplementary variable – for example, a satellite or airborne measurement of grass biomass in this pixel.and a supplementary measure of biomass. We can then plot the two variables together and fit a line, which might go through the zero point on both axes.

(this is a ratio) or might not (a regression), as shown in Fig. 1C. If we are lucky, the model will show that we can predict grass height with some success from the supplementary variable, effectively taking into account much of the variance. Now our statistical calculation of mean grass height can use the smaller difference between the measured and predicted values to calculate the variance of the estimate.

Ratio estimators are also widely used where sample units are of unequal size, notably in forest inventory. When counting ancient trees in several woodland compartments of unequal size, we can use the size of the compartment as the supplementary variable to reduce the variance of the total estimate.

## 2.6    Samples and Weights

Each of our individual sample measurements of grass height could be thought of as representing similar but unmeasured values at other points. Those with values close to the average will represent many similar points; those with more extreme values fewer points. If we knew the representativeness of each point, this could be incorporated into our strategic sampling by giving sample values different weights, with the most representative points given the greatest weight and consequently most influence on the parameter estimates.

The design of simple random sampling gives every point equal weight and relies on the principle that, on average, more representative values will turn up in the sample more often. However, in a small sample we can easily have too great or too small a proportion of extreme values by chance, with no means of compensating for this by giving them different weights in the calculations. Stratified random sampling can give points unequal weight if the number of points in a stratum is made proportional to the variance – meaning that in a low-variance stratum a few points have a greater individual weight in the analysis than many more points in a high variance stratum.

When there is a well-correlated supplementary variable (as in ratio/regression estimation) the weights for each sample value can now be estimated by looking at how representative the supplementary values are, if (as is usual) we know these values for every part of our AOS. We can even use these estimates to determine which points should be sampled, giving priority to those which are most useful in the analysis – while retaining an element of randomness so as not to introduce bias. This is the method of probability proportional to prediction (PPP) sampling, widely used in forestry inventory. These advanced sampling methods have great potential for reducing variance in the right situation.

## 2.7    Minimising location effort

So far the emphasis has been on getting the most from sample measurements. However, any field ecologist knows that precious time can also be spent locating sample points and marking them for future re-location. In this second section we consider how this is influenced by sample layout. Here again the reference method is simple random sampling.

## 2.8    Staking out random points

Until the recent availability of cheap global positioning systems (GPS or DGPS), random points could only be set out using tapes. This method becomes more time-consuming and less accurate as the area of interest increases, and impractical for any area larger than 50 m x 50 m. Here a systematic grid in which (almost) every point is a fixed, regular distance and bearing from the last point is far easier to set out.

GPS allow us to record locations of points and navigate to points (a process known as staking out) with a horizontal accuracy of between 5 and 0.5 m (even down to 10 mm or less in some very expensive systems used by surveyors). GPS allows staking out of points in the field in any order and at any scale. The same reference system is used for small sites and very big sites, with no cumulative error – so that the accuracy of the final point is, on average, as good as that of the first point.

With GPS it is possible to stake out a set of random points efficiently provided they are also visited in nearest-neighbour order rather than the order in which they are generated. The shortest path between a given number of random points is roughly 20% shorter than between the same number of points on a regular grid. Even though there is no exact solution for this problem (known in mathematics as the 'travelling salesman' problem), good approximate solutions can be found using a computer. It is even relatively easy to estimate a good enough path between a few tens of points by mimicking the pattern of the approximate solutions.

## 2.9    Regular grids as substitutes for random points

On small sites, regular grids are easy to lay out without using GPS. Their principal disadvantage for statistical sampling is theoretical: they give a design-unbiased estimate of the mean but not the variance. Ecologists worry about grids coinciding with repeated patterns such as furrows or tree spacing; but it is hard to find published examples where this has been a real problem. Data from grids of sample points are usually analysed as if they were from random points.

## 2.10   Cluster sampling

For large areas, grids and random sampling may become difficult as the costs and effort of getting to the points rises in proportion to the costs and effort of making measurements. Here, a more effective strategy is cluster sampling. A number of primary locations are chosen. At each of these, visits are made to a cluster of local sampling points. Even though the sample size is small – corresponding to the small number of primary locations – the secondary, clustered, measurements can be used to reduce the variance of the estimate at each point. This is as if the clustered measurements make the primary sampling points more representative of their locality and reduce the possibility of their having extreme values by chance. Once again the aim of strategic sampling is to reduce the variance of the estimate and get the most representative measurements.

More advanced strategies might be to select the primary sampling units for cluster sampling using lattice data. For very large areas, the most promising approaches use remote sensing methods to provide thematic classifications for stratification and planning cluster sampling.

## 2.11   Hypothesis testing

So far we have considered strategic sampling for parameter estimation, using the example of mean grass height. Instead, we can think of monitoring as hypothesis testing, where instead of estimating grass height, what we need to know is whether the height is above or below a critical value. This approach offers new opportunities for savings in fieldwork.

For statistical sampling methods the same approach can be used in parameter estimation and hypothesis testing. The only difference is in the way we specify our standard of quality. Rather than specifying confidence limits, we decide on the minimum *effect size* we want to detect and the acceptable rate of statistical type 1 (false difference) and type 2 (missed difference) errors. For example, we may set a target value for grass height of 0.5 m or taller; and specify an effect size of 10 cm, with a type 2 error rate of 20%. This means that if the actual mean grass height is 10 cm or more below our target value, our sampling must be good enough to detect this 80% or more of the time. We can also specify an acceptable type 1 error rate to reduce the chances of concluding mean grass height is below our target level by mistake. The analysis of how many samples are needed to meet these requirements is called statistical power analysis and can be done using a simple computer programme (see texts referred to in the final section). Of course, if the actual grass height is much taller or much shorter than the target value, we can find this out with very few samples.

## 2.12   Setting targets for each sample point

Linking conservation objectives with hypothesis testing opens up new possibilities for strategic sampling, including non-statistical methods. In particular, we can apply the approach to setting a target for individual points, and set a target for how many points must pass or fail. Because we are no longer concerned with estimating parameters such as the mean, we no longer have to complete our sampling to guarantee an unbiased result.

Supposing that instead of grass height we wanted to measure the number of grass species in a 1 m square plot at each sample point and estimate the mean number for the whole field. As ecologists we know this is a difficult task, especially if it involves looking for small bryophytes or distinguishing between vegetative grasses, with all species given equal weight in the calculation. The potential for minimising recording effort can be seen in several modifications of our recording method.

The first modification is to reduce the number of species we are interested in, perhaps by deciding not to record bryophytes or non-flowering grasses. Better still, we can come up with an abbreviated checklist of species to be looked for. A second modification is only to consider species if they are abundant in the plot or make more than a minimum amount of cover. In bogs we might record *Sphagnum* cover only if it is 5% or more. This relieves us of having to search endlessly for obscure plants; though of course there might still be a fine judgement to be made over whether an occurrence is just under or just over 5%. We can also group the observations of attributes, for example by setting a target for each sample point that is passed if five or more desirable species are observed. Once we have five we can stop looking and move on to the next point.

This is the opposite of what we did in the last section. Instead of optimising the choice of locations to maximise the weight of a small number of samples, or maximising recording at each location (cluster sampling is the extreme example), we are minimising recording effort in order to be able to record many locations quickly. Part 3 of this book looks at this type of approach in more detail.

## 2.13   Mapping using correlated points in a regular grid

Supposing we have a regular grid of points close enough together to be more similar to one another, on average, than to more distant points. We might assume that each point is representative of any part of the surrounding area closer to it than to any other point. If this is a reasonable assumption, the effect of visiting the grid of points is to map out the whole habitat in the AOS by visiting the entire population of recording units. If the measurements are continuous it is possible to test this assumption of spatial correlation by looking at the semivariance.

## 2.14   Sampling using gradients

The final method illustrates very well the use of contextual data. If a target is set for, say, 70 out of 100 points to meet some criterion, we can have the exact answer once we have visited between 30 points (if they are all fails) and 100 (if the failed points are scattered among those that pass).

Habitat survey often produces contextual information in the form of gradients. These might be gradients in quality or condition (Fig. 2A) or in habitat extent (2B). Our knowledge of gradients or other contextual information can tell us where to start recording to maximise the number of fails (Fig. 2C and 2E). Since we have included all points in the sample (even if we do not in practice visit them all), we do not have to visit them in an unbiased order. We could, for example, visit all the known patches where we think the points might fail first even if these are not found together.

This approach is in the nature of a wager that we cannot lose. If we are right, we finish the monitoring as early as possible. If we are wrong, the only consequence is some increase in the time needed to complete enough of the remaining grid of points to test the hypothesis. This method can be extended to missing out areas for sampling because we

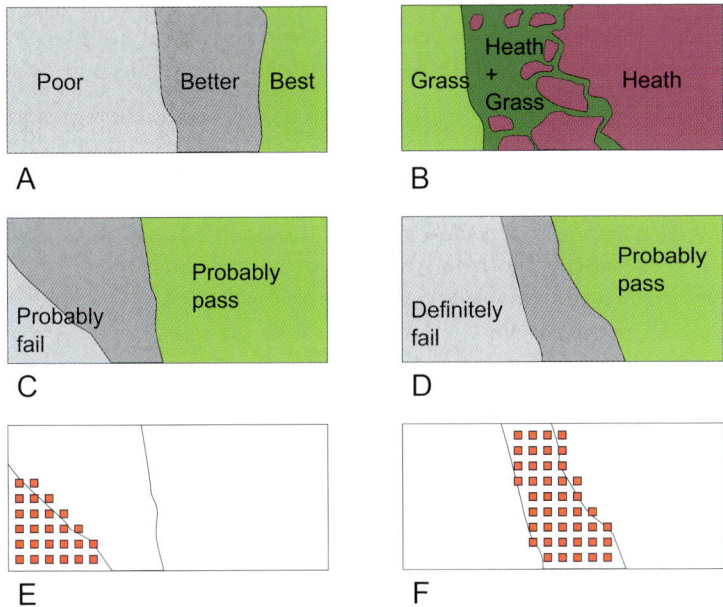

*Figure 5-2*: The use of gradients in sampling. Figs. 5-2A, 5-2C and 5-2D show gradients based on surveys or the subjective knowledge of the site manager, often based on the type of survey information shown in 5-2B (where the heathland is the feature of interest). The red squares in 5-2E and 5-2F represent sampling points, noting that the actual grid of points might be much more closely spaced.

are so certain they will pass or fail, and focusing sampling only on those areas where we are uncertain (Figs. 5-2D and 5-2F).

# 3.    CONCLUSIONS AND RECOMMENDATIONS

## 3.1    The choice of strategic sampling methods

This chapter has introduced a range of methods of strategic sampling, working from familiar classical probability-based sampling to pragmatic methods of monitoring small sites using a grid of correlated points, finishing once our questions are answered.

Classical sampling methods can be very sophisticated. In general, these aim to reduce the variance of estimates, taking advantage of contextual information to do so. As the size of the area of interest increases, the effort of finding and visiting points becomes more and more influential in the choice of methods. On small sites it is possible to visit a large number of point in a regular grid (which simplifies the location task). On larger sites we must either increase the spacing between points, in which case they are no longer spatially correlated and we are forced back into using classical statistical methods and may be better off using a random or stratified random layout. For larger sites still, samples must grouped together into clusters.

In the future the greater availability of supplementary measurements from remote sensing imagery will increase the importance of methods linking geostatistical and lattice data, notably ratio/regression estimators and PPP sampling.

Monitoring is part of a subjective process of decision-making by the site manager. Since this is site-specific, there is wide scope for the incorporation of contextual information to make new data collection as focused and quick as possible. This might involve choosing not to sample those areas which the site manager knows are bound to pass or fail the criteria set out in their conservation objectives. Once again the areas left may be small enough to sample using non-statistical methods.

On managed sites in cultural landscapes, condition and site management are closely linked. Unlike surveillance (Chapter 4), monitoring needs to collect just enough information to give the conservation manager confidence to make their decisions, and no more. To do this, we need to use strategic sampling.

## 3.2    Further reading

Good accounts of sampling for ecologists can be found in Krebs (second edition 1998) and a host of recent texts such as Elzinga *et al.* (2001). The slightly more technical publications by Schreuder *et al.* (2004) and edited by Sit and Taylor (1998) are both highly recommended, especially since both can be obtained as .pdf format files on the internet! Brewer (2002) gives excellent mathematical insights into sampling methods at a more advanced level.

## 4.    REFERENCES

Brewer, K (2002): Combined Survey Sampling Inference: Weighing Basu's Elephants. Hodder Arnold. 256 pp.
Elzinga, C.L., Salzer, D. W., Willoughby, J. W. & Gibbs, J. P. (2001): Monitoring Plant and Animal Populations: A Handbook for Field Biologists. Blackwell Science Inc. 368 pp.
Hurlbert, S. H. (1984): Pseudoreplication and the design of ecological field experiments. Ecological Monographs 54: 187-211.
Krebs, C. J. (1998): Ecological Methodology. Longman. 624 pp.
Schreuder, H T.; Ernst, R.; Ramirez-Maldonado, H. (2004). Statistical Techniques for Sampling and Monitoring Natural Resources. General Technical Report RMRS-GTR-126 of U.S. Department of Agriculture, Forest Service, Rocky Mountain Research Station, USA. 112 pp.
Sit, V. and B. Taylor (editors). (1998). Statistical Methods for Adaptive Management Studies. British Columbia Ministry of Forests, Forestry Division Services Branch Land Management Handbook No. 42. 148 pp.
Taylor, B. L. & Gerodette, T (1993): The uses of statistical power in conservation biology: the Vaquita and the Northern Spotted Owl. Conservation Biology 7: 489-500.

# PART III

## DEVELOPING  PROJECTS FOR MONITORING HABITATS

# CHAPTER 6

# DEVELOPING A HABITAT MONITORING PROJECT

CLIVE HURFORD

*Countryside Council for Wales, Plas Penrhos, Fffordd Penrhos, Bangor, Gwynedd, LL57 2BQ*
*clive.hurford@serapias.net*

## 1.    INTRODUCTION

This part of the book outlines a logical process for developing an efficient and reliable monitoring project. The process has several distinct phases: some of these are to do with making important management decisions; the others are related more to collecting monitoring data. Without making the necessary management decisions, it becomes more and more difficult to develop the monitoring project. The key steps in the process are:

1. Identifying the conservation priority on the site;
2. Incorporating existing knowledge from research;
3. Developing a conservation strategy;
4. Developing site-specific condition indicators;
5. Deciding where to monitor;
6. Collecting the monitoring data;
7. Using photography to support the monitoring data; and
8. Storing the monitoring data in GIS.

Steps 1-4 are essentially preparatory phases, while Steps 5-8 are implementation phases. Excepting the inclusion of a chapter that discusses ways to minimise observer error, this part of the book follows the sequence above.

Figures 6-1 and 6-2 show the sequence of management and monitoring actions associated with the maintenance and restoration phases of habitat management: these figures, adapted from Brown (2001), represent the practical application of the model described in Chapter 2. We should consider these management phases as separate exercises.

*C. Hurford & M. Schneider (eds.), Monitoring Nature Conservation in Cultural Habitats, 57–59.*
© 2007 *Springer.*

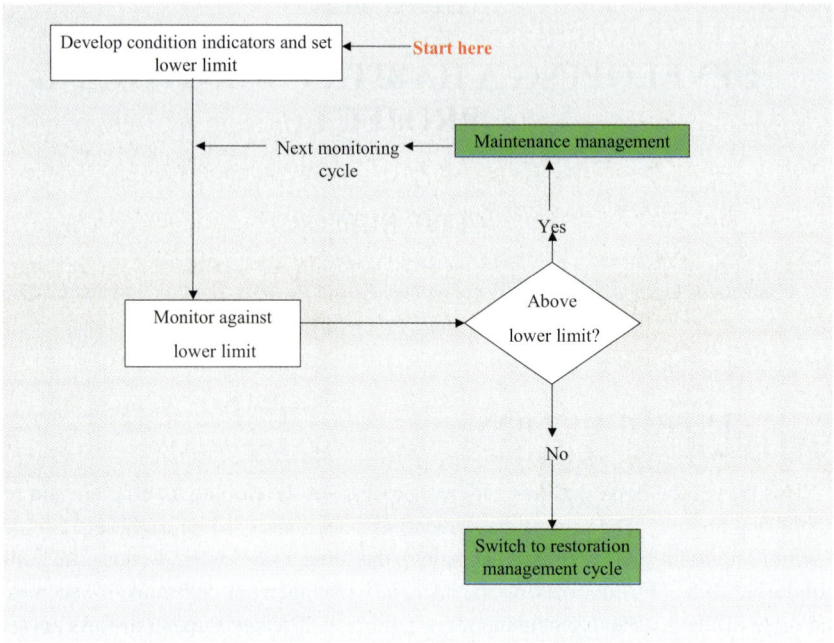

*Figure 6-1.* The monitoring and management cycle when the habitat is in optimal condition.

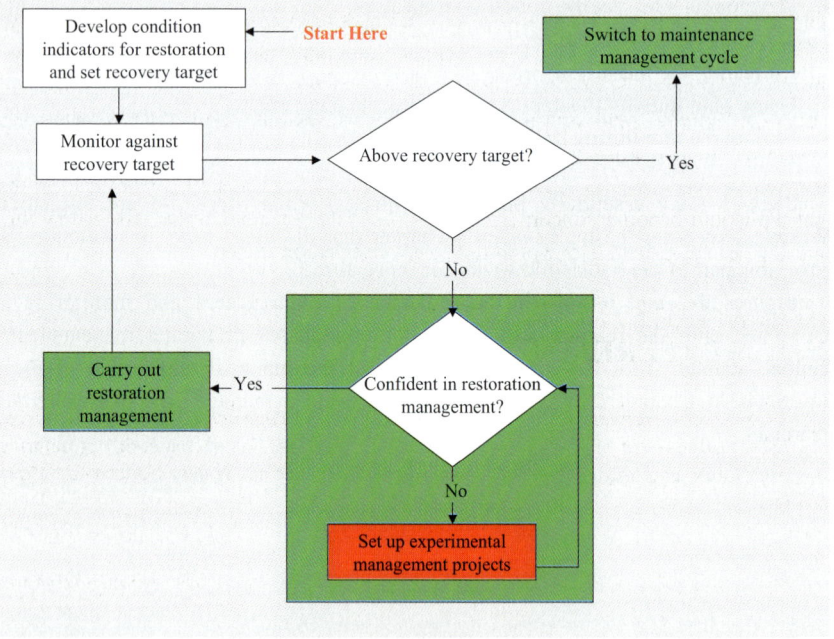

*Figure 6-2.* The monitoring and management cycle when a habitat is being restored.

*Photograph by Clive Hurford.*

*Figure 6-3.* Grazing exclosures, like this one near Cwmystwyth in central Wales, are an easy way of testing the potential for the restoration of heath vegetation. This exclosure had been erected only four years previously.

If we know, or suspect, that a habitat is in optimal condition then we monitor against a lower limit (Fig. 6-1): this identifies the point that we would become concerned that the habitat was degrading. If the habitat passes the condition indicator criteria, we stay on the maintenance management cycle, and remain there until the condition falls below the lower limit. When this happens, we switch to the restoration management and monitoring cycle (Fig. 6-2). We then monitor against, a recovery target until the habitat has been restored.

If we are uncertain of the condition before the first monitoring event, we should assume that the habitat or species is in optimal condition and set a lower limit for the point that we would become concerned. If the condition of the habitat falls below the lower limit, we can assume that it would not pass a restoration target.

## 2. REFERENCES

Brown, A. (2000). *Habitat Monitoring for Conservation Management and Reporting. 3: Technical Guide*. Life-Nature Project no LIFE95 NAT/UK/000821. Integrating monitoring with management planning: a demonstration of good practice in Wales. Countryside Council for Wales, Bangor.

# CHAPTER 7

# IDENTIFYING THE CONSERVATION PRIORITY

## Using limited resources to best effect

CLIVE HURFORD

*Countryside Council for Wales, Plas Penrhos, Ffordd Penrhos, Bangor, Gwynedd, LL57 2BQ*
*clive.hurford@serapias.net*

## 1.  INTRODUCTION

Conservation organisations are generally responsible for directly managing, or overseeing the management on, many sites of conservation interest. Without exception, the resources available for doing this are limited and it is essential that they are used wisely.  Despite this, I have seen only one management plan that that lists the key habitats and species on a site in priority order (Usher, 1973).  This plan, which was developed for Aberlady Bay in Scotland, made it absolutely clear where to focus the management and monitoring effort if resources were limited.

I suspect that this general lack of enthusiasm for formalised prioritisation in conservation stems from a) a reluctance to actively discriminate against anything that could be considered of any conservation value, and b) an inability to agree what the conservation priority should be.  Yet whenever we commit resources for management, inadvertently we are benefiting some habitats and species and discriminating against others.  This indirect and *ad hoc* form of prioritisation can lead to:

- Management discontinuity, arising from changes of conservation manager;
- Consistent discrimination against rare and threatened habitats; and
- Ineffective conservation management for every habitat and species on our sites.

Management discontinuity is a major problem for conservation as, given the option of where to focus the management effort, conservation managers will drift naturally towards their personal preferences.  There have been countless examples of management shifts as a result of changes in conservation manager.  Therefore, in the absence of a clear guidance for prioritising our conservation management and monitoring effort, we will only deliver effective conservation by chance.

*C. Hurford & M. Schneider (eds.), Monitoring Nature Conservation in Cultural Habitats, 61–64.*
© 2007 *Springer.*

## 1.1    The scale of the problem

In Wales alone, more than 6000 habitats and species of national and international importance have been identified on more than 1000 (mostly small and fragmented) protected sites. A medium-sized site of c.500 ha could be expected to host 30 or more habitats or species of international or national conservation concern. In practice, however, we do not have the resources to deliver effective conservation management or monitoring for even one priority habitat or species on each of these sites. Therefore, we must priorities our our resources to best effect, and plan to maintain biodiversity across sites at a national level – rather than within individual sites. The alternative is to risk conserving nothing.

## 2.    DEVELOPING A SYSTEM FOR PRIORITISING CONSERVATION MANAGEMENT AND MONITORING

Faced with so many threatened habitats and species and limited resources for management and monitoring, we need to develop a system that identifies a) the habitat or species of primary conservation importance on each site, and b) the sites of primary conservation importance.

One way we can do this is to develop a scoring system, based on an overview of the international and national conservation resource: this would dispassionately identify the conservation priority on each site.

The EC Habitats Directive (Council Directive 92/43) already provides an international overview: the rare and threatened habitats listed in Annex I of the Directive and the threatened species in Annex II. Within these annexes, the Directive draws attention to habitats and species that:

*   Are an international priority, wherever they are; and
*   Each member state has a special responsibility to conserve.

On top of this, in the UK, the Joint Nature Conservation Committee (JNCC) has introduced a grading system (A to D) that allows us to prioritise according to the status of the habitat or species on individual sites. In theory at least, this means that the sites designated under the Natura 2000 legislation should already have clear conservation priorities.

If we can provide a similar overview of the national resource, we will be able to use this to grade the habitats and species on our sites according to their international and national status. An example might be:

*   Grade 1 – internationally and nationally rare
*   Grade 2 – internationally and nationally rare but locally common
*   Grade 3 – internationally rare but nationally common
*   Grade 4 – widely distributed internationally and nationally but locally uncommon
*   Grade 5 – widely distributed and common throughout.

Table 7-1 shows an example of a weighted scoring system based on this type of information that could be used to identify the conservation priority on a site. The conservation managers would independently assess each of the habitats and species designated of conservation value on each site against the values attributed to international and national importance, taking account of how each interest feature contributes to the national resource. A similar scoring system could be developed for sites with habitats and species of regional importance.

*Table 7-1.* An example of a scoring system to identify the conservation management priority on protected sites in the UK. This example assesses the conservation value of the humid dune slack habitat at Kenfig NNR: a Natura 2000 Annex I habitat with two dependent Annex II species.

| Habitat conservation value assessment | | | | | |
|---|---|---|---|---|---|
| **Site: Kenfig NNR** | | | | | |
| **Habitat: Humid dune slacks** | | | | | |
| | | | | | |
| **Habitat designation** | **Value** | **Site Score** | **Dependent species designation** | **Value** | **Site Score** |
| International priority habitat and special UK responsibility | 10 | 0 | International priority species and special UK responsibility | 10 | 0 |
| International priority habitat | 9 | 0 | International priority species | 9 | 0 |
| Annex I habitat and special UK responsibility | 8 | 0 | Annex II species and special UK responsibility | 8 | 16 |
| Annex I habitat | 6 | 6 | Annex II species | 6 | 0 |
| SSSI habitat | 3 | 0 | SSSI species | 3 | 0 |
| | | | | | |
| **Area of habitat** | | | **Population size** | | |
| | | | | | |
| > 50% of national resource | 10 | 0 | > 50% of national resource | 10 | 10 |
| 26-50% of national resource | 8 | 8 | 26-50% of national resource | 8 | 0 |
| 11-25% of national resource | 6 | 0 | 11-25% of national resource | 6 | 0 |
| 6-10% of national resource | 4 | 0 | 6-10% of national resource | 4 | 0 |
| 1-5% of national resource | 3 | 0 | 1-5% of national resource | 3 | 0 |
| <1% of national resource | 1 | 0 | <1% of national resource | 1 | 1 |
| | | | Any of above - but no threat | 0 | 0 |
| **Habitat total** | | 14 | **Dependent species total** | | 27 |
| | | | | | |
| **Overall score = 41** | | | | | |

This type of scoring system not only identifies the management priority on each site, it also allows us prioritise the resource allocation at the site level, e.g. a site where the highest overall score for a habitat was 38 would be prioritised for resources ahead of a site where the highest overall score was 25.

*Photograph by Clive Hurford*

*Figure 7-1.* The entire UK population of Yellow Whitlowgrass *Draba aizoides* is restricted to one Natura 2000 site in South Wales. The species is not a high conservation priority, however, as it is well distributed over more than 10 km of the South Gower coastline and under no immediate threat.

This would ensure that the most important sites, at least, are resourced sufficiently well to deliver effective conservation management. This approach also ensures that we do not pour resources into sites that need restoration before we have secured the management on the remaining sites of high conservation value. Note that species that are not under threat do not register a population size score: this should ensure that sites hosting rare and threatened species are prioritised for management and monitoring ahead of sites hosting rare but less vulnerable species.

Although this approach to conservation management would represent a change of culture for many conservationists, by focusing our management effort on the most important and threatened habitats, the rare and threatened species naturally associated with those habitats will benefit as a by-product of the management.

## 3.      REFERENCES

European Commission, 1992. Council Directive 92/43/EEC of 21.5.1992 on the conservation of wild habitats and of wild fauna and flora. *Official Journal of the European Communities*: No L 206: 22.7.1992.
Usher, M.B. (1973) Biological Management and Conservation: theory application and planning. Chapman and Hall.

# CHAPTER 8

# INCORPORATING KNOWLEDGE FROM RESEARCH

CLIVE HURFORD

*Countryside Council for Wales, Plas Penrhos, Ffordd Penrhos, Bangor, Gwynedd, LL57 2BQ*
*clive.hurford@serapias.net*

## 1. INTRODUCTION

Information generated by research underpins many of our management and monitoring decisions. Some of this research provides background information on the habitats or species of interest and informs our selection of attributes for monitoring. This area of research is too broad to be covered in detail here, but its influence can be seen in each of the case studies towards the end of the book.

This chapter focuses on those research projects that have had a fundamental influence on both the underlying principles and practical application of conservation monitoring outlined in this book.

## 1.1 Key areas of research for conservation management and monitoring

We rarely have access to all the information that we would like to advise our management and monitoring projects, yet most of the knowledge that we need already exists. For the purposes of conservation management and monitoring, the key areas of research deal with:

- Habitat succession;
- Species co-existence;
- Understanding how species respond to disturbance and stress; and
- Habitat management practices.

*C. Hurford & M. Schneider (eds.), Monitoring Nature Conservation in Cultural Habitats, 65–72.*
© 2007 *Springer.*

Much of this information is either already available, or is relatively easy to collect, particularly for rare or threatened habitats and species. The following sections draw attention to the main sources of this knowledge.

## 2.       HABITAT SUCCESSION

The basic process of habitat succession was well documented by ecologists during the $20^{th}$ century. Some of the earlier works, e.g. Clements (1916) and Watt (1947), described the process of temporal change in vegetation, while more recent studies have focused on the underlying causes (Grime, 2001). Farmers and foresters, of course, have understood habitat succession, and how to control it, for thousands of years.

Habitat succession occurs in two basic forms: as primary and secondary succession (Grime *et al.*, 1988 and 1990). Primary succession occurs as a result of natural processes, and starts with the colonisation of bare substrates, such as rock or sand. Examples of primary succession would be the development of forest in the land-rise area along the Baltic coast (Fig. 8-1 and Fig. 8-2), or the development of sand dune vegetation from newly accreted sand. In the absence of disturbance, we would expect successional processes to run their course, driven by aggressive competitive species. A gross oversimplification of the process would see bare ground colonised by lichens and bryophytes, and subsequently develop into grassland, then scrub and finally woodland.

Secondary succession differs from this by originating from disturbances (such as fires, wind-throw and ploughing) to already established habitats. The critical difference is that these disturbances unlock nutrients that are present in the biomass, litter and soil: these nutrients are not readily available during the early stages of primary succession. Consequently, secondary succession is usually a much faster process than primary succession. For example, in sand dunes it can take in the region of 15 to 20 years for dune slack habitat to progress from newly accreted sand to the herb-rich phase of development favoured by Fen Orchids *Liparis loeselii*, whereas if you use close mowing to create open ground from mature slack vegetation, the sward will have closed again within c.5 years.

Traditionally, habitat management has been about arresting successional processes to deliver habitats of commercial value, e.g. for food, maintaining domestic stock, fuel, buildings and clothing. The current landscape of Europe has been shaped by these needs. Nature conservation played no part in the development of the countryside before the $20^{th}$ century and, with a few notable exceptions, has had a limited influence since.

In most habitats, farmers are striving to prevent the development of secondary succession. This is not always the case, however, as foresters will also take management control of habitats that have arisen through primary succession as soon as they become of commercial value. Using the primary succession forests as an example, the climax phase of the succession would be virgin Western Taiga forest (Svensson, 2002), an increasingly rare and ecologically valuable habitat. However, as soon *Pinus sylvestris* and / or *Picea abies* become dominant, the habitat tends to come under a regime of timber management.

Leaving to one side the fanciful notion of turning the clock back thousands of years to a time when Aurochs *Bos primigenius* grazed the forests and the UK did not have

*Photograph by Clive Hurford*

*Figure 8-1.* Lichens, including species of *Cladonia*, are often the first colonisers of rocky substrates.

*Photograph by Clive Hurford*

*Figure 8-2.* Stress-tolerators such as Sea Holly *Eryngium maritimum*, Sea Spurge *Euphorbia paralias* and Marram Grass *Ammophila arenaria* are early colonisers of freshly accreted sand in embryo dune habitats.

*Photograph by Clive Hurford*

*Figure 8-3.* Primary succession forest at Drivören, an exposed shore on the rising coastline of the Bothnian Sea. This photograph shows the relatively recently exposed rocky substrate, which was initially colonised by lichens, and subsequently by Grey Alder *Alnus incana* and Birch *Betula* spp., before Norway Spruce *Picea abies*. In places, these phases of the succession can be found within a 20-30 m of the sea (Svensson, 2002).

60 million human inhabitants, most conservationists would see the partial restoration of farmland and forest to something approaching pre-1920s conservation value as a reasonable conservation goal. The land management at that time was, albeit inadvertently, producing habitats of high conservation value that could support many species that have since suffered dramatic declines. However, this goal can only be achieved if we manage the habitats sympathetically for the species that we value. Removing the management disturbances altogether would simply open the way for aggressive competitive species to drive successional processes, and these species will not necessarily be native species. This management option would continue to discriminate against many of the habitats and species that we value and lead to further conservation losses.

For monitoring purposes, if we know that a habitat is not being actively managed, then our selection of condition indicators should focus on the species most likely to be lost, and the species most likely to replace them as a result of habitat succession.

## 3.      SPECIES CO-EXISTENCE

The most efficient and reliable monitoring projects focus on key groups of co-existing species. These co-existing species may be a) known associates of a rare species, or b) indicative of an important phase in habitat development. Many of the monitoring case studies in this book focus attention on a small group of species that we would expect to co-exist if the habitat (or species) was in optimal condition. The selection of these species was informed by a combination of published habitat descriptions and site-based surveys carried out specifically to determine which of the species in the published texts were present on the site being monitored.

Any publications that describe plant communities or habitats, e.g. Rívas-Martinez (1984), Polunin & Walters (1985), Rodwell (1991 *et seq.*), are underpinned by studies of species co-existence. For practical purposes, it is important to understand that these texts will have drawn together data from many different sites, and that any one site would be expected to host only a subset of the species mentioned in the text. Furthermore, the data sets will not have drawn on information from every site, so it is possible that some of the species that co-exist on your site may not even be mentioned: Chapter 11 suggests ways of overcoming these problems. These plant community texts summarise the information gathered through a vast number of research and survey projects, and (through the associated reference sections) can point you in the direction of research papers that are perhaps more relevant to your site.

If we are serious about delivering sites of high conservation value, then we must consider a) how the component species of a habitat co-exist during different phases of habitat succession and under different stresses and disturbances, and b) which species (and not just plants) should be associated with the habitat. We should not consider a habitat to be in optimal condition if the scarce plants, mammals, birds and invertebrates etc. that should be associated with the habitat are no longer present. A coherent management strategy will consider the potential for species co-existence on the site from the outset, and will be stronger for considering the key species as components of the habitats, and not as independent entities.

## 4. UNDERSTANDING HOW PLANT SPECIES RESPOND TO DISTURBANCE AND STRESS

We can build on our understanding of which species co-exist during different phases of habitat development by identifying the species that are most likely to respond first to adverse management impacts. This helps us to identify condition indicators that can be adapted to take account of local distinctiveness.

There are several publications that can help us to understand how plants respond to a range of different stresses and disturbances: those by Grime *et al.* (1988) and Ellenberg (1978, 1988 and 1991) are probably the most significant. Used together, these texts are a valuable aid to the selection of condition indicators for monitoring.

## 4.1 The C-S-R model

Grime *et al.* (1988) allocate most of the commoner British vascular plant species to a position within the C-S-R model. In this model, each species is classified into one of seven categories: three primary categories and four intermediates.  The primary categories are a) competitors, b) stress tolerators, and c) ruderals.

Grime also suggests that the factors that limit the amount of living and dead plant material in any habitat can be classified into two broad categories: stress and disturbance.

Stress is defined as phenomena that restrict photosynthetic production, e.g. temperature and shortages of light, water and available nutrients.  Disturbance is associated more with the partial or total destruction of the plant biomass by herbivores, pathogens, and management activities such as grazing, ploughing, mowing and burning, as well as through natural phenomena such as wind damage, frosting, fire damage, droughting and soil erosion (Grime, 2001).  In general terms:

- Ruderals exploit conditions of low stress and high disturbance, e.g. arable weeds;
- Stress tolerators exploit conditions of high stress and low disturbance, e.g. species that occupy densely shaded forest floors or highly acidic soils; and
- Competitors exploit conditions of low stress and low disturbance, e.g. species with a high growth rate and the capacity for vigorous lateral spread above and below ground.

Many species adopt more than one of these strategies.  For example, several canopy forming tree species are competitors when mature, but stress tolerators as seedlings and saplings (CS strategists).  Other species, including many perennial herbs, can exhibit traits of all three strategies (CSR strategists) depending to some degree on the environmental conditions.  The CSR strategists are very much a 'mixed-bag' of species, and for the purposes of conservation management at least, need to be classified further according to their primary strategy in habitats of high conservation value.  For example, *Dactylis glomerata* is classed as a CSR strategist, and this is understandable in as much as it can:

- Colonise open ground, e.g. roadsides, and behave as a ruderal species;
- Survive disturbances and stresses in established habitats, e.g. in droughted grassland; and
- Compete strongly by comparison with many other ruderals and stress tolerators.

However, in low disturbance situations, e.g. under-managed hay meadows, *Dactylis glomerata* becomes an aggressive competitor.  So, although it is accurately classed as a CSR strategist, in hay meadows *Dactylis glomerata* behaves primarily as a competitor.

Thinking in these terms is particularly useful when considering the management of cultural habitats, as the disturbances in these habitats are caused by human activities. Traditionally, management disturbance, e.g. through ploughing, grazing, cutting and burning, would have favoured species with the traits of ruderals or stress tolerators, and discriminated against competitors.  However, agricultural developments over the past century, particularly with respect to herbicide, pesticide and fertiliser applications, have given farmers far greater control of their land and produce.

For example, in the early 1900s, ploughing land created the opportunity for colonisation by ruderal species, whereas improved herbicide and pesticide applications now suppress that growth, and crops can develop unhindered by weeds.  Similarly, in hay meadows and pastures, where management was traditionally intended to discriminate against succession to competitive grasses and scrub, re-seeding with competitive strains of grass and the application of improved fertilisers has given competitive grasses the capacity for rapid growth: this has led to the exclusion of the stress tolerators that formerly co-existed with them. Few ruderals or stress tolerators have found a way of adapting to these advances in farming practices.

## 4.2    Ellenberg's indicator values

Over the period from 1978 to 1991, Ellenberg produced a series of publications on indicator values for the vascular plants of central Europe. This research complements Grime's work on the C-S-R model, by giving scores for the tolerance of each species to five key environmental stresses: light, moisture, pH (soil or water, as appropriate), nitrogen (as an indicator of soil fertility) and salt. These indicator values, which have been used extensively in central Europe and adjacent countries, have also been adjusted for the British flora (Hill *et al.* 1999).

This information gives us the ability to identify indicator species that are sensitive to the key stresses on our sites. For example, if our site is a mature forest on neutral soils, and our main concern is increasing acidity as a result of atmospheric deposition, we could collate a list of the species in our ground-flora and use the Ellenberg values to identify those species sensitive to increases in acidity but tolerant of shading. Species that meet these criteria would be potentially good indicators of acidification on that site.

## 5.    HABITAT MANAGEMENT PRACTICES

While this subject area is mostly beyond the scope of this book, it is important for conservation managers and monitoring specialists to understand how the vegetation and its associated species are likely to respond to different management practices. To some degree, the likely effects on the vegetation can be predicted by combining the knowledge of a) the management activity, b) the species that comprise the habitat, and c) the respective C-S-R strategies of these species with respect to disturbance.

Furthermore, there are many texts available to guide practical conservation management, e.g. Bullock & Pakeman (1997), Fletcher *et al.* (2001), Rowell (1988), Kirby (1992), Jones *et al.* (1996), and Sutherland & Hill (1995). These texts are complemented by other publications that offer a fascinating insight to how the current British countryside evolved, e.g. Rackham (1986).

## 6.    IN SUMMARY

Standard texts based on extensive research have provided us with the knowledge to recognise:
- The groups of co-existing species that form habitats;
- The likely direction of habitat succession in the absence of disturbance events;
- How individual species are likely to respond to disturbance events;
- How individual species are likely to respond to different environmental stresses;

- The types of management most likely to be applied to different habitats.

This information makes it possible to identify small suites of co-existing species that can be monitored efficiently and give reliable results. The remaining chapters in this part of the book outline a process for developing site-specific habitat monitoring projects, and the case studies illustrate how this knowledge has been applied at a site-specific level in monitoring projects on sites of international conservation importance.

# 7.    REFERENCES

Bullock, J. M. & Pakeman, R. J. (1997). Grazing of lowland heath in England: Management methods and their effects on heathland vegetation. *Biological Conservation* **79**: 1-13.

Clements, F.E. (1916). *Plant Succession. An Analysis of the Development of Vegetation.* Carnegie Institute. Washington.

Ellenberg, H. (1979). Zeigerverte von Gefässpflanzen Mitteleuropas. *Scripta Geobotanica* **9**: 1-122.

Ellenberg, H. (1988). *Vegetation Ecology of Central Europe, 4th Edition.* Cambridge University Press. Cambridge.

Ellenberg, H., Weber, H.E., Düll, R., Wirth, V., Werner, W. & Paulisson, D. (1991). Zeigerverte von Pflanzen in Mitteleuropa. *Scripta Geobotanica* **18**: 1-248.

Fletcher, A., Wolseley, P. & Woods, R. (2001). *Lichen Habitat Management.* British Lichen Society. London.

Grime, J.P., Hodgson, J.G. & Hunt, R. (1988). *Comparative plant ecology: a functional approach to common British species.* Unwin Hyman. London.

Grime, J.P., Hodgson, J.G. & Hunt, R. (1990). *The abridged comparative plant ecology.* Chapman & Hall. London.

Grime. J.P. (2001). *Plant Strategies, Vegetation Processes and Ecosystem Properties (2nd Ed.).* John Wiley & Sons. Chichester.

Hill, M.O., Mountford, J.O., Roy, D.B. & Bunce, R.G.H. (1999). *Ellenberg's indicator values for British plants.* ECOFACT Volume 2. London Department of the Environment, Transport and the Regions.

Jones, P.S. *et al.* (eds). (1996). *Studies in European coastal management.* Samara Publishing Ltd, Cardigan.

Kirby, P. (1992). *Habitat Management for Invertebrates: a practical handbook.* Royal Society for the Protection of Birds. Sandy.

Polunin, O. & Walters, M. (1985). *A Guide to the Vegetation of Britain and Europe.* Oxford University Press. Oxford.

Rackham, O. (1986). *The History of the Countryside.* J.M. Dent & Sons Ltd. London.

Rivas-Martínez, S., Díaz, T.E., Prieto, J.A.F., Loídi, J. & Penas, A. (1984). *Los Picos de Europa: la vegetacion de la alta montaña de Cantabrica.* Ediciones Leonesas. León.

Rodwell, J.S. (1991 *et seq.*). *British Plant Communities. Vols.1-5.* Cambridge University Press, Cambridge.

Rowell, T.A. (1988). *The Peatland Management Handbook.* Research and Survey in Nature Conservation, No.14. Nature Conservancy Council, Peterborough.

Sutherland, W.J. & Hill, D.A. (1995). *Managing Habitats for Conservation.* Cambridge University Press. Cambridge.

Svensson. J. (2002). *Succession and Dynamics of Norway Spruce Communities on Gulf of Bothnia Rising Coastlines.* Doctor's Dissertation. ISSN 1401-6230, ISBN 91-576-6323-8.

Watt, A.S. (1947). Pattern and process in the plant community. *Journal of Ecology* **35**: 1-22.

# CHAPTER 9

# DEVELOPING A CONSERVATION STRATEGY

CLIVE HURFORD

*Countryside Council for Wales, Plas Penrhos, Ffordd Penrhos, Bangor, Gwynedd, LL57 2BQ*
*clive.hurford@serapias.net*

## 1.    INTRODUCTION

For monitoring purposes, we should think of our rare and threatened habitats and species as vulnerable patients exposed to a life threatening illness, and approach the problem with the same mindset as a doctor checking either for signs of ill health or for complications during recovery. After all, we will already know, from previous research:

- How to recognise if there is a problem with the habitat;
- The likely cause of any problems;
- The potential for restoring the habitat; and (if it can be restored)
- The management needed to restore it.

Ultimately, we need monitoring to provide us with an early warning when the condition of a habitat or species starts to deteriorate, or with evidence of its recovery through restoration management. We do not need to resort to complex statistical tests to provide this information, any more than a doctor does when monitoring the recovery of a patient. We simply need to look for the right signs in the right places.

Before we can do this, however, the conservation manager must decide what their management is aiming to achieve and where. The worst-case scenario is that we do not have enough information to make these management decisions, in which case, we have to go out and collect it. Knowing where the habitat is and what state it is in is a fundamental prerequisite for a responsible conservation management strategy. We will not be able to plan or carry out appropriate management without this information.

In this chapter, we describe the process for producing a habitat quality map. This type of survey has proved to be an effective tool for facilitating well-informed conservation strategies on small, medium and large sites of conservation value, and is a valuable aid to monitoring.

*C. Hurford & M. Schneider (eds.), Monitoring Nature Conservation in Cultural Habitats, 73–77.*
© 2007 *Springer.*

## 2.        PRODUCING HABITAT QUALITY MAPS

A habitat quality map is the product of a 'one-off' survey designed to provide a general impression of the extent and distribution of the key habitat/s on a site, and to indicate how much of that habitat is of high conservation value. With this information we can develop a monitoring strategy that focuses on critical areas. Because we can be flexible over the amount of data we collect, this type of survey is relatively efficient at gathering the essential information. For example, we could:

- Map the distribution of the key habitat on our site and allocate each habitat patch to a pre-defined habitat quality class; or
- Map the distribution of the broad habitat types on our site but map quality classes for only the key habitat; or
- Provide quality definitions for all of the habitats on our site and map the distribution of each habitat class.

Often, we will already have a map or image that shows the approximate distribution of the broad habitats on the site, if this is the case, we can simply revisit the areas of key habitat and add a layer of quality information. Even without a habitat map, we could obtain a remote image of the site and eliminate areas that are definitely not the key habitat. As most conservation sites are in the medium size range (400 to 800 ha) or smaller, habitat quality surveys should not be a major drain on resources.

## 2.1    The survey method

The first phase of a habitat quality survey is critical, and involves generating site-specific definitions of the key habitat states: this requires familiarity with the key habitat, good botanical field skills, and an understanding of successional processes and likely management impacts. If you are unfamiliar with the site or habitats, then you should seek help to identify and define the appropriate habitat classes. One way of generating these definitions is to visit the site and describe the vegetation from stands of the habitat that are visually different in terms of species composition, structure, or both. Eventually, you will reach the point where virtually all the vegetation on the site can be allocated to the habitat classes you have defined. Table 9-1 shows the habitat class definitions for the dune slack vegetation at Kenfig Pool: similar definitions were generated for the dune grassland.

After defining the habitat quality classes, the rest of the survey is straightforward: you simply take the habitat quality definitions and an aerial photograph out on site and assign each habitat patch to the appropriate habitat and habitat class.

The habitat quality map in Fig. 10-1 shows the approximate extent and distribution of the habitat quality classes for dune grassland and dune slack vegetation at Kenfig Pool, a Natura 2000 site of ca. 450 ha in south Wales. The conservation management priority at Kenfig is the successionally-young humid dune slack vegetation, which supports two internationally rare species: the Fen Orchid *Liparis loeselii* (Figure 9-2) and the Petalwort *Petalophyllum ralfsii*. However, as virtually all of the sand dune vegetation at Kenfig is considered to be of international conservation value, we decided to map the

*Table 9-1.* The site-specific habitat quality classes for dune slack vegetation at Kenfig Pool.

| Vegetation type | Definitions of dune slack quality at Kenfig |
|---|---|
| **Embryo dune slack** | Some 25-50% open ground present in the immediate area of an active blow-out, *Salix repens* occurring in distinct clonal patches, with *Carex arenaria* an obvious associate and with either *Sagina nodosa* or *Juncus articulatus* present within 2 m of any point in the stand |
| **Successionally-young slack** | A mosaic of patchy bare soil with thalloid liverworts and low, closed vegetation, with patches of moss cover, mostly of *Campylium stellatum* or *Calliergon cuspidatum*, but bryophytes not forming a dense mat. *Salix repens* can be abundant but not canopy forming and grasses should be generally scarce. At least two of, *Carex viridula, Juncus articulatus, Anagallis tenella, Samolus valerandi* and *Eleocharis quinqueflora* should be present within 2 m of open soil, and *Liparis loeselii* may also be present with other orchids, e.g. *Epipactis palustris, Dactylorhiza praetermissa, D.incarnata* |
| **Orchid-rich slack** | Little or no bare soil evident, though orchids, e.g. *Dactylorhiza praetermissa, D. incarnata, Epipactis palustris* and *Gymnadenia conopsea* are patchily common. *Salix repens* can be canopy forming and *Calliergon cuspidatum* can form dense mats in places. At least two of *Holcus lanatus, Poa subcaerulea, Pyrola rotundifolia* and *Galium palustre* should be present within 2 m of any point, while *Phragmites australis, Calamagrostis epigejos* and *Molinia caerulea* can be evident, none of these will be dominant |
| **Species-poor wet slack** | Either as above, but species-poor with few orchids, or *Salix repens* co-dominant with *Carex nigra*, typically with a dense cover of *Hydrocotyle vulgaris* under the *Salix repens* |
| **Dry slack** | A drier, species-poor, slack type, where *Salix repens* forms a shrubby canopy with *Holcus lanatus* and *Festuca rubra* notable among the associates, prone to invasion by *Betula* and taller *Salix* shrubs e.g. *S. cinerea* or *S. caprea*. Slacks where grasses such as *Festuca rubra* and *Elymus repens* are locally co-dominant with the *Salix repens* should be placed in this category |
| **Brackish slack** | Stands of *Juncus maritimus* present |
| **Single species stands** | *Calamagrostis epigejos* or *Phragmites australis* forming dense stands or *Molinia caerulea* tussocks dominant. |

quality of each habitat across the whole site. Two students from University of Wales Swansea carried out the habitat quality survey at Kenfig, as part of their MSc theses (Aubrey, 1997, and Besley, 1997). This survey took approximately three weeks to complete: for comparison, it took an experienced botanist two summers to complete a plant community (NVC) survey of the site. It is worth noting that a similar exercise carried out in dune habitats at Whiteford Burrows the following summer required a different set of habitat class definitions (Cooper, 1998).

The map shows that, while there was no shortage of fixed dune grassland or humid dune slack vegetation at Kenfig, only small fragmented patches of successionally-young dune slack habitat were suitable for colonisation by either *Liparis* or *Petalophyllum*. We already knew (from aerial photographic surveillance) that no new areas of open sand were being created and that the dunes were in an advanced stage of stabilisation. Therefore, the options were limited, the site manager could either a) attempt to recreate new areas of successionally-young slack vegetation in the parts of the site that have matured, or b) attempt to arrest the succession in the remaining habitat patches with a view to gradually increasing their extent.  In the event, he decided on the latter option, i.e. to secure what was already there and attempt to build on it.  If he had taken the first option, he risked losing the remaining fragments of successionally-young slack vegetation while attempting, perhaps unsuccessfully, to recreate the habitat elsewhere.

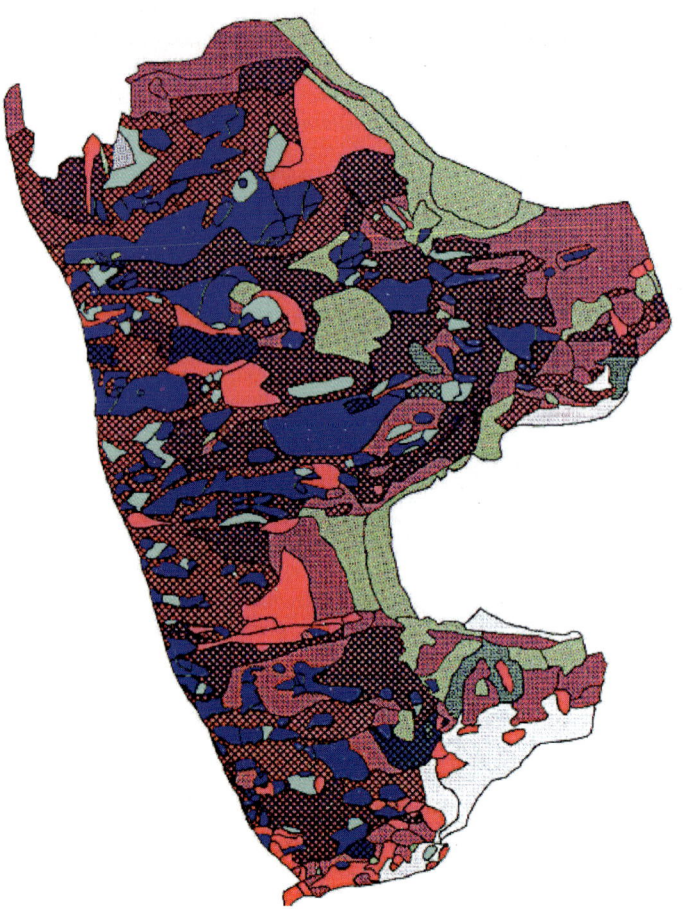

*Figure 9-1.*  A habitat quality map of Kenfig Dunes in South Wales.  The successionally-young slack vegetation is the light turquoise colour on the map, while the more mature phases of dune slack development are represented by the darker blue polygons.  The successionally-young grassland is mapped as bright red.

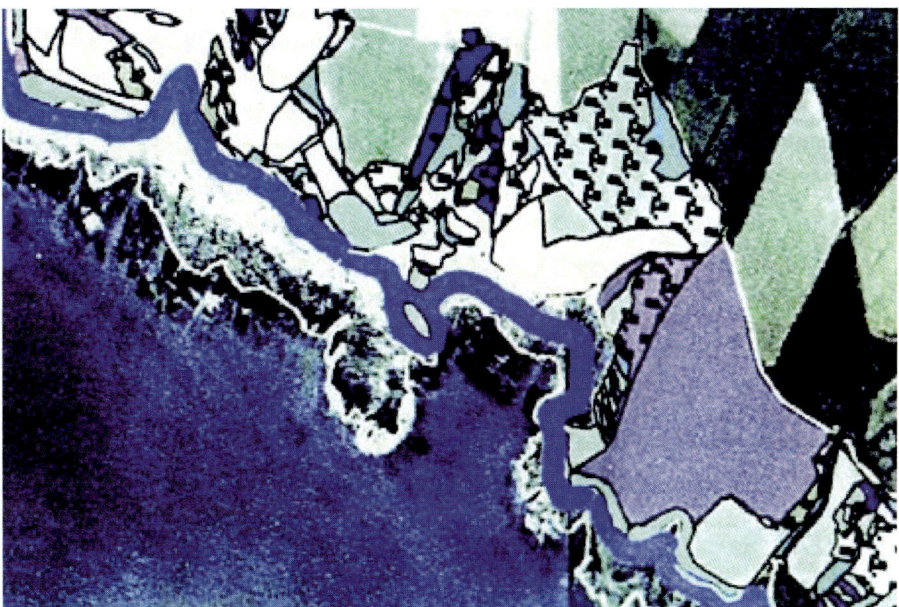

*Figure 9-2.* In a similar habitat quality mapping exercise on the Gower Peninsula in south-west Wales, Walker (1999) and Somers (1999) mapped the various classes of grassland vegetation directly onto aerial photographs.

## 3.    DISCUSSION

The habitat quality survey of Kenfig was designed for one purpose and one purpose only: to facilitate a conservation management strategy for the site. We made no real attempt to map the precise extent of and distribution of the various habitat classes; our only aim was to gain an impression of the distribution and relative proportions of the key habitat classes. On this basis, we would recommend not attempting to derive habitat class measurements from the map, or using the map as any sort of baseline for future comparison. The map had served its purpose as soon as the conservation strategy was in place.

## 4.    REFERENCES

Aubrey, M. (1997). Habitat mapping and TWINSPAN analysis of dune grasslands at Kenfig National Nature Reserve. MSc Thesis, University of Wales Swansea.

Besley, K. (1997). Habitat mapping and TWINSPAN analysis of dune slacks at Kenfig National Nature Reserve. MSc Thesis, University of Wales Swansea.

Cooper, A. (1998). Can generic keys be applied to map the condition of humid dune slack vegetation on dune systems in South Wales? MSc Thesis, University of Wales Swansea.

Somers, L. (1999). Can Common Standards Monitoring be used to assess the condition of limestone grassland at the West Glamorgan Nature Reserve at Port Eynon? (MSc thesis). University of Wales Swansea.

Walker, K. (1999). Can Common Standards Monitoring be used to assess the condition of limestone grassland in the South Gower NNR at Rhosilli part of the SAC? (MSc thesis). University of Wales Swansea.

# CHAPTER 10

# MINIMISING OBSERVER ERROR

Increasing the reliability of a monitoring project

CLIVE HURFORD

*Countryside Council for Wales, Plas Penrhos, Ffordd Penrhos, Bangor, Gwynedd, LL57 2BQ*
*clive.hurford@serapias.net*

## 1.    INTRODUCTION

As the monitoring result has a direct influence on the way that a site is managed, a monitoring project must provide the same result regardless of who does the monitoring, and that result must be the right result.

To achieve this level of consistency, we must minimise the opportunities for observer bias during the data collection phase of the monitoring project. In practice, this means giving careful consideration to the respective merits of recording cover, structure, abundance and frequency: these are the measurements that we are most likely to use for habitat monitoring. The following sections use the results from sampling trials to show where observer bias is most likely to arise, and to illustrate ways of minimising it.

## 1.1    Background information on the sampling trial data

Over the period from 1996 to 2004, members of the CCW monitoring team carried out a series of multiple-observer sampling trials to assess the degree of observer bias attached to measures commonly used in vegetation survey and surveillance projects. These sampling trials were carried out in a variety of habitats, and by observers with varying degrees of field experience, e.g. university students, conservation site managers, habitat specialists and professional field surveyors.

During the sampling trials, a series of observers assessed the same attributes at fixed sample points. The number of observers involved in each sampling trial ranged from eight to 20. The main measures that we tested were 1) estimates of percentage cover, 2) recording cover / abundance scores using the Domin scale; 3) assessing the vegetation against cover targets, and 4) and recording species frequency.

*C. Hurford & M. Schneider (eds.), Monitoring Nature Conservation in Cultural Habitats, 79–92.*
© 2007 *Springer.*

These results from these trials build on information gathered during an exercise carried out by the English Field Unit (Leach & Doarks, 1991), which looked at the degree of variation between experienced vegetation surveyors when compiling a species list (and abundance scores) for two fixed quadrats (see 3.1).

The data sets that I have used to demonstrate observer bias are typical of the results obtained from the sampling trials: the only modification I have made is to remove one outlier data set from each set of results. In every case, the inclusion of the outlier data set would have increased the range of observer variation in the results, not reduced it. Over time we have come to realise that you can train three out of four people to be vegetation surveyors: the fourth person should be actively discouraged from pursing this career path.

## 2. RECORDING ESTIMATES OF VEGETATION COVER

If you decide to record estimates of vegetation cover in your monitoring project, then you must consider which form of cover assessment to use; there are several options. The commonest forms of cover assessment in current usage are subjective measures:
- Estimates of percentage cover;
- The Domin cover scale; and
- The Braun-Blanquet cover scale.

Two other options are pin-frame recording and estimating against cover targets. Some researchers favour pin-frame recording, because it is an objective measure with good statistical properties. The method is too time consuming, however, to be used for monitoring large areas of vegetation. Conversely, cover targets have been used in many CCW monitoring projects, and this form of assessment is covered in the sections below.

### 2.1 Estimates of percentage cover

Ecologists have been uncomfortable with the use of straightforward estimates of percentage cover for some time. This discomfort gave rise to the Braun-Blanquet and Domin cover scales, which are assessed in 2.2.

The main problems with recording percentage cover estimates arise simply because it is a subjective measure, and can be influenced by a number of variables, e.g.:
- The familiarity of the observer with the habitat or species being assessed;
- The size of the area of search;
- The complexity of the vegetation; and
- The structure of the vegetation.

To demonstrate the degree of observer bias and subjectivity attached to recording estimates of percentage cover, I have used data from a sampling trial where the attribute being assessed, ericoid cover, was easy to identify and easy to see. The habitat was

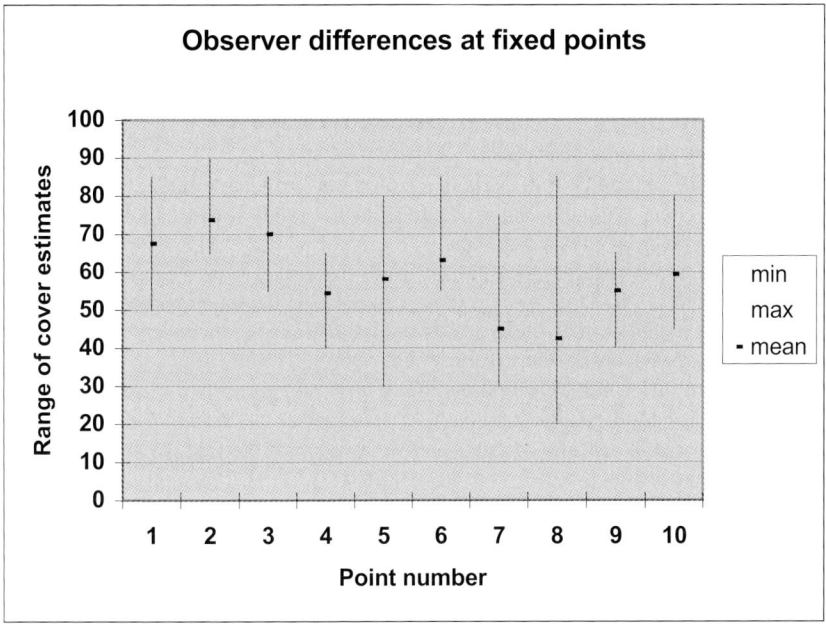

*Figure 10-1.* The results from a sampling trial to test the range of variation between observers estimating the percentage cover (at intervals of 5%) of ericoids at fixed points in blanket bog vegetation. The mean range of uncertainty was 36%.

blanket bog, which is naturally species-poor, and the species-group comprising the attribute (i.e. *Calluna vulgaris*, *Erica tetralix* and *Empetrum nigrum*) could not be confused with anything else in the sample area. In addition, the recorders were all experienced field surveyors or habitat specialists with a professional interest in that habitat type. Fig. 10-1 shows the results from this sampling trial, where the seven observers assessed the percentage cover of ericoids within a 1m-radius area of search at the same ten sample points. We marked the location of each sample point with a numbered cane.

The data set in Fig. 10-1 shows that, on average, the difference between the lowest and highest cover estimate at each sample point varied from 15% to 65%, with a mean difference of 35%. Bearing in mind that I have already removed an outlier data set, these results should be a concern for anybody recording percentage cover estimates for surveillance or monitoring purposes. In my experience, these results are not atypical of the results from multiple-observer field trials. In fact, I have seen many that are less well aligned, particularly when the attribute being assessed was either difficult to measure, such as bryophyte cover or bare ground, or difficult to separate from similar species in the search area, such as a species of grass or sedge.

The range of uncertainty associated with recording percentage cover estimates has been known, or suspected, for many years, which raises the question of why we persist with the measure. From a monitoring perspective, we cannot base habitat condition assessments on a measure where we have to ignore cover shifts of ca. 35% (either side of our estimate) to accommodate observer error: changes of that magnitude would have a dramatic impact both on the conservation value of a habitat and on the species that are associated with it

## 2.2    Domin and Braun-Blanquet cover scales

Strictly speaking, the Domin scale is a cover and abundance scale, but this is a moot point as, in practice, both the Domin and Braun-Blanquet scales (Table 10-1) are used primarily for recording vegetation cover.

Both of these cover scales have been used extensively in vegetation survey and surveillance projects, the Domin scale mostly in the UK, Braun-Blanquet more in Europe. The initial impression is that these scales reduce the scope for observer error, with the Braun-Blanquet scale being more robust than Domin.  Both scales were developed to describe vegetation for survey purposes and, within an appropriate classification system, both can do that to good effect.

However, problems arise for monitoring from two areas: the first is related to the process that the surveyors use to arrive at their cover class, the second is related to the implications attached to making an error.  I will deal with the problems that arise through the recording process first.

When a vegetation surveyor is using either of these cover scales, they initially estimate the percentage cover of the species in question, and then translate that estimate into the appropriate cover class.  The problem here, as we saw in the sampling trial results for percentage cover estimates (2.1), stems from the accuracy of the original cover estimate, which varies as a result of observer bias.

*Table 10-1.* The Domin and Braun-Blanquet scales.

| | Domin scale | | Braun-Blanquet scale |
|---|---|---|---|
| + | A single individual.  No measurable cover | + | Less than 1% cover |
| 1 | 1-2 individuals.  No measurable cover. | 1 | 1-5% cover |
| 2 | Several individuals, but less than 1% cover | 2 | 5-25% cover |
| 3 | 1-4% cover | 3 | 26-50% cover |
| 4 | 5-10% cover | 4 | 51-75% cover |
| 5 | 11-25% cover | 5 | 76-100% cover |
| 6 | 26-33% cover | | |
| 7 | 34-50% cover | | |
| 8 | 51-75% cover | | |
| 9 | 76-90% cover | | |
| 10 | 91-100%cover | | |

The second problem comes in to play when the original cover estimate is at or near the boundary between two cover classes, as then the observer is forced to reconsider whether the cover is above or below that boundary before allocating a cover class to the species. In our experience, in marginal situations, where the cover of a species is close to a boundary between two cover classes, the chance of two observers allocating the species to the same cover class is no better than 50:50.  Unfortunately, if you are using the Domin scale, you are never far away from a boundary between cover classes.  In one respect, this suggests that the Braun-Blanquet scale is better adapted for monitoring than the Domin scale, because there are less cover classes and therefore fewer boundaries. This is not necessarily true, however, because the implications attached to making an error are greater. Table 10-2 shows the results of multi-observer sampling trials designed to look at observer differences using the Domin scale to assess vegetation cover.

For both sampling trials, we fixed the location of four quadrats (two 1 x 1 m quadrats and two 50 x 50 cm quadrats) and asked the participants to record a Domin score for each of the attributes listed in the table. A different set of observers was used to record the Domin scores at each quadrat.

The results from these trials show that all three surveyors agreed on the Domin score in only two (10%) of the 20 assessments (and in one of those the attribute was absent). By contrast, the surveyors recorded three different Domin scores in seven (35%) of the 20 assessments. Also of significance, 50% of the assessments differed by more than one Domin score, with a maximum difference of five Domin scores recorded for one assessment.

If these results are an accurate reflection of observer variation using the Domin scale, and the results from other sampling trials suggest that they are, then even if there was no change in the vegetation cover at the sample points, during a repeat monitoring exercise 50% of the assessments would differ by more than one Domin score as a result of observer error. So if we want to use Domin for recording vegetation cover, we have to accept that there is a 50% chance at every sample point that there will be observer error of at least two points on the Domin scale. What are the implications of this for conservation management?

*Table 10-2.* A comparison of Domin scores from two different sampling trials. Trial 1 involved professional ecologists and Trial 2 involved university students. Observers 1, 2 and 3 recorded the Domin estimates in fixed quadrats 1-4. Trial 1 was carried out in mire vegetation and Trial 2 was in dune grassland. The blocks of data highlighted in red draw attention to assessments that differed by more than one Domin score.

| Attribute | Quadrat 1 | | | Quadrat 2 | | | Quadrat 3 | | | Quadrat 4 | | |
|---|---|---|---|---|---|---|---|---|---|---|---|---|
| | 1 | 2 | 3 | 1 | 2 | 3 | 1 | 2 | 3 | 1 | 2 | 3 |
| **Trial 1** | | | | | | | | | | | | |
| **Sphagnum cover** | 4 | 6 | 8 | 10 | 8 | 9 | 5 | 5 | 5 | 7 | 8 | 8 |
| **Grass cover** | 4 | 5 | 7 | 0 | 0 | 0 | 4 | 4 | 5 | 0 | 1 | 0 |
| **Trial 2** | | | | | | | | | | | | |
| **Moss cover** | 6 | 7 | 7 | 7 | 5 | 6 | 7 | 5 | 6 | 5 | 4 | 4 |
| **Grass cover** | 4 | 5 | 4 | 4 | 4 | 5 | 7 | 5 | 5 | 9 | 10 | 8 |
| **Bare sand** | 7 | 6 | 7 | 8 | 5 | 6 | 2 | 0 | 0 | 2 | 0 | 0 |
| | | | | | | | | | | | | |

Most of the differences occurred within the range of Domins 4 to 8. So at the bottom end of that range, Domins 4-6, we would have to ignore any changes of cover from 4% to 33% because there would be 50% chance that there had been no change at all. Similarly, in the range from Domins 6 to 8, we would have to ignore changes in cover from 33% to 80% for the same reason. Cover changes of this magnitude are not an early warning of change: again, they would have a dramatic impact on the conservation value of any habitat. Unless we are prepared to accept this level of uncertainty, we cannot use the

Domin scale for monitoring the condition of vegetation. Furthermore, it is worth noting that the data set collected by the professional ecologists in Trial 1 was no better aligned than the data set collected by university students in Trial 2.

We have not tested the Braun-Blanquet scale, but there is no reason to believe that we would not see a similar pattern. If anything, the effects of observer error using the Braun-Blanquet scale would be even greater, because the cover bands are wider. If we had to accommodate observer differences of only one point on the Braun-Blanquet scale, for example from 2 to 3, then we would have to write-off a change from 5% to 50% cover when there may have been no change at all. Consequently, neither of these cover scales are appropriate for monitoring habitat condition.

## 2.3    Cover targets

The last estimate of vegetation cover that we have tested extensively is the assessment of cover targets, also known as 'cover pseudospecies'. These are used in multi-variate statistical analyses, such as TWINSPAN, to help to separate vegetation types with a similar species composition. To the best of my knowledge, however, cover targets had not been used in vegetation monitoring projects before 1996, when we began to look at measurability issues as part of EU/CCW Life Project (Brown, 2000; Hurford & Perry, 2000).

The concept underpinning the use of cover targets is that if we can identify the point at which the cover of the competitive or dominant species starts to impact on the occurrence of the more sensitive stress tolerating species, we only have to assess whether the cover of the competitors is greater or smaller than that value.

*Table 10-3.* The results from sampling trials to assess the scale of observer variation when recording cover targets. Observers 1, 2 and 3 assessed whether the vegetation cover of the attributes was greater than or less than 20% in fixed quadrats 1-4. If the cover of the attribute was borderline (at or around 20%), the recorders were told to record it as greater than 20%. Trial 1 was carried out in mire vegetation: Trial 2 was in dune grassland. The blocks of data highlighted in red draw attention to inconsistent assessments.

| Attribute | | Quadrat 1 | | | Quadrat 2 | | | Quadrat 3 | | | Quadrat 4 | | |
|---|---|---|---|---|---|---|---|---|---|---|---|---|---|
| | | 1 | 2 | 3 | 1 | 2 | 3 | 1 | 2 | 3 | 1 | 2 | 3 |
| <20% / >20% | | | | | | | | | | | | | |
| Sphagnum cover - Trial 1 | | > | > | > | > | > | > | > | < | > | > | > | > |
| Grass cover        - Trial 1 | | > | > | > | < | < | < | < | > | < | < | < | < |
| Moss cover        -Trial 2 | | > | > | > | > | > | > | > | > | > | < | < | < |
| Grass cover        -Trial 2 | | < | < | < | < | < | < | > | > | > | > | > | > |
| Bare sand        -Trial 2 | | > | > | > | > | > | > | < | < | < | < | < | < |
| | | | | | | | | | | | | | |

The pitfall to this approach, of course, is that you cannot just pull this value out of the air. You must have a good understanding of your site and the habitat that you are monitoring before you can decide what this cover value should be. However, an experienced field recorder should be able to obtain this value from a relatively quick survey exercise (see Chapter 11).

Table 10-3 shows the results of cover target assessments carried out by the same groups of recorders that carried out the Domin sampling trial (Table 2). I have used this data set to illustrate that, although the surveyors struggled to achieve any level of consistency recording Domin values, the same set of recorders could achieve a high level of consistency when assessing cover targets.

The results in Table 10-3 show that the surveyors provided consistent assessments in 18 of the 20 assessments (90%). This level of consistency compares favourably with the trial results for percentage cover estimates and Domin scores. These results are in keeping with those from other sampling trials, though it is not unusual for surveyors to achieve 100% consistency. This level of consistency has a price, however, because the results do not give any indication of the actual cover of the attribute, they only tell you whether it is obviously more than, or less than, the cover target. This is not a particularly high price, however, if you take account of potential impact of the observer error associated with recording estimates of percentage cover.

By focusing attention on a single boundary, cover target assessments are relatively straightforward, and will deliver consistent results, until you are on a site where the cover of the attribute is consistently close to the cover target. There are two ways to avoid an inconsistent monitoring result in this situation. ? You can:

1. Introduce a decision rule stating that, for example, 'the grass cover at the sample point must be obviously less than 20% cover: if you have to stop to think about it, then the point must fail'; and

2. Use the cover target as one of a suite of co-occurring attributes (see Chapter 11) that must all pass before the sampling point can pass. For example, you can state that before a sample point can pass, 'four positive indicator species must be present, the grass must form <20% cover, and all of the negative indicator species must be absent'.

We regularly employ both of these precautionary measures in projects developed to monitor Natura 2000 habitats. The most damaging error that we can make in a monitoring project is to say that a habitat is in good condition when it isn't. Both of these measures discriminate against making that mistake. If we are going to make an error, we will err on the side of caution, which is how it should be on sites of high conservation value.

## 3.     RECORDING SPECIES PRESENCE AND ABUNDANCE

The section looks at the observer variation associated with estimating species diversity and abundance. Management plans often include an aim to 'maintain or increase the diversity (or biodiversity) of a habitat or site'. This section looks at the options that are available for measuring this, and at the levels of observer variation associated with some of the more common recording methods that are used. The sections below include assessments of species diversity, species frequency, and species abundance. Where available, we have used sample trial data to inform our recommendations.

## 3.1    Assessing species diversity

During the 1980s, the English Field Unit carried out a sampling trial designed to assess observer variation in recording species diversity and abundance in grassland vegetation (Leach & Doarks, 1991). This trial involved 14 experienced grassland surveyors, who were asked to independently record all of the species in two fixed quadrats, one 1 x 1 m quadrat and one 10 x 10 m quadrat.

The results from this trial showed that, in the 1 x 1 m quadrat, the most successful surveyor found 73% of the species recorded in the quadrat, while on average the surveyors recorded only 63% of the species. As might be expected, the detection rate in the 10 x 10 m quadrat was considerably lower, with the most successful surveyor recording only 63% of the species and the average detection rate falling to 55%.

These results suggest that if we are interested in recording changes in species diversity, then we have to be prepared to live with observer variation of □ 30%. I fear that we have to accept that we will never have this level of information for our sites (particularly if we take account of the diversity of other species groups associated with the habitats, e.g. invertebrates) and that we must learn to live without it.

To some degree, information on species diversity is 'nice to know', in as much as we have never had this level of information before, and this has not stopped us managing habitats. I have known it prevent conservation managers making management decisions, but the habitat was still being managed of course, albeit passively. In truth, we will probably never know (except maybe in very species-poor habitats) the true diversity of species on a site. The only practical alternative is to record reliable indicators of diversity.

## 3.2    Assessments of frequency

Frequency is an objective measure that uses 'presence or absence' data to assess how frequently an attribute is present in a set of samples: this figure is typically expressed as a percentage. For example, you could assess the frequency of otter spraints along a stretch of river by dividing it into 20 sections and simply recording whether otter spraints were found in each section. If otter spraints were found in 15 sections, they occurred at a frequency of 75%.

*Table 10-4*. The results from sampling trials to test the effects of observer bias on species frequency data. Trials 1 and 2 involved students from University of Wales Swansea; Trial 3 involved professional ecologists. The blocks of data highlighted in red draw attention to inconsistent assessments.

| Attribute | Quadrat 1 | | | Quadrat 2 | | | Quadrat 3 | | | Quadrat 4 | | |
|---|---|---|---|---|---|---|---|---|---|---|---|---|
| | 1 | 2 | 3 | 1 | 2 | 3 | 1 | 2 | 3 | 1 | 2 | 3 |
| **Trial 1** | | | | | | | | | | | | |
| *Calluna vulgaris* | + | + | + | + | + | + | + | + | + | + | + | + |
| *Erica tetralix* | + | + | + | + | + | + | + | + | + | + | + | + |
| *Trichophorum cespitosum* | + | + | + | + | + | + | + | + | + | + | + | + |
| *Juncus squarrosus* | + | + | + | + | + | + | + | + | + | + | + | + |
| *Sphagnum sp.* | - | - | - | - | - | - | - | - | - | - | - | - |
| | | | | | | | | | | | | |
| **Trial 2** | | | | | | | | | | | | |
| *Cerastium semidecandrum* | + | + | + | + | + | + | + | + | + | + | + | + |
| *Erophila verna* | + | + | + | + | + | + | - | + | - | + | + | + |
| *Hornungia petraea* | - | - | - | - | - | - | - | - | - | + | + | + |
| *Saxifraga tridactylites* | + | + | + | + | + | + | + | + | + | + | + | + |
| *Peltigera canina* | + | + | + | + | + | + | + | + | + | + | + | + |
| | | | | | | | | | | | | |
| **Trial 3** | | | | | | | | | | | | |
| *Molinia caerulea* | + | + | + | - | - | - | + | + | + | - | + | + |
| *Potentilla erecta* | + | + | + | - | - | - | + | + | + | - | - | - |
| *Andromeda polifolia* | - | - | - | + | + | + | - | - | - | + | + | + |
| *Narthecium ossifragum* | + | + | + | + | + | + | - | - | - | + | + | + |
| *Rhynchospora alba* | - | - | - | + | + | + | - | - | - | + | + | + |
| | | | | | | | | | | | | |
| | | | | | | | | | | | | |

The quality of a habitat can be assessed in much the same way. When we have defined how to recognise good quality habitat, we can monitor to assess how frequently the vegetation meets the definition: there are several examples of this in the habitats case studies. The reliability of the result, however, will depend on the measurability of the attributes that we assess at each monitoring point.

We have tested the ability of surveyors to record species frequency on many occasions, and Table 10-4 shows the results from three sampling trials. A minimum of 12 surveyors participated in each trial, including many with little or no previous experience of vegetation recording or of the species that we selected for the exercise.

To help to overcome these problems, before each trial we spent c. 30 minutes training the surveyors to identify the relevant species. To ensure that the surveyors searched intensively for the species, we subdivided each of the fixed quadrats into 16 cells, and asked the surveyors to search for the presence of each species in each cell. The main purpose of the exercise, however, was to test the ability of the surveyors to detect the presence of each species in each of the four fixed quadrats.

The sampling trial results in Table 10-4 show that, overall, the surveyors agreed on the presence or absence of the selected species in 58 of the 60 assessments (97%), and that the level of consistency within each trial never dropped below 95%. The results of Trial 2 were particularly impressive, as most of the participants were students with backgrounds in either marine biology or zoology, and four of the five species being assessed were diminutive spring annuals.

These results suggest that, given a short period of training in species identification, even inexperienced surveyors can achieve consistent results when assessing the presence or absence of a small number of species at a sample point.

## 3.3    Assessments of abundance

Abundance differs from frequency by referring to the number of individuals of a species present, as opposed to the mere presence or absence of the species (see Fig. 10-2).

The commonest scale for assessing species abundance, in the UK at least, is the DAFOR scale, where D = Dominant, A = Abundant, F = Frequent, O = Occasional and R = Rare. Unfortunately, there does not appear to be a standard definition of these categories, so you will come across various interpretations of the DAFOR scale in different publications.

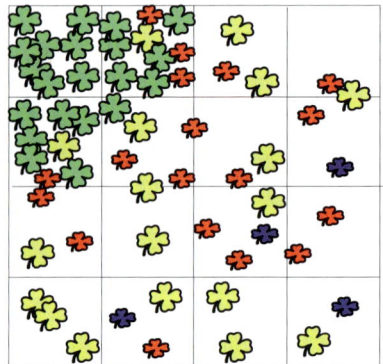 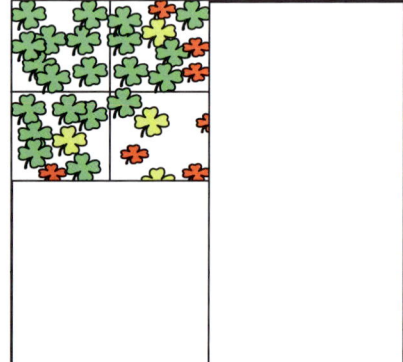

*Figure 10-2*. Frequency and abundance: the green plant (clustered in the top left corner of the 1 x 1 m quadrat on the left) is present in only four of the 25 x 25 cm cells, giving a cell frequency of 25%, but it has an overall abundance of 21 in the quadrat. The scarcer blue plant is also present in four of the cells and shares the same cell frequency score as the green plant (25%), but the blue plant has an overall abundance of only four. Note how, in the smaller quadrat on the right, it is no longer possible to see an increase in the frequency of the green plant as it now has frequency of 100%, nor could we record a decline in the blue plant as it is no longer present in the sample area. The size of the area of search is critical if we are using frequency measures for monitoring.

A sampling trial carried out by the English Field Unit (Leach & Doarks, 1991) found wide variation between experienced grassland surveyors using a form of the DAFOR scale. For example, the DAFOR assessments for *Lotus corniculatus* (a relatively common and easily identified species) ranged from absent to abundant between surveyors recording the same patch of vegetation. Leach and Doarks concluded that while compiling a species list with DAFOR may be a good way for a surveyor to become familiar with a habitat, it is probably of little value as a monitoring method.

Straightforward counts are an alternative form of abundance recording, though these are generally reserved for species monitoring projects. However, counts can also be used to good effect in habitat monitoring projects, particularly for dealing with negative indicator species. For example, the Heath Rush *Juncus squarrosus* is a regular component of wet heath vegetation, but it can also be an indicator of over-grazing. So while we might expect to find the occasional plant of *Juncus squarrosus* in a stand of wet heath, we might be concerned if we were seeing small clusters of plants scattered throughout it. One way to deal with this is to set an upper limit for the density of plants that we are prepared to tolerate within the area of search at your monitoring points. For example, we could set an upper limit of no more than five plants of *Juncus squarrosus* within a 1 m radius of a monitoring point.

The approach could be applied equally well to aggressive species such as Bracken *Pteridium aquilinum* in dry heath, or Common Reed *Phragmites australis* in transition mires. Counts can also be used in habitat monitoring projects as an alternative to subjective vegetation cover estimates. This is particularly true in broad-leaved woodland, where we can use densities and ratios to provide reliable monitoring results for most of the important structural attributes (Chapter 27).

*Photograph by Clive Hurford*

*Figure 10-3*. Heath Rush *Juncus squarrosus*, here with Ling *Calluna vulgaris* and Mat Grass *Nardus stricta*, responds positively to over-grazing in heath vegetation.

As a general rule, if we can assess the attributes using simple objective measures, then we should use them, because they are much less prone to observer bias than subjective assessments.

# 4.    ASSESSING VEGETATION HEIGHT

Vegetation height can be an important attribute of a habitat, particularly in grassland vegetation.   Many of the species associated with grassland have specific structural requirements: several species of butterfly, small mammals, and fungi spring readily to mind.  If vegetation height is an important attribute of the habitat, then we should include upper and / or lower limits, as appropriate, in the condition indicator table.   There are three methods commonly used for measuring vegetation height in ecological studies: direct measures; sward sticks; and drop discs.

All of these can be used to good effect, and the methods and their relative strengths and weaknesses are outlined below.  Stewart *et al.* (2001), carried out an evaluation of these methods, and concluded that all three were easy to use and delivered consistent results with negligible observer bias.

## 4.1    Direct measures

To record vegetation height using direct measures, we place a card (or hand) lightly on the vegetation at the point where ca. 80% of the vegetation is growing at or below that height.  We then take a reading of this height on a ruler (Hodgson *et al.*, 1971).  This is the most subjective of the three methods, as the observer has to decide where to place the card.  The qualifier 'ignoring tall stalks' improves the likelihood of consistency between recorders.

Direct measures are well adapted for recording the fine scale 'micro-heterogeneity' that some invertebrates require in short swards.  The design of sward sticks and drop discs rule them out as practical alternatives if micro-heterogeneity is an issue.

## 4.2    Sward sticks

For the sward stick method we use a 45 cm metal rule, with 0.5 cm graduations, with a sleeve supporting a 2 x 1 cm piece of clear Perspex.  The rule is held vertically, and the sleeve lowered until the Perspex touches the first piece of green non-flowering vegetation: we read the measurement from the rule at this point (Barthram, 1986).

The sward stick samples the smallest area of vegetation of the three methods and therefore gives the most variable results.   Its advantage is that, by taking several measurements at each sample point, we can detect structural heterogeneity at each sample point.  On the negative side, the sward stick is less well adapted for measuring variation in short vegetation than direct measures  (Stewart *et al.*, 2001).

## 4.3   Drop discs

The drop disc method (Holmes, 1974) is simple but effective: we simply let a disc (which is a standard size and weight) slide down the measuring stick from a height of 1.5 m until it rests on the vegetation at the sample point. We then take the reading of the vegetation height from where the disc is resting against the measuring stick. The measuring pole is marked at 1 cm intervals.

In medium to tall swards, the drop disc is recommended as the best method for measuring productivity and the effects of herbivory: it is also recommended for use in agri-environment schemes. The disadvantages are associated with the relatively large surface area of the disc, which makes it less well suited for detecting fine-scale heterogeneity, particularly in low swards. However, if the vegetation height data from a drop disc is combined with species frequency data, then it may well be possible to assume micro-heterogeneity at a monitoring point.

The drop disc is certainly the least subjective and simplest of the methods for measuring vegetation height, and would be more than adequate for assessing vegetation height in most situations. Unfortunately, drop discs are not produced commercially and have to be constructed by the surveyor. The disc itself should have a diameter of 30 cm and weigh 200 g, with a central slot or hole for sliding down the measuring stick. As the critical features of the disc are surface area and weight, the disc can be made from various materials, even cardboard secured with sticky tape. The main difficulty in making these discs is achieving the correct weight, as cardboard tends to be too light and plywood too heavy. However, with cardboard discs we can keep adding tape until we reach a weight of 200 g (this has the added bonus of waterproofing the cardboard), while with plywood discs you can drill holes in the disc to reduce the weight (it is best to varnish the disc first though, as a layer of varnish will push the weight back up). A 1.5 m or 2 m rule will suffice for a measuring pole in most grassland habitats.

Perhaps the most important point from a monitoring perspective is that these methods are not interchangeable - they provide different results. The golden rule therefore is not to change methods during the course of a monitoring project: choose the method and persist with it.

## 5.   IN SUMMARY

In this chapter we have discussed various forms of field assessment available for monitoring the quality of a habitat, concentrating on three principal components: vegetation cover, species composition, and vegetation height.

The results from multiple-observer sampling trials have indicated that the most reliable measures for monitoring habitats are presence and absence data; simple counts of abundance; and using a drop disc to record vegetation height.

As a general rule, we should try to avoid using estimates of vegetation cover in a monitoring project unless absolutely necessary. If we decide that it is essential, then we should monitor against cover targets. The results from sampling trials suggest that if we set up a monitoring project where the result can depend solely on estimates of vegetation cover, then the reliability of the monitoring result will be compromised by unacceptable levels of observer bias.

For this reason, we should think carefully about what we need to know about the vegetation that we are monitoring before deciding how to monitor it. If we consider, within any broad habitat type, which examples of a habitat we regard to be of high conservation interest, and why, we will probably begin to focus on those with a good representation of stress tolerating species (Chapter 8). These species will become scarcer as the more competitive species achieve dominance. This suggests that, in most cases at least, it is actually the presence of the stress tolerators (and associated species) that dictates the conservation value of the habitat, rather than the cover of the potentially dominant competitors. If we accept this, then the most efficient and reliable approach to monitoring the condition of a habitat is to focus on the frequency (or abundance) of the stress tolerating associate species, and not the cover of the dominants.

## 6.  REFERENCES

Stewart, K.E.J., Bourne, N.A.D. & Thomas, J.A. (2001). An evaluation of three quick methods commonly used to assess sward height in ecology. *Journal of Applied Ecology 2001*: **38**: 1148-1154.

Barthram, G.T. (1986). Experimental techniques – the HFRO sward stick. *Biennial Report of the Hill Farming Research Organisation 1984-85* (ed. M.M. Allcock), pp. 29 – 30. Hill Farming Research Organisation, Penicuik, Midlothian, UK.

Brown, A. (2000). *Habitat Monitoring for Conservation Management and Reporting. 3: Technical Guide*. Life-Nature Project no LIFE95 NAT/UK/000821. Integrating monitoring with management planning: a demonstration of good practice in Wales. Countryside Council for Wales, Bangor.

Holmes, C.W. (1974). The Massey grass meter. *Dairy Farming Annual*, pp.26 – 30. Massey University, Palmerston North, New Zealand.

Hodgson, J., Taylor, J.C. & Lonsdale, C.R. (1971). The relationship between intensity of grazing and the herbage consumption and growth of calves. *Journal of the British Grassland Society*, **26**: 231 – 237.

Hurford, C. & Perry, K (2000). *Habitat Monitoring for Conservation Management and Reporting. 1: Case studies*. Life-Nature Project no LIFE95 NAT/UK/000821. Integrating monitoring with management planning: a demonstration of good practice in Wales. Countryside Council for Wales, Bangor.

Leach, S.J. & Doarks, C. (1991). *Site Quality Monitoring methods and approaches (with particular reference to grasslands)*. Project No.135. English Field Unit, Nature Conservancy Council, Peterborough.

# CHAPTER 11

# IDENTIFYING SITE-SPECIFIC CONDITION INDICATORS FOR HABITATS

## Adding efficiency to reliability

CLIVE HURFORD

*Countryside Council for Wales, Plas Penrhos, Ffordd Penrhos, Bangor, Gwynedd, LL57 2BQ*
*clive.hurford@serapias.net*

## 1.     INTRODUCTION

This chapter addresses the problem of how to define the management aims for a habitat in concise, unambiguous and measurable terms. This will ensure that a) the land manager is clear about what we want the management to achieve and b) we will be able to obtain a reliable monitoring result. If we get this right, then with a minimum of training, anyone should be able to look down at the vegetation at their feet and say a) whether they are standing in the key habitat, and if so, b) whether the habitat is in a state of high conservation value. By this stage in the development of a monitoring project, we should have:

- Identified the key habitat on our site;
- Identified the major threats to the key habitat; and
- Developed a conservation strategy for the key habitat.

Chapter 10 drew attention to the most reliable forms of data capture, and suggested that a reliable habitat monitoring project should focus on collecting frequency or abundance data, possibly combined with some measure of structure, such as vegetation height. The general recommendation was to avoid recording estimates of vegetation cover if at all possible, and to use cover targets if some form of cover estimate was essential. We must bear this in mind when we are developing the habitat definitions for monitoring.

*C. Hurford & M. Schneider (eds.), Monitoring Nature Conservation in Cultural Habitats*, 93–103.
© 2007 *Springer.*

## 2.     DEVELOPING CONDITION INDICATORS FOR HABITATS

We use the term 'condition indicators' to describe the suite of attributes and targets that we have selected to define when a habitat (or species) is in optimal condition. In effect, the condition indicators are a form of ecological shorthand to help us recognise when the key habitat is in a state of high conservation value. Typically, the condition indicators, which should be applied at the management unit level, will comprise:

- A target for the overall extent of the broad habitat;
- A target for the extent of good quality habitat; and
- Unambiguous definitions for both the broad habitat and good quality habitat.

The condition of a habitat is typically determined by its extent, species composition, structure and physical integrity. However, if the physical integrity is damaged, this will be manifested through habitat loss, changes in species composition and changes in structure. Therefore, for practical purposes, these are the three critical attributes that we should use to guide our definitions of habitat quality in a condition indicator table. The following sections provide guidance on how to develop a set of site-specific condition indicators for a habitat.

### 2.1     Setting targets for the extent of a habitat

In most cases, we do not know the exact extent of the habitats on our sites, as all of our area estimates are derived from habitat maps that incorporate an unknown magnitude of error. This error originates from personal interpretations of ambiguous habitat definitions and imprecise mapping methods (Chapter 3).

Against this background, unless the habitat boundary is a) unambiguous, e.g. bordered by fences, hedgerows or walls, or b) can be reliably identified on remote images, e.g. forests, expressing targets in hectares is meaningless, as we have no way of knowing whether our target would represent an increase or decrease on what is already there. Furthermore, it is not good enough for us to declare that we want to maintain the current extent of a rare and threatened habitat, if we do not know a) where the habitat is, b) how much habitat there is, and c) whether there is the potential for expanding the area of habitat. This is refusing to accept management responsibility.

The practical answer to setting targets for habitat extent involves using survey information to identify a) areas where we <u>know</u> the key habitat is present, and b) areas that have the potential to be the key habitat. A habitat quality map can provide this information (Chapter 9).

When we have access to the relevant survey data, we can make an informed decision about how much of the habitat we want to be present on our site and where. We can then develop a management strategy for maintaining (and perhaps increasing) the extent of the habitat at its current locations, and for restoring the habitat in the areas that we have identified for expansion. After transferring this information onto a 1:10 000 map (or remote image), we can set up a monitoring project that will feed back information on whether the management aims are being achieved in the key parts of the site, e.g. those areas most vulnerable to loss or degradation, or those targeted for habitat expansion.

Before setting targets for the habitat across the whole of the site, we should decide what we want the management to produce in each management unit, and set appropriate targets for delivering this. There are several good reasons for approaching target setting in this order, but principally because:

- The land managers will need this information. If we do not tell the land managers what we want them to deliver in their respective management units, they can only deliver it by chance; and
- Our targets for the site as a whole will be founded on a carefully considered management strategy.

When we have been through this process, the target (or limits) for the extent of the habitat in the condition indicator table for the site should refer back to the map showing the desired extent. This map should be an integral component of the site management plan (Nature Conservancy Council, 1989).

## 2.2    Setting targets for the quality of a habitat

When developing habitat definitions for management and monitoring we must remain constantly aware of the problems caused by observer error and ambiguity. In the first instance, this means steering clear of using estimates of vegetation cover. A logical alternative is to turn our attention away from the 'competitors', and focus more on detecting the effects that increases in cover will have on the 'stress tolerating' species (Chapter 8). These are the species most likely to decline as the cover of competitive species increases.

I first became aware of the scale of observer error associated with recording cover estimates in the mid 1990s when, after years of working as an individual, I joined a small team of vegetation surveyors in a project to demonstrate monitoring in Natura 2000 habitats (Hurford & Perry, 2001). As a matter of good practice, we would check periodically to see whether our recording was standardised. Whenever we did this, we found that the range of variation between our cover estimates was great enough to undermine confidence in the monitoring result. In other words, it became apparent that the monitoring result was determined more by who did the monitoring than by the condition of the vegetation.

Consequently, we arranged a multiple-observer sampling exercise at Whiteford Burrows (a local sand dune system) to explore the alternatives to recording estimates of vegetation cover. This exercise, which involved MSc students from University of Wales Swansea, was designed to:

- Ascertain how the species composition of dune grassland changed as it succeeded from vegetation with patches of bare sand to closed grassland;
- Examine changes in the frequencies of stress-tolerating species in response to invasion by competitors; and

- Assess the variation between observers recording straightforward estimates of vegetation cover, cover-abundance scales, and species frequency (see Chapter 10).

In preparation for this exercise, we visited the site beforehand and recorded a series of vegetation quadrats in grassland at different stages of succession, e.g. in habitat with patches of bare sand and annual species present; in vegetation with less sand and a high cover of bryophytes; in vegetation with low bryophyte cover but a greater cover of higher plants etc. We then examined the data set to ascertain which species disappeared from the quadrats as the succession progressed. After doing this, we selected five or six species that were relatively easy to identify, and used these in the exercise with the students. The species we selected were spring annuals, simply because we did the trials at the end of a teaching module scheduled for late March or early April, when these were the only species in flower. During the course of the exercise, we used fixed 1 x 1 m and 50 x 50 cm quadrats (each sub-divided into 16 cells) to record the frequency of each species at a series of locations with increasing in vegetation cover. On examining the results from the quadrats, it was no great surprise to find that:
- In general, the frequencies of the annual species declined in keeping with the gradual loss of bare sand and bryophyte cover; and
- The declines in frequency were generally steeper in the 50 x 50 cm quadrats than in the 1 x 1 m quadrats.

It was surprising, however, to find that no individual species emerged as being more sensitive than the others, and that there was no obvious pattern to the scale of the declines. In one quadrat Species A would suffer the greatest decline in frequency, in the next quadrat Species F would suffer the greatest decline, and so on. Occasionally, a species would manage to maintain its frequency as the succession progressed, but again, we could not predict which species, as most managed to do it one quadrat or another.

It was only on entering the data in spreadsheets that a pattern began to emerge. It then became apparent that as the succession progressed, so the assemblage of species broke down, and that this started as soon as the bare sand began to disappear. So, whereas all seven of the species could co-exist within an area of 50 x 50 cm in the earliest stages of the succession, only five of the species could co-exist at the next stage. Again, we could not predict which of the seven species would disappear, but we could predict that at least two would. This exercise proved extremely valuable because the results suggested that:
1. The presence of an assemblage, rather than the presence of individual species, could be used to monitor increases in vegetation cover and provide an early warning of successional shifts;
2. The area of search was important, because if all seven species could co-exist within a relatively small (50 x 50 cm) area, then the individual species in the assemblage would take longer to disappear from a 1 x 1 m area of search; and
3. Even inexperienced recorders, with less than 30 minutes training in species identification, could reliably detect the presence or absence of the assemblage.

If these results were repeated on other sites and in other habitats, it meant that we could reliably predict, and detect, a decline in diversity from the earliest onset of the successional process, and long before there we could reliably detect changes in vegetation cover and structure.

Subsequently, we arranged an MSc project to test whether a similar pattern emerged on a different dune system: at Merthyr Mawr Warren in South Wales. Figure 11-1 shows an output from this project (Mathebola, 1999). The project was designed to test the effects of increases in grass cover and height on an assemblage of seven stress-tolerating species associated with the open, sandy habitats on the site. These species were Thyme-leaved Sandwort *Arenaria serpyllifolia,* Eyebright sp. *Euphrasia* sp., Dove's-foot Crane's-bill *Geranium molle,* Autumn Hawkbit *Leontodon autumnalis,* Bird's-foot Trefoil *Lotus corniculatus,* Wild Thyme *Thymus polytrichus* and Wild Pansy *Viola tricolor.* We identified this assemblage of species using the same process employed at Whiteford Burrows. However, because the field recording was carried out in midsummer, the assemblage comprised species that had replaced the spring annuals in the open sandy vegetation.

The results conformed to the pattern that we had seen in the previous exercise at Whiteford Burrows, i.e. that when grasses formed <5% of the vegetation cover and the grass height was <5 cm, the vast majority of the quadrats contained six or seven of the species in the assemblage, but when the grass cover was estimated to be in the 5-50% range, all except one of the quadrats contained five of the species or less.

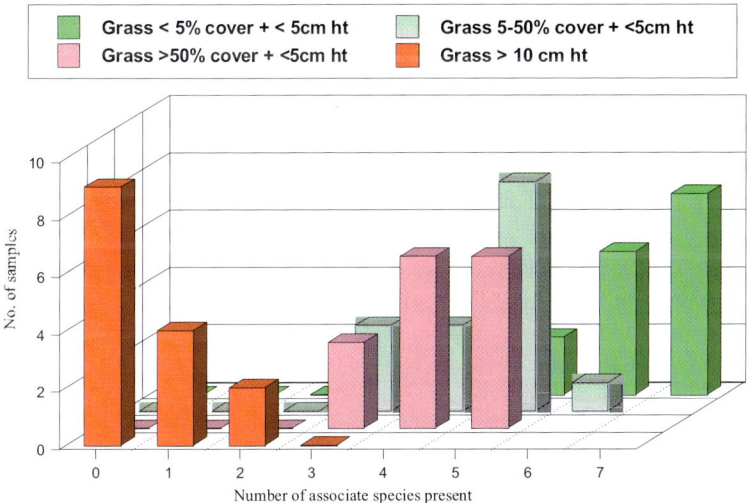

*Figure 11-1.* The results of an experiment to test the effects of grass cover and height on an assemblage of annual species at Merthyr Mawr Warren in south-east Wales.

In the light of these results, we began testing the approach in other habitats, with a view to identifying indicator assemblages that could define when these habitats were in optimal condition. We found, almost without exception, that whenever the quality of the vegetation was associated with a successional phase of development, we could identify an assemblage of species that would break down as soon as it came under pressure from competitive species. The only exceptions were upland heaths, which are naturally species-poor, and woodlands (see Chapters 27 and 28).

## 2.3    Identifying site-specific indicator assemblages

Indicator assemblages often comprise a small number of stress-tolerating species that we expect to come under pressure if the equilibrium shifts in favour of competitive species. The whole purpose of the assemblage is to provide an early warning of habitat change or degradation, which will allow us to make a prompt management response before the conservation value of the site is compromised. The approach can also be applied in reverse. If we are in a restoration phase of management, the reappearance of the species forming the assemblage can give us confidence that our habitat is recovering, while a return to widespread co-existence on the site will tell us when the balance has been restored. Critically, however, we have found that reliable indicator assemblages are site-specific.

### 2.3.1    The process for identifying a site-specific indicator assemblage

If we are familiar with our site and the species composition of the key habitat, then identifying a site-specific indicator assemblage should be straightforward. We have two basic scenarios, a) sites that are in optimal condition where we want an early warning of degradation, and b) sites that are degraded where we want evidence of recovery. To detect early warnings of degradation, we should visit the site and identify three key habitat states:

- Habitat patches that we consider to be of high conservation value;
- Habitat patches that we consider to be of low conservation value; and
- Habitat patches that are showing signs of moving from high conservation value to low.

Consider a pasture with areas of species-rich grassland over much of the site, but with ranker, species-poor grassland along one edge as a result of eutrophication from a neighboring field. Somewhere between these two extremes, we will find vegetation that is still of interest, but not as species-rich as the vegetation well distanced from source of the eutrophication: this is the vegetation that should reveal the early warning indicator assemblage.

We can distil the assemblage by recording a small number of samples in each of the three habitat states and noting:

- All of the species present in each quadrat;
- The approximate cover of the dominant species;
- The approximate cover of bare ground (if appropriate); and
- The height of the vegetation.

After tabulating these data, we should arrange the quadrats in columns from left to right in terms of declining conservation value. This will draw attention to the species most likely to disappear as the cover of the competitors, or the height of the vegetation, increases (see below).

## 2.3.2    Data from Blanches Blanques in Jersey

The data set in Table 11-1 was collected from successionally-young dune grassland at Blanches Blanques in Jersey. This phase of dune grassland development was identified as the conservation priority due to the rare plant species associated with it: these included Dwarf Pansy *Viola kitaibeliana*, Early Sand-grass *Mibora minima* and Sand Crocus *Romulea columnea*.

Collecting the data for this exercise took no more than two hours. During this time we selected eight quadrat locations: four in open vegetation with >20% bare sand or moss cover, and four in vegetation that was still successionally-young, but with >50% cover of grass. Initially, we compiled a species list for each 1x1 m quadrat and then noted cover estimates for bare sand, moss and grass, and the height of the vegetation.

A total of 40 species was recorded in the eight quadrats, with a mean of 16.5 species recorded in Quadrats 1-4 and a mean of 15.75 species in Quadrats 5-8. Significantly, nine species occurred only in Quadrats 1-4; these were Rue-leaved Saxifrage *Saxifraga tridactylites,*Parsley-piert *Aphanes arvensis,* Buck's-horn Plantain *Plantago coronopus,* Lesser Hawkbit *Leontodon taraxacoides,* Common Stork's-bill *Erodium cicutarium, Mibora minima,* Procumbent Pearlwort *Sagina procumbens*, Sand Cat's-tail *Phleum arenarium* and Cat's-ear *Hypochaeris radicata.* These are mostly stress-tolerating annual plants with an obligate requirement for open ground; they will begin to disappear as conditions become more suitable for competitive species.

It is impossible to predict which of these 'early warning' species will disappear first as the area of open sandy patches diminishes, but it is clear from the data set that the annual plant assemblage breaks down as this starts to happen. However, our experience from other sites suggested that some of these species can persist in a low closed sward, and are not generally restricted to open sandy habitats, notably *Hypochaeris radicata, Leontodon taraxacoides*, and *Sagina procumbens*. Conversely, Common Whitlowgrass *Erophila verna* and *Viola kitaibeliana* are species that are strongly associated with open sandy habitats, and, although they were recorded in Quadrats 5 and 6, this was only because they were persisting in very small patches of bare sand. So, on the basis of the data collected at Blanches Blanques, and taking account of what we already knew of the species, a site-specific indicator assemblage for Blanches Blanques would comprise *Aphanes arvensis, Erophila verna, Erodium cicutarium, Mibora minima, Phleum arenarium, Plantago coronopus, Saxifraga tridactylites* and *Viola kitaibeliana.*

If we now examine the individual quadrat data for the co-existence of these eight species, by looking down the columns in the spreadsheet, at least four of them were present in each of Quadrats 1-4. No more than two of them were present in Quadrats 5 and 6, and none was found in Quadrats 7 and 8. This information, perhaps, but not necessarily, coupled with a positive cover target for bare sand or bryophyte cover and a negative target for vegetation height, could be used to define dune grassland of high conservation value at Blanches Blanques (Table 11-3).

*Table 11-1.* A data set collected from different quality dune grassland states at Blanches Blanques in Jersey. The crosses indicate that the species was present in that quadrat.

| Species name | Quadrat no | | | | | | | |
|---|---|---|---|---|---|---|---|---|
| | 1 | 2 | 3 | 4 | 5 | 6 | 7 | 8 |
| *Saxifraga tridactylites* | + | + | | + | | | | |
| *Galium verum* | + | + | + | + | + | + | + | + |
| *Sedum acre* | + | + | + | + | + | + | + | + |
| *Thymus polytrichus* | + | | + | + | + | + | + | + |
| *Cerastium sp.* | + | + | + | + | + | | + | + |
| *Euphorbia portlandica* | + | | + | + | + | + | | |
| *Viola kitaibeliana* | + | + | + | + | + | + | | |
| *Aphanes arvensis* | + | | + | + | | | | |
| *Plantago coronopus* | + | | | | | | | |
| *Leontodon taraxacoides* | + | + | | + | | | | |
| *Leontodon autumnalis* | + | | | | | + | | |
| *Erodium cicutarium* | + | + | + | + | | | | |
| *Verónica arvensis* | + | | | | | | + | |
| *Erophila verna* | + | + | + | + | + | | | |
| *Mibora mimima* | | + | | + | | | | |
| *Geranium molle* | | + | | + | + | | + | + |
| *Sagina procumbens* | | + | | | | | | |
| *Phleum arenarium* | | + | | | | | | |
| *Luzula campestris* | | | + | | + | + | + | |
| *Trifolium dubium* | | | + | | | + | | |
| *Vicia sativa* | | | + | | | + | | |
| *Hypochaeris radicata* | | | | + | | | | |
| *Ranunculus bulbosus* | | | | | + | + | | + |
| *Lotus corniculatus* | | | | | + | + | + | |
| *Centaurium erythraea* | | | | | | | + | + |
| *Vicia lathyroides* | | | | | | | | + |
| | | | | | | | | |
| **Species total** | 15 | 15 | 18 | 18 | 18 | 14 | 15 | 16 |
| Bare sand | 55 | 35 | 0 | 0 | 1 | 0 | 0 | 0 |
| Grass cover | 3 | 35 | 30 | 35 | 90 | 90 | 70 | 80 |
| Moss cover | 30 | 30 | 20 | 65 | 5 | 3 | 5 | 15 |
| Vegetation height | 2.6 | 2.5 | 4.0 | 3.5 | 4 | 4 | 5.5 | 6 |

This leaves the conservation manager only one decision: how much of the dune grassland would have to be in this state for the habitat to be considered to be in optimal condition? There is no universal answer to this question. The conservation manager simply has to decide when to take management control of the habitat if the dynamic processes slow down or stop.

*Photograph by Clive Hurford.*

*Figure 11-2.* Dwarf Pansy *Viola kitaibeliana* is associated with bare sand at Blanches Blanques.

*Photograph by Clive Hurford.*

*Figure 11-3.* Early Sand-grass *Mibora minima* is another species strongly associated with open sandy habitats at Blanches Blanques.

It is always worth bearing in mind that as long as we have enough successionally-young vegetation on a site, we will always have the potential for the later seral phases.

This approach to identifying site-specific condition indicators can be applied equally well in more stable, culturally managed, habitats.  For example, the main threat to a grazed fen meadow could be drying out as a result of drainage.  In this situation, we can record vegetation data from the stands of high conservation value on our site and from areas that are showing signs of drying, and compare the data sets in much the same way as we did for the phases of dune grassland development.  The early indicators for drying out will be amongst those species present in the data collected from the stand of high conservation value but absent from data collected in the drier vegetation.  We can also refer to texts like Ellenberg (1999) to identify which of these species are likely to be the best indicators, i.e. least tolerant of drying out.

Similarly, if drying out is a perceived threat but there is no evidence of it yet, we can collect species data from the fen meadow vegetation and use Ellenberg's text to identify which of the species on our site are most likely to decline if drying out becomes a problem in the future.

Finally, if we include a species that is either a) difficult to find or b) difficult to identify in an indicator assemblage, it is likely to compromise the monitoring result.  We can avoid this by excluding these species and reducing the number of species that must co-exist at the monitoring points.  So instead of asking for at least five of eight species to be present, we would want four of seven.

# 3.     STRUCTURING A CONDITION INDICATOR TABLE

When we have identified our condition indicators, we should ensure that they are available in a concise and user-friendly format.  Tables 11-2 and 11-3 illustrate a way of doing this.

*Table 11-2.* The basic structure of a condition indicator table.

| Condition indicators | Statement of intent here | |
|---|---|---|
| Habitat extent | Lower limit | Refer to areas identified on maps or images for maintaining or restoring broad habitat |
| Habitat quality | Lower limit | State proportion of broad habitat to be in a state of high conservation value here |
| Site-specific habitat definitions | | |
| Broad habitat name here | Concise site-specific definition of broad habitat here | |
| Habitat class name here | Concise site-specific definition of habitat class here | |

The template illustrated in Table 11-2 can contain all of the essential information to guide both the management strategy and the monitoring project.   If the conservation priority is a species, we simply add another row at the top of the table stating the lower limit for the population size on the site.  The lower limit for the distribution of a species can be incorporated with the lower limit for population size, e.g. >50 individuals present in each of Areas A, B, and C (refer back to a map or remote image to identify these areas).

The information in Table 11-3 is concise and unambiguous. It clearly states the overriding management aim for the site, and contains sufficient detail to inform the monitoring. Chapters 12 and 13 describe how this information can be incorporated in an efficient and reliable monitoring project.

The site-specific-habitat definition for successionally-young dune grassland focuses on the co-existence of an indicator assemblage within a relatively small area of search (a 50 cm radius). This means that a) the recorder will soon become familiar with the appearance of the required habitat class and b) it will be possible to teach the land manager how to recognise the habitat class.

Finally, it is good practice to provide the rationale behind the selection of the condition indicators, and to store this with the site management plan and details of the monitoring project. This information will be important for future site managers responsible reviewing the monitoring project.

*Table 11-.3.* An example of a completed condition indicator table. This example incorporates the critical information distilled from the data collected in the dune grassland habitat at Blanches Blanques (Table 11-1).

| Condition indicators | The dune grassland habitat at Blanches Blanques will be in optimal condition when: | |
|---|---|---|
| **Habitat extent** | **Lower limit** | Extent of dune grassland habitat outlined on the map in Figure 1 of the site management plan (1998 version). |
| **Habitat quality** | **Lower limit** | >20% of the dune grassland vegetation is in a successionally-young phase of development |
| **Site-specific habitat definitions** | | |
| **Dune grassland vegetation** | Vegetation growing on a sandy substrate that is dry throughout the year. *Ammophila arenaria* is locally dominant and likely to be present within any 10 m radius. Trees and scrub absent. | |
| **Dune grassland vegetation in a successionally young phase of development** | Vegetation where more than three of *Aphanes arvensis, Erophila verna, Erodium cicutarium, Mibora minima, Phleum arenarium, Plantago coronopus, Saxifraga tridactylites* and *Viola kitaibeliana* are present within a 50 cm radius. | |

# 4. REFERENCES

Hill, M.O., Mountford, J.O., Roy, D.B. & Bunce, R.G.H. (1999). *Ellenberg's indicator values for British plants.* ECOFACT Volume 2. London Department of the Environment, Transport and the Regions.

Hurford, C. & Perry, K (2000). *Habitat Monitoring for Conservation Management and Reporting. 1: Case studies* Life-Nature Project no LIFE95 NAT/UK/000821. Integrating monitoring with management planning: a demonstration of good practic ein Wales. Countryside Council for Wales, Bangor.

Mathobela, P.C. (1998). *Can vegetation structure and height be used as an indicator of condition in dune grassland habitats?* MSc Thesis. University of Wales Swansea.

Nature Conservancy Council. (1989). *Site management plans for nature conservation: a working guide.* NCC report. Peterborough.

# CHAPTER 12

# WHERE TO FOCUS THE MONITORING EFFORT

CLIVE HURFORD[1] & DAN GUEST[2]

*Countryside Council for Wales, Plas Penrhos, Ffordd Penrhos, Bangor, Gwynedd, LL57 2BQ*

*clive.hurford@serapias.net*[1]
*D.Guest@ccw.gov.uk*[2]

## 1.     INTRODUCTION

For all but the smallest areas of habitat, it is both impractical and inefficient to attempt to monitor all of the vegetation.  Our alternatives are a) to take a random sample from across the whole habitat and use statistical inference to draw conclusions about its overall condition (Chapter 5) or b) to monitor in selected areas and use logic (or our knowledge of the inter-relationship between different parts of the habitat) to infer the condition elsewhere.  This chapter outlines the factors that can guide our decisions on where to monitor if we choose the latter approach.

Before deciding where to focus our monitoring effort, we should not only have a map showing the present distribution of the habitat, but also a map showing desired distribution of the habitat and unambiguous definitions of the desired habitat states.  This much is true whether we are planning to use a statistical or selective method.     It is important to remember however, that irrespective of the size of the site, land management will be applied at the level of the individual enclosure or management block.   A landscape-scale conservation project can only succeed if each individual landowner knows what the management of their land is expected to deliver and where.  The same is true of monitoring projects.    If the aim of the monitoring is to feed back into management, then we must be able to interpret the results at the scale of the individual management units.

This is a problem for monitoring projects based on random sampling methods because, unless there are enough randomly distributed samples within each management unit, the monitoring cannot provide feedback to the individual site managers.   For example, if we apply a random sample across the whole area of habitat, our monitoring result is only likely to indicate the overall condition of the habitat.  If the sampling area comprises two or more management units, we are unlikely to have enough samples in any single management unit to derive a statistically significant result.  Therefore, we will not be able to provide feedback to any land manager on the condition of the habitat on their

*C. Hurford & M. Schneider (eds.), Monitoring Nature Conservation in Cultural Habitats*, 105–118.

land. This will rule out an informed management response, which is the whole point of the monitoring in the first place.

The practical alternative is to monitor in carefully selected areas of our site, and use what we know about the habitat and site management to infer the condition of the habitat elsewhere. This approach focuses on ecological significance as opposed to statistical significance.

## 1.1    General principles of selective monitoring

In projects that adopt a selective approach to monitoring, only carefully chosen areas of the habitat are monitored. The overall condition of the habitat is then deduced on the basis of the known relationship between the monitoring plots (which are best regarded as reference points to inform management) and the rest of the habitat. This focused approach to monitoring allows us to draw conclusions about the condition of relatively large areas of habitat without demanding excessive amounts of fieldwork. It does, however, require that the conservation manager has a good understanding of the habitat, the site and the factors most likely to impact on it. A selective monitoring approach is appropriate in a wide range of situations, including:

- Medium and large areas of habitat;
- Habitats where the likely direction of change, and the areas likely to be most affected, are predictable (most cultural habitats fall into this category);
- Habitats where some areas are intrinsically of higher conservation value than others;
- Habitats that include areas which are inherently more fragile or more vulnerable to change than others; and
- Habitats in sub-optimal condition where we can predict which areas will be last to be restored given the anticipated programme of restoration management.

## 1.2    Assumptions that we can make if we adopt a selective approach

Irrespective of the habitat type, there are a number of scenarios where the results from a small number of carefully located monitoring plots will allow us to infer the condition of the key habitat in every management unit on large multiple-owner sites. For example:

1. If our habitat quality map (see Chapter 9), or our knowledge of the site's management history, indicates that most of our key habitat will be in poor condition, we can focus our monitoring on those management units where the habitat patches have the best chance of passing the criteria in our condition indicator table. If the key habitat in these management units is found to be in sub-optimal condition, we can reasonably conclude that it will be in poor condition in the other management units. The coastal heath case study (Chapter 17) demonstrates how this approach can be applied in practice.

2. If our habitat quality map suggests that most of the habitat will be in optimal condition, we can focus our monitoring on those management units where the habitat is most likely to fail to meet the criteria in our condition indicator table. If the habitat

in these management units is found to be in optimal condition, then it is unlikely that the condition of the habitat in the other management units will be sub-optimal.

These assumptions allow us to give conservation managers reliable feedback on the condition of the habitat on their sites without intensive monitoring activity. As a consequence, the manager will be able to develop a strategy for bringing the habitat in each management unit under an appropriate management regime. This will take time to implement, and monitoring plots can be established in each management unit as the restoration (or maintenance) management becomes operational. This allows us to increase our monitoring coverage on a 'need to know' basis. If the habitat in a management unit is in sub-optimal condition and we have not done anything to change the situation, we will not need monitoring to tell us the condition of the habitat. The same is true if we have initiated restoration management that we know will take decades to deliver the desired outcome. In this case, however, we may want to revisit the monitoring plots periodically for reassurance that the habitat is recovering.

Essentially, we need to establish monitoring plots as reference points for the conservation manager when we take management control of a site. If we are investing resources in conservation management we need evidence that the management is having the desired effect.

## 1.3 Monitoring the desired habitat condition

As the over-riding aim of our monitoring is to assess whether or not the habitat is in optimal condition, we should focus on where we want the habitat to be, rather than where it is at present. For example, consider a small upland site, which supports a range of habitat types grading from broad-leaved woodland on lower ground, through dry heath on the west-facing slopes, into dry acidic grassland on the summit and gentler east-facing slopes where the grazing pressure is highest (Fig. 12-1).

If our management aim is for the dry heath to expand into the adjacent acid grassland, monitoring only within the current areas of dry heath will never tell us if we have achieved this. Conversely if the overall aim is to allow the re-establishment of a natural tree-line at the expense of some areas of heath, assessing the condition of the dry heath in these areas would be a waste of resources, as we should be monitoring against criteria for woodland. Consequently, we must take account of what we want the management to achieve before deciding where to monitor.

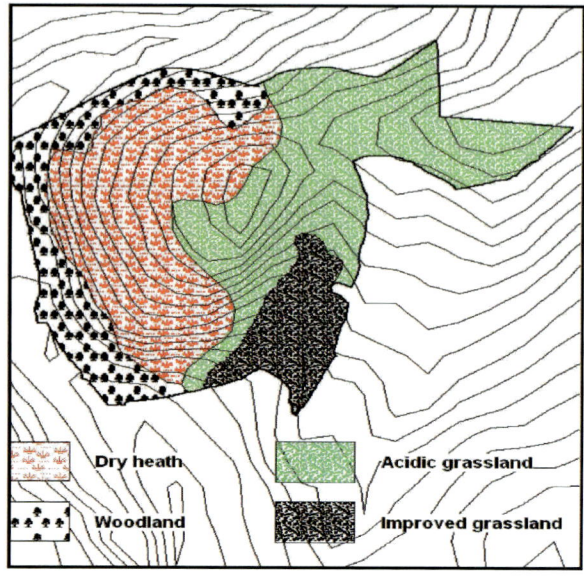

*Figure 12-1.* A diagram of an upland site for dry heath vegetation.

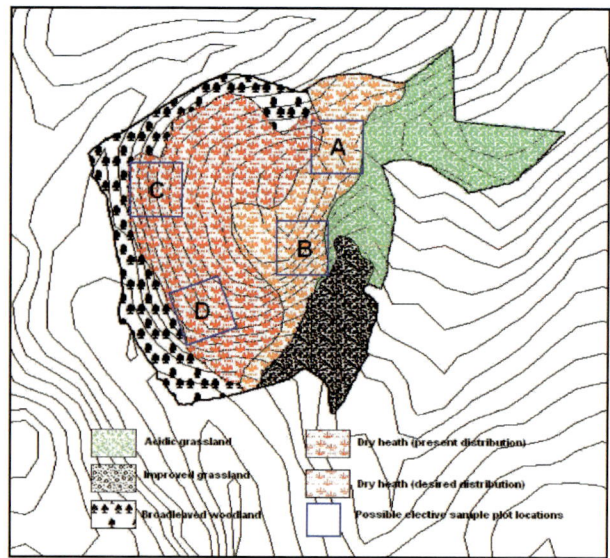

*Figure 12-2.* A diagram showing the planned expansion areas for the heath, and the location of selectively positioned monitoring plots.

# 2. TAKING ACCOUNT OF THE MANAGEMENT HISTORY

Choosing where to monitor is a key decision: if we select the wrong areas we risk getting the wrong monitoring result, which could result in inappropriate management and habitat loss or degradation. Furthermore, we will be wasting valuable resources by committing future recorders to long days in the field when a shorter more focused visit would suffice. Understanding the management history of our site can have a major influence on our approach to monitoring it.

Most habitats of conservation value can be allocated to one of the following three categories:

1. Habitats subjected to regular and intensive management disturbances, typically on some form of rotation, e.g., intensively managed dry heaths, arable land, and drainage ditches;
2. Semi-permanent habitats that are subjected to regular management disturbances, e.g. meadows, pastures and forest; and
3. Naturally formed habitats, e.g. dunes, fens, raised bogs and lakes, impacted by direct or indirect disturbances as a result of human activities.

Habitats in Categories 1 and 2 have a long history of cultural management and their condition will be inextricably linked to how they are managed. Developing monitoring projects in these habitats should be relatively straightforward, though we need to be aware of the management rotation for habitats in Category 1.

Habitats of high nature conservation value that do not fit into these categories should probably be considered wilderness and are beyond the scope of this book. Perhaps our only interference with these increasingly rare habitats should be to a) give them the highest level of protection, b) buffer them from the effects of human activities, and c) provide the opportunity for expansion. If the value of the habitat is simply its ability to be self-sustaining, we should not have any expectations of it. As such, research and surveillance programmes are more appropriate in wilderness situations than monitoring.

## 2.1 Land subjected to regular and intensive management disturbances

Of the habitats generally considered to be of high conservation value, only arable land, dry heaths, drainage ditches and perhaps coppiced woodlands fit readily into this category. These habitats are subjected to intensive management on a rotational basis. Because of the extreme nature of the management, e.g. regular ploughing or burning, we can expect the habitats that are produced to be relatively homogeneous, at least in terms of vegetation structure. In theory, this means that within any management unit, we should be able to monitor any part of the habitat and get the same monitoring result. If this holds true, we can also monitor any part of the habitat and assume that the vegetation outside the monitoring plot will be in the same condition. This assumption is more likely to hold true, however, for dry heath vegetation and woods than for arable habitats or drainage ditches.

## 2.1.1    Dry heaths

In the UK much of the lowland and sub-montane heath is intensively managed by a combination of light grazing and rotational burning or mowing.   This management appears to present difficulties for monitoring, as each individual patch of heath changes character as it develops from one growth phase to the next.

A selective approach to monitoring dry heath relies on us knowing the management cycle on our site and using this knowledge to focus our monitoring effort.  We should know, for example, how recently different patches of the heath have been burnt or mown, and we can use this information to predict the development stage of each patch.  On this basis, we might predict that patches burnt less than five years ago will be in the pioneer phase of development, patches burnt between five and ten years ago in the building phase and so on.  We also know that the pioneer and degenerate phases of development are poorly suited to monitoring, as the character of the heath changes relatively rapidly in these growth phases. The building and mature stages are more stable, however, and sufficiently well advanced to show the effects of intensive management.

One strategy would be to focus the monitoring on the building or mature heath growth phases, and use our knowledge of the management to identify these stands.  This would reduce the monitoring effort in each cycle and ensure that the entire habitat was monitored over a ten-year period.   By the end of this period, we will have identified those management units where the condition of the habitat was sub-optimal, which will allow us to focus the monitoring on these stands in the future.

*Photograph by Clive Hurford.*

*Figure 12-3.* The effect of intensive sheep-grazing on heath vegetation is clearly illustrated by the contrasting habitats on either side of the fenceline.

## 2.1.2    Arable habitats

Although arable habitats are also subjected to intensive management on a regular rotation, our approach to the monitoring has to be slightly different to the one that we described for heath vegetation. This is because the traditional crop rotation on arable land involves a complete change of habitat, from cereals, to roots and temporary leys and, while the end product is predictable, it will vary according to the stage in the rotation cycle. Generally, we would focus the monitoring effort on the cereal stage of the rotation, as most conservation value is attached to this; it is a waste of resources to monitor fields that are under a ley. This presents us with two options:

1. We can carry out the monitoring in keeping with the rotation, i.e. monitor in permanent plots and only in years when the fields with the plots are planted with cereals; or
2. We can assess whichever fields are planted with cereals in the year scheduled for monitoring.

On balance, it is probably better to select the first option, as this will allow direct comparison, though you could put together a good case for adopting either approach.

We can be selective about where we place our monitoring plots in the cereal fields, as many arable weeds of high conservation value have a low tolerance of shading and could not survive within the main crop. Therefore, if the weeds are not present in the field margins (perhaps because of herbicide applications – or because there is no obvious field margin) we can safely assume that they will not be anywhere else. The case study in Chapter 18 provides an example of monitoring arable fields.

## 2.2    Semi-permanent habitats subjected to regular disturbances

The commonest habitats in this category are grasslands and broad-leaved woodland (excluding coppice). The management disturbances in these habitats are regular, but not intensive, as plant biomass is only partially removed. These habitats are similar in that one of the major threats to their conservation interest is when the management becomes irregular or stops. We deal with woodland separately in Part 5 of the book.

### 2.2.1    Meadows and pastures

These are probably the most straightforward habitats to monitor, as each management unit will be subjected to a uniform level of management intended to deliver a predictable end product. All we have to do is check that the end product is of sufficiently high conservation value. One or two carefully located monitoring plots will normally suffice to inform management, though these may need to be situated in areas perceived to be threatened by offsite activities (see 2.3.1), e.g. areas susceptible to nutrient enriched run-off from adjacent fields. The major threats to these habitats (on sites under regulated conservation management at least) are overgrazing or neglect, both of which will result in

*Photograph by Clive Hurford.*

*Figure 12-4.* This south-facing chalk grassland near Lulworth Cove in Dorset, England is an important site for Adonis Blue *Polyommatus bellargus* butterflies.   A monitoring plot situated along this eastern edge of the main breeding area would reveal a) whether the grazing intensity was maintaining or increasing the area of abundant Horseshoe Vetch *Hippocrepis comosa* (the Adonis Blue foodplant) and b) whether Tor-grass *Brachypodium pinnatum* (in the top right of the photo) was reducing the area of suitable breeding habitat. This plot, combined with a second situated along the western margin of the breeding area, should be sufficient to inform the management of the site.

an impoverished flora, where sustained neglect will ultimately lead to habitat loss through successional processes.


## 2.3    Naturally formed habitats impacted by direct or indirect disturbances

In general, these habitats pose the greatest problems for monitoring, primarily because we rarely consider them to be actively managed.  It is easy, however, to think of examples that have been subjected to intensive management, e.g. raised bogs that have been cut for peat.  How we approach the monitoring in these habitats will depend entirely on site-specific factors.  If the habitat is, or has been, managed in the past, we should treat it no differently to habitats with a history of cultural management.  Alternatively, if there is no history of management, the monitoring should focus on the parts of the site most likely to degrade as a consequence of damaging off-site influences.  The following sections highlight ways of applying a selective approach in these situations.

## 2.3.1    Selective monitoring in areas most vulnerable to modification

In this approach to selective monitoring, the parts of the habitat that are considered to be most susceptible to adverse modification are identified and monitored using selectively located plots.   These areas tend to be either:

- The areas closest to a localised or directional threat, e.g. areas of broad-leaved woodland adjacent to a block of forestry where the principal concern is infestation by conifers; or
- The parts of the habitat that are inherently more susceptible to modification or loss, e.g. early successional stages of habitat development in dynamic habitats, as these will be the first areas to be lost to succession in more stable systems, or bog pools within natural hummock-hollow complexes on raised bogs, as we know that this microhabitat is particularly sensitive to drying out.

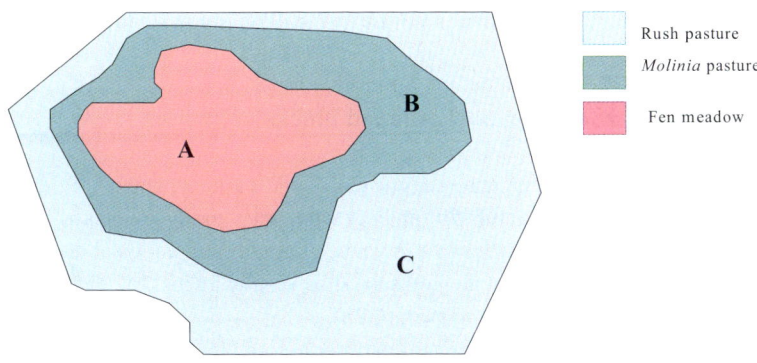

*Figure 12-5.* A diagrammatic representation of a single management unit, including an area of fen meadow (Habitat 6410).   Sampling the whole management unit (Areas A-C) is potentially misleading here as the value we place on the component communities differs. Replacement of part or all of the fen meadow by acidic rush or *Molinia* pasture would be unacceptable and any monitoring strategy must reflect this. A selective approach might concentrate on the fen meadow habitat, as this reflects the much higher value we attach to this vegetation type.

## 2.3.2    Selectively monitoring the areas of highest conservation value

Habitats are seldom of uniform conservation value: typically some areas are and always will be of much higher value than others.   These areas are often prioritised for management, and as such should be the principal focus of our monitoring effort.

For example, consider a marshy grassland site, which comprises a mixture of acidic rush and Purple Moor Grass (*Molinia caerulea*) pasture and Purple Moor Grass dominated fen meadow (Fig. 12-5). Of these plant communities, the fen meadow vegetation has the highest conservation value, as it represents an internationally important habitat (Habitat 6410) listed in Annex I of the EU Habitats and Species Directive.

In this case, our strategy may be to focus our monitoring effort on the stands of fen meadow. The rationale for adopting this approach would be as follows:

1. As the fen meadow vegetation is the most important component of the broad habitat, maintaining it or restoring it to optimal condition must be the principal aim of the conservation management;
2. If the management is successful in maintaining or restoring fen meadow vegetation then, in broad terms, we must accept whatever condition it delivers for the other areas of marshy grassland in the same management unit.

## 2.4    Some difficult cases

Selective monitoring is not always straightforward, but then neither is any other reliable form of monitoring.  Every site is different, and will pose a different set of problems.  None of these have yet proved insurmountable, though they have proved difficult at times.  In the following sections, we outline some of the more frequently encountered problems, and their solutions.

### 2.4.1    Dealing with irregularly shaped blocks of habitat

With the exception of some enclosed meadows and pastures, habitats rarely come in regular blocks that fit conveniently and exactly into easily relocated rectangular monitoring plots.  Therefore, setting up a permanent monitoring plot that covers the whole of the key habitat, and nothing but the key habitat, is usually impossible without the use of a high accuracy Global Positioning System (GPS).

One alternative is to expand the area of our monitoring plot to encompass the whole of the key habitat, accepting that we will also be recording vegetation that is not the key habitat and never will be. This does not present a major problem for monitoring, as the time spent recording at points outside our target feature is likely to be minimal (we simply record them as 'other habitat' and move on).  If we choose to do this, however, we must recognise the presence of non-target habitat within the plot and adjust the target in our condition indicator table to accommodate it. To do this accurately we need to know what proportion of the plot is potential key habitat.  So, for example, if we were looking at our dry heath site (Fig. 12-2), we may want to record points landing in a) dry heath habitat and b) dry acidic grassland with suppressed ericoids, as the combination of these equates to potential dry heath vegetation.

This information should be gathered in the first monitoring cycle, when at each sample point we ask first 'are we in our target or potential target habitat?' and second 'if so, is the vegetation of the desired quality?' When we know more accurately how much of our sample plot comprises potential key habitat, we may then want to revisit our condition indicator table and revise our original target.

Another alternative is to place smaller monitoring plots a) within the main block of habitat and b) in the area of potential habitat. Provided that the quality of the habitat in the main block of habitat was being maintained, and the vegetation in the area of potential habitat was being restored, we could safely assume that our management was having the desired effect elsewhere in that management unit.

## 2.4.2    Dealing with small-scale mosaics

As a general rule, our approach to monitoring small-scale mosaics of different habitats is similar to our approach to monitoring in plots that include areas of non-key habitat, except that we focus the monitoring on the habitat component that is either a) most sensitive to the perceived threats or b) that is of the highest conservation value - whichever is most appropriate on that site.

If we have a habitat mosaic with a long history of cultural management, e.g. calcareous grassland and scrub, we might want to maintain both of these component habitats because of the associated bird and plant interest. The major threat to this mosaic would probably be neglect, which would result in an increase in the extent of the scrub at the expense of the calcareous grassland. Given what we understand of successional processes, we are unlikely to see a reduction in the extent of the scrub unless we actively remove it. Therefore, if we focus the monitoring on the extent and quality of the calcareous grassland component, and include scrub saplings as a negative attribute in our condition indicators, this will provide us with an early warning of scrub invasion. We could increase our confidence further by selectively locating our monitoring plots in areas adjacent to the larger stands of scrub: if the grassland was being maintained in these areas we could assume that the same would be true elsewhere in that management unit.

We could adopt a similar approach to monitoring small-scale mosaics in naturally formed habitats, such as raised bogs. The raised bog surface typically comprises a complex mosaic of 'hummock and hollow' vegetation, where the hollow vegetation is represented by bog pool species. The commonest threats to raised bogs are drainage and peat cutting, which can both result in drying out at the bog surface. The earliest indication of this will be a reduction in the extent and distribution of the bog pool vegetation. Hence, in this situation we could set a lower limit for the extent of bog pool vegetation on the bog surface, and focus our monitoring on this component of the mosaic. Given our existing knowledge of the raised bog habitat, if the bog pool component of the vegetation is well distributed we can assume that the hummock vegetation will be present too. The converse is not true: the bog could become dominated by hummock vegetation if the surface began to dry.

If we decided to use monitoring plots in this fragile habitat, we should probably monitor near the central area of the raised dome and at locations near the edge of the dome, as these areas are likely to be the first to show signs of drying (Hurford & Perry, 2000).

## 2.4.3    Dynamic habitats

Dynamic habitats that have a cyclical, as opposed to directional, pattern of succession, present a number of problems for monitoring. This is particularly true if there is no history of cultural management and if the conservation value of the habitat is associated with the ongoing creation of new habitats, e.g. bare sand on sand dunes or hollow vegetation on raised bogs. The greatest threat to these habitats is that the dynamic processes will either slow down or stop, leading to a shift from cyclical succession to directional (see Fig. 12-6). In the first instance, this will result in an inevitable decline in the early phases of habitat development. Therefore, in dynamic habitats, the successionally-young vegetation should be prioritised for monitoring.

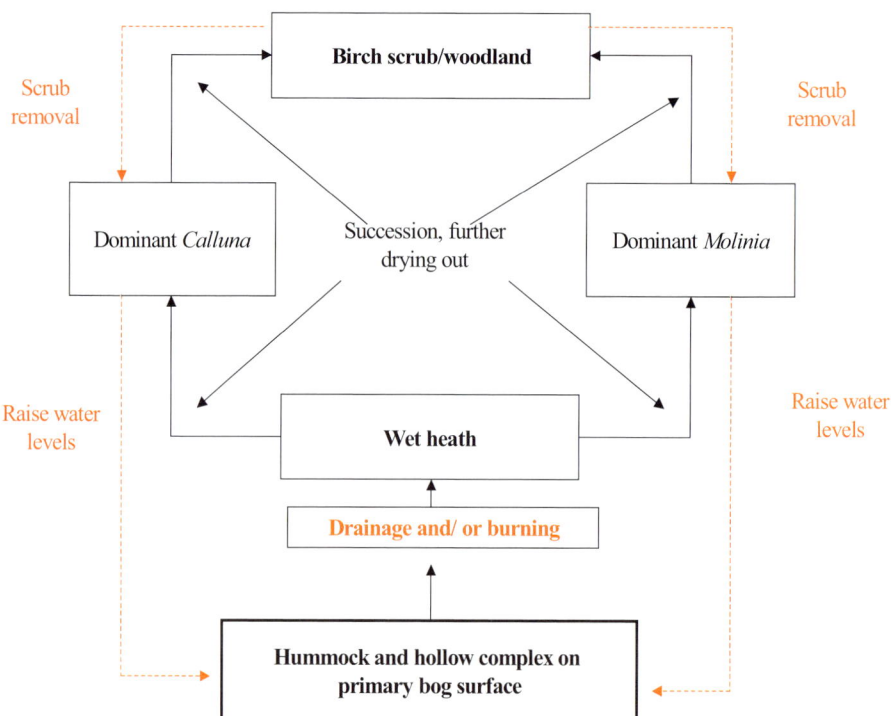

*Figure 12-6.* A simplified diagram showing how primary raised bog vegetation changes from the cyclical succession of hummock and hollow creation to directional succession after drainage or burning (adapted from Heathwaite & Gottlich, 1993). The red lines indicate management responses and point to the likely effects.

On sites that are still dynamic, monitoring in permanent plots could lead to an inaccurate monitoring assessment, as the monitoring would simply reveal what we already know about cyclical succession, i.e. that the successionally-young vegetation will develop from one seral stage to the next. This is not a problem, of course, as long as new habitat is being created elsewhere on the site.

Therefore, we must abandon the concept of monitoring in fixed plots and follow the key habitat as it moves around the site. Before each monitoring event, we should review the suitability of the existing monitoring plots and, if necessary, set up new plots to accommodate the changing distribution of the habitat. It is, however, important to note that we do not need to go through an expensive and time-consuming survey phase in advance of repeat monitoring visits, as we can use our existing survey information or remote images to determine where to look for the key habitat.

For example, consider the monitoring of Fen Orchid, *Liparis loeselii*, habitat. We know that in Wales this species is restricted to the early and early to mid-successional stages in humid dune slack development. A habitat quality map showing the distribution of this habitat type and embryo dune slack will provide us with the information we need to identify monitoring areas in the first instance, and allow us to predict where suitable habitat is most likely to develop in the future. This allows us to focus our search in future monitoring cycles. In the longer term, we can also use remote images to detect areas of newly deposited sand.

In the event of the dynamic processes stopping, we will have to undertake active management if we want to maintain the early phases of development. In this situation, the monitoring should treat the site the same as any other where the condition is linked to the management, i.e. the monitoring should focus on the areas prioritised for management: there will be no point in looking anywhere else for successionally-young habitats.

## 2.5    A balanced approach to selective monitoring

In deciding where to place our monitoring plots, we should identify a series of plots which, taken together, will be capable of establishing the condition of the habitat in any given set of circumstances. So, in our example of grazed upland heath, we may want to establish a series of four plots, two (A and B) in the acidic grassland and two more (C and D) in the heath adjacent to the areas of woodland (Fig. 12-2). We can probably assume that if our management is restoring the heath in the area of acid grassland, it will also be maintaining the existing habitat, in the short to medium term at least.

Returning to Fig. 12-2, Plots A and B will tell us when the heath has successfully reached its desired extent, while plots C and D will provide us with confidence that the lower areas of heath have not become overrun with trees and scrub. We need not, however, visit all of our plots each time we monitor the site: we should use our existing knowledge to visit and assess only those plots which we believe will give us a clear result first. So, in our upland heath example, in the first monitoring cycle we may choose to visit only plots A and B, but as time passes and these plots approach good condition we may then want to turn our attention to all four plots, finally perhaps focussing only on plots C and D as under-management and associated scrub invasion becomes the principal threat in our newly restored heath.

Finally, for practical purposes, we do not need to establish all monitoring plots from the outset. In the first instance, we could set up plots A and B, and then wait until we had restored the extent of the heath before turning our attention to invasion by trees and scrub.

## 2.6    The importance of periodic reviews

Although selective monitoring is an extremely efficient and reliable way of assessing habitat condition, we can all make mistakes, and we can all misread situations. Therefore, it is important that we do not rely blindly on selectively placed sampling plots in habitat monitoring. Both before and during repeat monitoring events, we should revisit the rationale underlying the selection of the monitoring plot locations to check that it still holds true.

Furthermore, we can often bolster our confidence through combining evidence from selective monitoring with evidence from remote images. For example, if we return to consider the area of marshy grassland described in Fig. 12-5, the selective monitoring plot may provide us with confidence that the key area of fen meadow is being maintained, while aerial photograph interpretation can provide us with confidence that the associated areas of *Juncus* and *Molinia*-dominated pasture have not been invaded by bracken or scrub or undergone other forms of gross change.

## 3.    REFERENCES

Heathwaite, A.L. & Gottlich, K.H. (1993). Mires: process, exploitation and conservation. Wiley, Chichester.

Hurford, C. & Perry, K (2000). *Habitat Monitoring for Conservation Management and Reporting. 1: Case studies* Life-Nature Project no LIFE95 NAT/UK/000821. Integrating monitoring with management planning: a demonstration of good practice in Wales. Countryside Council for Wales, Bangor.

# CHAPTER 13

# COLLECTING THE MONITORING DATA

CLIVE HURFORD[1] & DYLAN LLOYD[2]

*Countryside Council for Wales, Plas Penrhos, Ffordd Penrhos, Bangor, Gwynedd, LL57 2BQ*
*clive.hurford@serapias.net[1]*
*Dy.Lloyd@ccw.gov.uk[2]*

## 1.    INTRODUCTION

During the data collection phase of a monitoring project we have to make decisions relating to:
- When to carry out the monitoring;
- The size and shape of our monitoring plots;
- How to collect the monitoring data; and
- How to re-find the monitoring plots.

The following sections draw attention to the issues that we need to consider during the course of making those decisions.

## 2.    WHEN TO CARRY OUT THE MONITORING

As a general rule, we should plan the monitoring for the period when the key attributes, e.g. the indicator assemblage of co-existing species and negative indicator species, are most visible. The safest period to monitor in many habitats is at the end of the growth period, when we have the option of recording species in flower, leaf or seed. In some habitats we will have a choice: in successionally-young dune grassland, for example, we could use an assemblage of spring annual species to monitor early in the season, or an assemblage of summer annuals to monitor later. If our monitoring result is likely to be influenced by apparent changes in vegetation cover, however, the end of the growth period is probably the only time that we can monitor without the result being compromised by fluctuating seasonal growth patterns <u>and</u> observer error.

119

*C. Hurford & M. Schneider (eds.), Monitoring Nature Conservation in Cultural Habitats, 119–128.*
© 2007 *Springer.*

*Figure 13-1.* Clockwise from left, Greater Butterfly Orchid *Platanthera chlorantha*, Bitter Vetch, *Lathyrus montanus* and Wood Bitter-vetch *Vicia orobus*. These species have all proved to be particularly sensitive to agricultural improvements and suffered steep declines in UK grasslands. However, any monitoring project where these species were present would have to take account not only of the peak flowering period, it would also need to take into account the timing of the hay cut and grazing regime.

In other habitats, the management cycle may dictate the monitoring period: hay meadows, for example, must be monitored before the hay is cut. The only hard and fast rule is to think carefully about when is the best time to carry out the monitoring.

# 3. THE SHAPE AND SIZE OF THE MONITORING PLOT

The general rule is to keep everything as straightforward as possible. The time spent trying to find a poorly marked monitoring plot can far outweigh the time recording in it. On this basis, one large plot is always a better option than several small ones, and square or rectangular plots are preferable to more complex shapes. In most situations, the default should be a square plot. Rectangular plots, however, increase the efficiency of the monitoring a) in habitats that form a narrow band, e.g. transitional fens, b) in habitats associated with linear features, e.g. flushes, and c) when we are monitoring the boundary vegetation between habitats.

Unless we are using a high accuracy (sub 50 cm) Global Positioning System (GPS), and we know that similar equipment will be available for repeat monitoring events, it is best to keep the shape of the monitoring plot simple.

The size of the monitoring plot should be determined by a) the spatial patterning of the vegetation when it is in optimal condition and b) the number of points that we want to record.

In more homogeneous habitats, e.g. upland heaths, we would use 200 x 200 m plots; whereas in fine-scale mosaics, e.g. the surface of raised bogs, we would use 50 x 50 m plots. In most other habitats, e.g. meadows and pastures, we would use 100 x 100 m plots. If we were using rectangular plots, we would try to ensure that the plot covered the same area as a square plot in the same habitat. We have never used plots larger than 200 x 200 m.

There are two exceptions to these guidelines: a) when the key habitat occupies a very small area of the site and there is little scope for expansion, in which case we would monitor all of it, and b) when we are using a high accuracy GPS and attempting to map the condition of the entire area of habitat and potential habitat.

# 4. COLLECTING THE MONITORING DATA

Our preferred method of data collection involves recording the condition indicators at points on a systematic grid (Fig. 13-2). This monitoring method, which evolved through the pursuit of consistent and reliable results that can feed back critical information for habitat management, involves systematically saturating the monitoring area with recording points, and works within the spatial patterning of the vegetation.

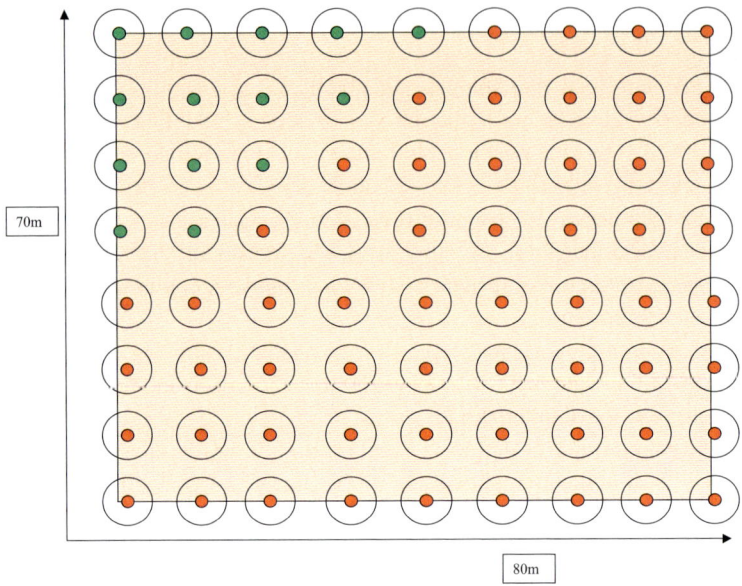

*Figure 13-2.* The systematic layout of monitoring points in a 70 x 80 m plot: the distance between the points is 10 m. The closed red and green circles represent the 50 cm radius area of search for checking the co-existence of the positive indicators, while the open black circles represent a 3 m radius area of search for negative indicators. After monitoring, we can colour code the points to show where the vegetation passed the criteria in the condition indicator table (green) and where it failed (red).

By focusing the monitoring on:
- The presence of a site-specific suite of co-existing positive indicator species;
- The presence of a small number of site-specific negative indicators; and
- Recording within the spatial patterning of the habitat,

the monitoring result will confirm that our management is achieving its aims, provide an early warning of degradation, or provide evidence of recovery.

Often, a habitat that has been managed inappropriately will be homogeneously in poor condition, and we would get the same monitoring result whether we recorded points every 10 m or every 50 m, as almost all of the points would be expected to fail to meet the criteria in the condition indicators. However, our management aim may well be for 75% of the habitat to be in optimal condition, which could be patchily distributed: our monitoring points must be sufficiently frequent to detect this level of heterogeneity.

For this reason, we recommend that, even in relatively homogeneous habitats, points should be recorded at least every 20 m, while in more heterogeneous habitats it is better to record every 5 m or 10 m. If we are uncertain, it is better to err on the side of caution and over-sample than to risk recording too infrequently to detect changes in the condition of the vegetation. By applying a selective approach, we have made so many savings in efficiency that we should be able to justify an extra hour or two in our monitoring plots to be certain of obtaining a reliable result.

If the distance between our monitoring points is within the spatial variation of the habitat in the plot, then we should achieve a similar percentage pass rate regardless of where we start to record. This means that, during repeat monitoring events, we do not have to revisit the same monitoring points to have confidence in the result. This is important if you do not have access to sub-50 cm accuracy GPS, or if you are monitoring a fragile habitat and do not want to set up a regular monitoring grid for fear of damaging the vegetation. One advantage of using a high accuracy GPS is that it allows us to revisit the same points during repeat monitoring cycles. This gives the conservation manager the opportunity to detect patterns of recovery or decline.

## 4.1    Recording at the monitoring points

When we have decided a) where and when to carry out the monitoring, b) the size and shape of the monitoring plot, and c) the distance between the monitoring points, all that remains is to record the co-occurrence of the attributes in the condition indicator table (Chapter 11) at the monitoring points.

This should be straightforward. Typically, the condition indicator table will combine positive and negative indicators. As negative indicators tend to be relatively easy to see, e.g. *Pteridium aquilinum,* the most efficient sequence for recording at each point is to check for the negative indicators first. If the point fails, note the reasons why it failed, and then move on to the next point: there is no need to look for the positive indicator assemblage, the point will fail regardless of whether it is present or not. There may, or may not, be advantages (for surveillance purposes at least) attached to checking for all of the attributes at each point, but it will take longer and it will not affect the monitoring result.

To determine the condition of the vegetation in the monitoring plot, we would check the lower limit for habitat quality in the condition indicator table, which would be expressed in terms similar to '>75% of the grassland has to be species-rich', and if more than 75% of our monitoring points passed the criteria for species-rich grassland we would consider the habitat to be in optimal condition. Sampling trial results, using recorders with varying levels of experience, have indicated that the difference in the percentage pass rate differs by less than 5% between recorders using this method.

If we want to adopt a precautionary approach to accommodate this error, we can introduce a rule whereby if the monitoring result is within 5% of the lower limit (or upper limit or target – as appropriate) the habitat should be considered to be in sub-optimal condition. This option is certainly worth consideration.

Finally, some notes of caution. The very least that we should do is test the vegetation for each of the negative indicators at every monitoring point, as there may be more than one reason for a habitat being in sub-optimal condition. If we check only the first negative attribute on the recording form, we risk failing to detect the underlying cause of the problem. This will subsequently result in an inappropriate management response, and a continuing decline in conservation value. In UK woodlands, for example, we could note

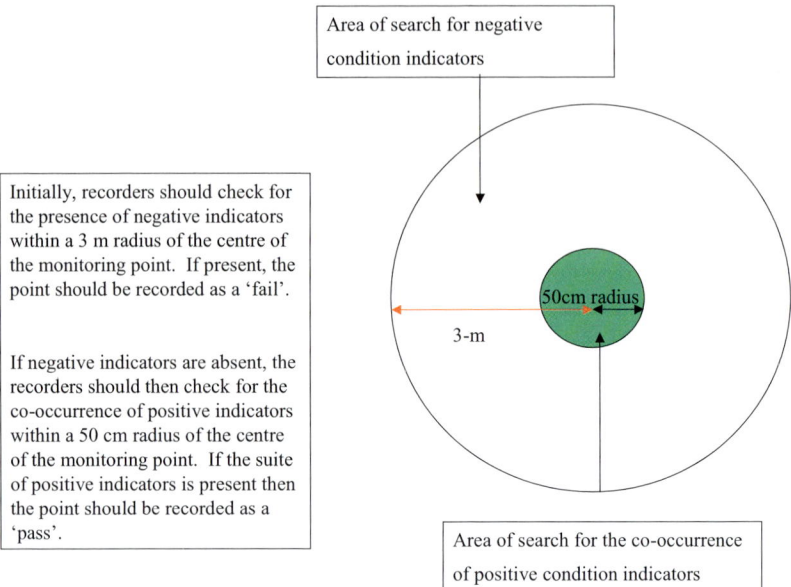

Area of search for negative condition indicators

Initially, recorders should check for the presence of negative indicators within a 3 m radius of the centre of the monitoring point. If present, the point should be recorded as a 'fail'.

If negative indicators are absent, the recorders should then check for the co-occurrence of positive indicators within a 50 cm radius of the centre of the monitoring point. If the suite of positive indicators is present then the point should be recorded as a 'pass'.

50cm radius

3-m

Area of search for the co-occurrence of positive condition indicators

*Figure 13-3.* A diagram of an individual monitoring point. The areas of search for the positive and negative indicators will vary according to the habitat type: the 50 cm radius and 3 m radius areas of search at this point would be appropriate for monitoring most meadows and pastures.

that there were no gaps in the canopy, but fail to note the number of viable non-native Sycamore *Acer pseudoplatanus* saplings waiting for gaps to be created. In this situation, going out and thinning the wood could exacerbate the problem, not solve it.

Secondly, if we decide to record all of the attributes at each monitoring point, we must search as diligently at points that have already failed as we would at points where the result was still uncertain. In multiple-observer sampling trials, we have noted a tendency for inconsistencies to arise in the attribute data at points that have failed because of an attribute that appeared early on the recording form. This did not affect the consistency of the overall monitoring result between recorders, nor the consistency of the percentage pass rate between recorders (because as long as the result of the point remained uncertain, the recorders maintained a high level of recording effort), but it did create inconsistency between recorders in the data for individual attributes. This suggests that we can have confidence in the overall monitoring result, confidence in the percentage pass rate, and confidence in detecting the underlying reasons for a habitat being in sub-optimal condition. This is sufficient information to guide the conservation management, and the results from our sampling trials suggest that we should resist the temptation to push the data any further.

# 5.      ENSURING THAT WE CAN RE-FIND THE MONITORING PLOT

In most cases, there are advantages to ensuring that we can re-find the corners of a monitoring plot on repeat monitoring visits, particularly if we are monitoring in boundary areas between habitats.   This will, without exception, increase confidence in the monitoring result.   There are three ways that we can do this:
- Use a high accuracy GPS (accurate to <50 cm);
- Use a lower accuracy GPS combined with underground markers; or
- Use triangulation methods.

In cases where we choose to use underground markers, thin plates of aluminium attached to the top of short lengths of untreated timber will give a strong signal on standard metal detectors.  These should be buried 5-10 cm below the surface.  Aluminium gives a strong signal on standard metal detectors.   Alternatively, aluminium or stainless steel tubing has also proved reliable (Hurford *et al.*, 2001).

## 5.1      Using Global Positioning Systems (GPS) to record the location of monitoring plots

Before the mid 1990s, relatively few habitat monitoring or surveillance projects survived beyond the first attempt to repeat them, primarily because the monitoring areas could not be re-found.  The advent of GPS has provided us with the ability to eliminate this problem in all but the most sheltered habitats and locations, e.g. broad-leaved woodland and gorges.

There are two long-standing satellite navigation systems: the USA GPS system and the Russian GLONASS system.  Within Europe there is another system, EGNOS, currently under deployment, with the long-term aim of establishing the 'Galileo' system: Europe's own global satellite navigation system.  The EGNOS system should enable European users to determine their position to within 5 m.

Many types of navigation equipment use these satellite systems, and new kits arrive on the market regularly.  These commercially available kits differ in their ability to collect and  process  position  information.   For  example,  the  lower  accuracy  kits  provide uncorrected code phase data; medium accuracy kits allow post-processed differential correction; while high accuracy kits use either real-time correction or the more demanding carrier processing methods.  As you would expect, the cost of the equipment increases with its accuracy.  For most conservation monitoring work, the critical issue is being able to obtain differential correction.

Currently, if we require position information with <10 m accuracy, then we need to use equipment that allows differential correction of the position fixes.  Even with the differential correction ability, however, there is no guarantee that the predicted accuracy levels will be achieved in the field, as the local topography, e.g. mountains and forest canopy, can have a significant effect on signal availability.

When deciding which type of GPS kit to use, we should consider:

- The level of accuracy required;
- The resources available for equipment purchase and maintenance;
- The resources available for supporting computer hardware and software;
- The resources available for data management; and
- How we intend to use the data in the future.

Even if we are using high accuracy equipment in optimal conditions, if precise relocation is critical, e.g. plot locations, we should safeguard our data by supporting the GPS positional fixes with permanent markers, location notes and photographs. This supplementary information should be stored, together with a record of the make, model and configuration of the GPS kit, with the GPS position data.

*Table 13-1.* A basic overview of terrestrial navigation equipment.

| GPS receiver type | General horizontal accuracy expected for 95% of position fixes in good conditions[1] (m). | Differential correction ability | Comments |
|---|---|---|---|
| Recreational GPS[2] | >10 m | Some | Sold for walking and other leisure pursuits |
| General navigation | >10 m | Most | Vehicle navigation on roads |
| Low specification mapping kits | 3-5 m | Yes | As used by some UK conservation workers. Precise relocation requires the use of underground markers. |
| High specification mapping kits | 1-2 m | Yes | Backpack systems – used where precise relocation is essential and the substrate is unsuitable for underground markers. |
| Geodetic surveying | < 1 cm | Yes | Specialist market equipment. |

[1] Accuracies typically degrade by a factor of 2-3 under forest canopy.
[2] Some recreational GPS are also EGNOS enabled. These potentially offer improved accuracy over similar non-EGNOS enabled kits.

## 5.2    General advice for recording GPS points in difficult locations

If we want to record GPS information in difficult terrain or habitats, e.g. deep gorges or forests, first we should investigate the manufacturer's specifications to ascertain a) the best approach for that type of GPS and b) the likely performance we can expect. We could also plan GPS work in deciduous forests for winter, when canopy cover is minimal, and time our visit for when the greatest number of satellites is available: this information can be obtained through satellite planning software. In woodland, an external antenna, held as high as possible, will improve our chances of success, though these tend to be available only for the more expensive GPS kits.

Whichever model we are using, it should be turned on for 10 minutes before entering the habitat, as this enables a new almanac to be recorded. Thereafter, if our GPS co-ordinates are generated by individual position fixes, we should leave the kit at each location for c. 10 minutes. If we record for a shorter period than this our position fixes could be subject to multi-path bias. It is also worth noting that we are unlikely to improve the accuracy if we record for much longer periods in one location.

Finally, in difficult terrains, we should be prepared to use offsets, particularly when attempting to record at the base of a tree or rock face. We may even have to accept that using GPS is not a practical option on our site and revert to using traditional triangulation methods.

## 5.3 Using triangulation methods to re-find plots

Before the increased availability of GPS, triangulation methods were used as an aid to re-finding monitoring plots. This still holds true if the recorder does not have access to a GPS. Critically, the method depends on the availability of semi-permanent and easily recognisable fixed points, e.g. isolated trees, telegraph poles, field corners and rock outcrops. These fixed points should be photographed and marked on a 1:5 000 OS map of

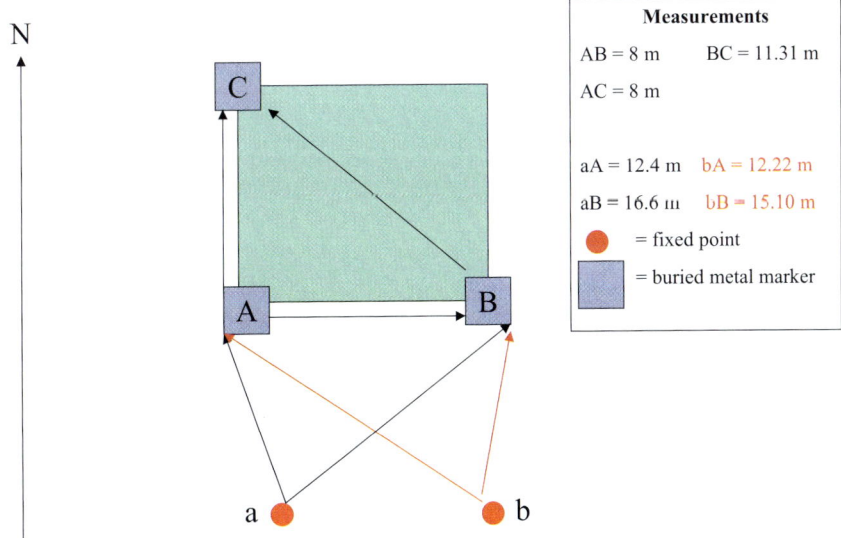

*Figure 13-4*. A diagram showing how triangulation can be used to re-find a monitoring plot.

The triangulation method is simple:

- Use temporary markers, i.e. bamboo canes, to mark the corners of the monitoring plot.
- Measure the distance from each of the two fixed points to each of the two nearest corners of the monitoring plot (Fig. 13-4).
- After noting these distances (and any supplementary information to aid the relocation of the fixed points), replace the temporary markers with permanent underground markers.

In habitats where underground markers are not an option, e.g. grassland on thin soils, it should be possible to re-find a plot from the tape intersects, though in this situation we would need to record the distance from our fixed points to all four corners of the plot. Our ability to recover the plot would then depend entirely on being able to get back to the fixed points, though this is equally true when we are using underground markers. The underground markers simply provide an additional layer of confidence that we have found the correct locations.

Ultimately, our ability to re-find a monitoring plot will depend on a) the clarity of the directions to our fixed points, b) the safekeeping and availability of the directions, and c) the continued presence of the markers. It is good practice to replace the plot markers during each monitoring cycle.

## 6.    REFERENCES

Hurford, C., Jones, M. & Brown, A. (2000). *Habitat Monitoring for Conservation Management and Reporting. 2: Field Methods*. Life-Nature Project no LIFE95 NAT/UK/000821. Integrating monitoring with management planning: a demonstration of good practice in Wales. Countryside Council for Wales, Bangor.

# CHAPTER 14

# THE SUPPORTING ROLES OF ON-SITE PHOTOGRAPHY

CLIVE HURFORD

*Countryside Council for Wales, Plas Penrhos, Ffordd Penrhos, Bangor, Gwynedd, LL57 2BQ*
*clive.hurford@serapias.net*

## 1.    INTRODUCTION

The importance of on-site photography is often understated in texts on surveillance and monitoring, perhaps because ecology courses tend not to promote the camera as a valid research tool. Consequently, there are relatively few historic photographs showing the structure or composition of habitats on sites of conservation interest. This is a problem, because we are not very perceptive when it comes to noticing the early signs of habitat succession; we tend not to notice it until a habitat has changed beyond recognition. Perhaps  this is because, on the one hand, if we visit an area regularly we adjust to the changes as they happen (because the vegetation looks very much the same from one day to the next). While on the other hand, if our visits are infrequent, our memories fade and become unreliable.

Photographs can help to overcome many of the uncertainties associated with assessing change over time, and for this reason, we recommend photographing the key habitats on a site during each monitoring cycle. In some cases, for example when assessing cliff ledge vegetation, photography may form a fundamental component of the monitoring. In other cases, it will be a means providing support for the monitoring result. Either way, photographic evidence will offer an additional layer of information (and confidence) to the conservation manager.

*C. Hurford & M. Schneider (eds.), Monitoring Nature Conservation in Cultural Habitats, 129–136.*
© 2007 *Springer.*

A series of photographs taken over time can provide evidence that the condition of a habitat has changed, and can be far more convincing than any data set. It is worth noting, however, that photographs are not particularly good at providing an early indication of change (unless a project is set up specifically for that purpose). They tend to reveal change after it has happened.

This chapter provides an overview of the options available for 'on-site' photographic surveillance, and offers guidance on how to set up these projects. The chapters in Part VI of the book look at how remote images, e.g. generated by aerial photography and remote sensing, can contribute to a monitoring project. On-site photographic methods tend to fall into one of the following categories:

- Photomonitoring, i.e. photographic panoramas taken from vantage points;
- Fixed-point photography, i.e. single photographs taken from vantage points; and
- Quadrat photography, i.e. a photographic record of species cover in recording frames.

## 2.    PHOTOGRAPHIC EQUIPMENT

At present, we recommend using a traditional 35 mm Single Lens Reflex (SLR) camera fitted with a standard fixed focal length lens. We also recommend using a 'neutral' 35 mm colour transparency film with a small grain size: Fuji Provia 100 or Kodachrome 64 are films generally considered to provide true colour representation and will retain detail when enlarged.

We realise that, with the 'digital revolution' gathering momentum, these recommendations could soon become outdated. At present, however, only the very best (and most expensive) digital SLR cameras can provide the detail of images captured on 35 mm film. And we must also take into account issues related to image processing and image storage before committing to a digital archive. For example, if we use digital images to record habitat change, then we should store the 'RAW' image, and not an image that been in any way manipulated or compressed. This presents two problems:

- Only the more expensive digital cameras have the capacity to save RAW images; and
- RAW images have relatively large file sizes (typically in the region of 10 Megabytes each).

The size of RAW files creates a problem for image storage; any network laden with large numbers of them will soon slow down. The option is to store the images on CD ROM media. However, we also face the prospect of changes in storage media in the future, which will result in further time spent transferring the images from CD-ROM to the new format: already CD-ROM is being replaced by DVD and memory sticks. Therefore, in 2005 at least, the simplest, and probably most reliable, option at present is to persist with 35 mm film.

*Photograph by Clive Hurford.*

*Figure 14-1.* A fixed-point photograph showing the habitat in the vicinity of permanently marked surveillance plot at Kenfig NNR, the blackboard identifies the plot number.

## 3. PHOTOMONITORING

A formal method for photomonitoring was developed in the late 1980s (Hope Jones, 1994a & 1994b). The underlying premise was simple: to provide a photographic record of sites by creating a panoramic snapshot of the vegetation from carefully selected vantage points. This involved noting the location of each vantage point and taking a series of photographs that, when placed alongside each other, would provide a 360° record of the vegetation. The general recommendation was to carry out a preliminary site visit to identify the best vantage points.

### 3.1 The method for photomonitoring

A summary of the method for photomonitoring follows:
1. Carry out the photomonitoring on an overcast day with 'flat light', and at the time of year when the habitat of interest is most easily recognised. Photographs taken on bright sunny days can produce aesthetically pleasing photographs, but their value can be undermined by shadows in early morning or late afternoon, and by bleaching during strong sunlight.
2. Check that the camera has a film in it and set the camera for the correct film speed;
3. Set the camera to 'aperture priority' and set the aperture to 11 (this should ensure that all of the vegetation in the image is in focus).

*Photographs by Clive Hurford*

*Figure 14-2.* The top photograph shows the most important dune slack for the internationally rare hepatic *Petalophyllum ralfsii* at Merthyr Mawr in late autumn 1994. The bottom photograph shows the same slack in February 2001. The loss of open ground, and increases in the cover of grasses and Creeping Willow *Salix repens* are clearly evident. By 2001, *Petalophyllum* was unable to compete in the dune slack habitat and was restricted, almost exclusively, to the footpaths.

4. Use a new film for each site, and use the first frame of the film to identify the name of the site, the vantage point location (e.g. the OS grid reference or a numbered cross-reference to a location marked on a map); and the date. We often wrote these details in chalk on a small blackboard, and then photographed the blackboard.

5. Use the second frame of the film to photograph the tripod in position at the vantage point. This will be a valuable reference during repeat exercises.

6. Place the camera on the tripod to eliminate blurring as a result of 'handshake'; this will be a problem if you are using a relatively slow film speed on an overcast day.

7. Take three photographs: one at the recommended shutter speed, one at one stop above it and another at one stop below it: this will minimise the risk of your photographs being under-exposed or over-exposed, either of which will compromise the value of the exercise.

8. Repeat this for each field of view (which should slightly overlap the previous field of view) until you have a complete photographic record of the vegetation over 360°.

9. Rewind the film and remove it from the camera.

10. Move on to the next vantage point, put a new film in the camera and repeat steps 2-9.

After developing the films, you should a) select the best-exposed set of transparencies from each vantage point, b) place them in slide-holders, c) label the slide holder with the site, date and vantage point location, and d) store the transparencies under appropriate conditions. This should be a cool, dark and fireproof location.

## 4.    FIXED-POINT PHOTOGRAPHY

The conservation value of a habitat usually depends on its structure and species composition: as these attributes will determine which species, other than plants, are associated with the habitat.  However, recording the structure of the vegetation is problematic using traditional recording methods, and fairly obvious structural changes can go unnoticed.  Fixed-point photography can provide clear evidence of changes in habitat structure that are less easy to detect in data sets.

Fixed-point photography is similar in many respects to photomonitoring.  Typically, however, it is associated with a set of surveillance data, and is set up to provide an additional layer of confidence with respect to structural changes in the sampling area.

## 4.1    The method for taking fixed-point photograph

A summary of the method for fixed-point photography follows:
1. Plan the photography to correspond with the surveillance sampling.
2. Make sure that the camera has a film in it; set up the camera for the correct film speed.
3. Set the camera to 'aperture priority' and set the aperture to 11.
4. Use a new film for each site, and use the first frame of the film to identify the name of the site, the fixed-point location, and the date.

5. Use the second frame of the film to photograph the tripod in position at the fixed-point.
6. Place the camera on the tripod.
7. Take three photographs at each fixed point, one at the recommended shutter speed, and then two more, bracketing either side of the recommended shutter speed.

Fixed-point photography is more widely practised than photomonitoring, primarily because it is relatively easy to set up and execute. There is a trade off, however, as we will have a record only of selected parts of a site, as opposed a more general overview.

## 5.      QUADRAT PHOTOGRAPHY

We have used quadrat photography primarily to provide a visual record of the vegetation cover in a quadrat. Multi-observer field trials have shown consistently that cover assessments are prone to unacceptable levels of variation between observers (Chapter 10). If we take photographs of the quadrats during each recording cycle, however, we will have a reference point that can help to overcome the uncertainty attached to vegetation surveillance data.

### 5.1     The method for taking quadrat photographs

The quadrat photographs should be taken after setting up the surveillance plot, but before starting to record the data. The success of this method depends entirely on the ability to precisely re-find the quadrats (Chapter 13).

Use a 35 mm SLR camera fitted with a 28 mm lens: this will allow you get the whole of a 1 x 1 m quadrat within the frame of the photograph. Use a colour-neutral film with a small grain size, i.e. with an ISO no greater than 100.

1. Make sure that the camera has a film in it; that the camera is set up for the correct film speed (100 ISO);
2. Set the camera to 'aperture priority' and set the aperture to 8: this should ensure that all of vegetation in the image is in focus.
3. Within each photograph, include a blackboard stating the site, plot number, quadrat number and date, and try to ensure that it is possible to see where the quadrat is positioned on the measuring tape. All of these things will help to ensure that repeat exercises are success.
4. Take all photographs from the northernmost edge of the quadrat: this will ensure that repeat photographs will not be affected by shadow (in the northern hemisphere at least).
5. Take three photographs at each fixed point, one at the recommended shutter speed, and then two more, bracketing either side.

*Photographs by Clive Hurford.*

*Figure 14-3.* These photographs show the rate of development in successionally-young dune slack vegetation at Kenfig NNR. The top photograph was taken in late autumn 1994, while the lower photograph shows the same quadrat in February 2001. In 1994, 39 plant species were present in the 1 x 1 m quadrat, including 62 thalli of the Annex II species *Petalophyllum ralfsii.* In 2001, the increase in *Agrostis stolonifera* had reduced the number of species present in the quadrat to <10, and the habitat was no longer suitable for *Petalophyllum*, which was absent. These results reflected the situation at Merthyr Mawr dunes (Fig. 4-2), which suggests that the South Wales populations of *Petalophyllum* are under serious threat.

Quadrat photographs will not necessarily allow you see the frequency of individual species in your surveillance quadrats, unless the species is particularly large and obvious, but they will allow you to determine whether there has been a significant increase in grass cover or bare ground for example, both of which are notoriously prone to observer error in data sets.

## 6.      PHOTOGRAPHY AS AN AID TO INTERPRETATION

Many of the photographs in this book are taken directly from surveillance or monitoring projects, mostly taken as aids to interpretation.  Interpretative photographs have a number of important uses, particularly in relation to:
*   Facilitating management decisions from site managers;
*   Providing feedback to site managers;
*   Training in habitat recognition; and
*   Illustrating the effects of successional processes.

If we want to explain to conservation managers why the habitats on their site are not in optimal condition, then photographs of the habitat when it is in optimal condition alongside photographs of the habitat as it actually is on their sites can save a lot of time.

## 7.      REFERENCES & BIBLIOGRAPHY

Bower, B. (1993).  Lens, Light and Landscape: the Art and Technique of Scenic Photography. David & Charles.  Newton Abbot.
Hope Jones, P. (1994a). *Photomonitoring: a procedure manual*. (CCW Contract Report 94/2/1): Countryside Council for Wales.
Hope Jones, P.  (1994b). Photomonitoring on sites of wildlife interest in Wales.  British Wildlife: Vol.6, No 1, pp. 23-27.

# CHAPTER 15

# INTEGRATING DATA IN GEOGRAPHICAL INFORMATION SYSTEMS

PAUL PAN

*Custom GIS Limited, 62 Llanishen Street, Cardiff, CF14 3QD, UK*
*paulpan64@yahoo.co.uk*

## 1.      INTRODUCTION

In this part of the book, we have examined not only the requirements for developing an environmental monitoring project, but also a number of key case studies in the UK, Sweden, Finland and Germany.  Several different layers of data were collected during the course of these monitoring projects, including one or more of field survey maps, remote images, surveillance quadrat data, species distribution data, Global Positioning System (GPS) point data, monitoring pass/fail data, monitoring plot co-ordinates, monitoring point co-ordinates and fixed-point photographs.  The increasing availability of medium and high accuracy GPS and GIS has made it possible to develop projects that integrate these site-based data and make them available at the push of a button.  Therefore we intend to draw these chapters to a fitting conclusion by discussing the integration of field data using modern Geographic Information Systems (GIS).

A few enthusiasts have described GIS as a panacea for applications of a geographical nature. Others hail it as the biggest breakthrough since the invention of maps many centuries ago. The use of GIS for environmental monitoring is not new, however, GIS have been around in different forms for many decades, though early GIS systems suffered from a lack of functionalities and were difficult to use. Moreover, their applications were restricted to a privileged few in the academic or research communities. These shortcomings were compounded by limitations in the capacity of early computer hardware.

It was not until the early 1990s that the tide began to turn in the favour of the practitioners and many robust and usable GIS began to emerge in the open commercial market. Advances in core computer technologies such as analytical algorithms, user-interfaces, microprocessors and memory storage were essential to the successful application of GIS.

*C. Hurford & M. Schneider (eds.), Monitoring Nature Conservation in Cultural Habitats, 137–145.*
© 2007 *Springer.*

To many of us, however, the key question remains – "How can we apply GIS to best effect in an environmental monitoring project?" A successful application of GIS in environmental monitoring is usually characterised by the presence of three distinctive components:

- A robust database design;
- Translation of data into the database; and
- Analysis of data to support environmental decision-making.

These three components form an integral part of the pre-fieldwork planning, fieldwork data collection and post-fieldwork analysis stages of environmental monitoring, respectively.

This chapter focuses on issues relating to the integration of data for conservation and environmental monitoring projects in a data system. It utilises Kenfig Dunes of South Wales (as discussed in Chapter 32), as a case study. A modern and widely used GIS has been selected for illustration purposes: this is ESRI's ArcView GIS version 3.2[1]. Another popular GIS used by many environmentalists is MapInfo GIS[2]. It is worth noting that files generated in ArcView can be converted for use in MapInfo and *vice versa*.

It is beyond the scope of this chapter to discuss the detailed functionalities of any specific GIS, though some of the most commonly used functions will be discussed in their generic terms.

## 2.        DEFINING THE PROJECT OBJECTIVES

From the outset, we must define clearly the objectives of the project. We need to ask ourselves what we want to achieve with GIS? For example, do we want to use GIS to manage our data? Or do we wish to model the growth of certain types of vegetation in the nature reserve for the next decade? The answers to these questions will determine not only the design of the database and data collection methods, but also the technical expertise required. The latter dictates whether the environmental monitoring project can be run entirely by conservation practitioners, or whether external GIS experts should be brought in.

For the purpose of this discussion, we will use our GIS to provide a system for managing environmental monitoring data for Kenfig Dunes in south-east Wales (Fig. 16-1).

## 2.1     Data Requirements

Once the objectives of the project have been defined and agreed, we need to produce a statement of data requirements. The principal function of this statement is to help to identify the data sets required to support the delivery of the objectives. Existing data sets are researched, and if necessary, the collection of new data is proposed.

---

[1] ArcView GIS is a product of the Environmental Systems Research Institute, Inc.
[2] MapInfo GIS is a product of the MapInfo, Inc.

The following is a typical checklist for existing data sets:
- Is it fit for the intended purposes?
- Is it available in digital (computer) format?
- If it is available in digital format, is it compatible with the GIS system?
- If it is not available in digital format, how much effort is required to translate it into digital format?
- Has the existing data any mappable geographical reference (e.g. National Grid Reference)?

The following is a typical checklist for new data sets:
- What geographical accuracy is required?
- How do we capture the geography in the field, e.g. do we use a GPS?
- Apart from geography, what other environmental data do we need to collect?
- Will automatic data loggers be used?
- Will photographs be taken in the field?

At a glance, this can appear to be a daunting task, and it certainly is if you are setting up a GIS for your organisation for the first time. Fortunately, many of the larger conservation organisations already employ (or have access to) GIS specialists and many existing data sets are already available in GIS-ready computer formats. In the case of our Kenfig Dunes GIS, the following data sets have been identified:

Existing data sets:
- Base maps – Ordnance Survey maps 1:50 000 and 1:10 000;
- Habitat maps – Hand-drawn maps from previous surveys.

New data sets:
Habitat survey includes:
- Geography and locations (using GPS);
- Habitat classifications (using 1 x 1m quadrats);
- Field photography (scanned film images).

Remote sensing survey includes:
- Using Compact Airborne Spectrographic Imager (CASI - see Chapter 30).

Topographical survey includes:
- Using Light Detection and Ranging (LiDAR) for capturing the topography of the study area as individual height data points.

# 3.    DATABASE DESIGN

A robust database design is vital to any environmental programme in which data may need to be held within the system for a long period of time and used for various monitoring and surveillance purposes. Fortunately, the database requirements for simple environmental monitoring projects are relatively straightforward. In most cases, a relational database that arranges data in two-dimensional tables is more than sufficient.

## 3.1    Translation of data into the database

This involves not only geographical data such as GPS locations, but also non-geographical data such as quadrat data sheets and habitat photographs taken in the field. The following section describes a number of generic tasks for constructing a GIS to manage environmental monitoring data. ESRI's ArcView GIS (version 3.2) is employed to illustrate the various generic functions used. It must be emphasised that it is not the author's intention to produce a comprehensive user-guide of ArcView GIS. Other commercial GIS systems provide similar, if not identical, functions.

This section is divided into two sub-sections. The first section describes some of the more basic tasks that can be undertaken by relatively inexperienced conservation practitioners, whilst the second describes tasks that require a higher level of technical expertise.

## 3.2    Basic data handling

### 3.2.1    Step 1 – Importing base maps

One of the most essential data sets for GIS projects is the base map.  In the UK, the Ordnance Survey (OS) supplies base maps at different scales under various agreements. The Pan-Government Agreement (PGA) allows all government departments and certain public bodies unlimited access to these base maps.  Other organisations may have to purchase them from the OS.

Amongst all the base maps, perhaps the most widely used products are the 1:50,000 scale maps in colour raster format. These colour raster base maps are usually supplied as individual files in .TIFF file format. Each of these .TIFF files represents a map tile covering a pre-defined area of some 400 $km^2$.

To display these map tiles in the GIS, a geo-referencing file must be created for each map tile. Fortunately, the OS has recognised the need to interface their base maps with GIS, and supplies these geo-referencing files through their website. In the case of .TIFF map tile files, the associated geo-referencing files have a .TIFW extension.  The command to import the base maps into ArcView GIS is

"View – Add Theme". Then "Image Data Source" under the "Data Source Type".

## 3.2.2 Step 2 – Importing other geometrically corrected images

The process for importing other geometrically corrected images is similar to that for OS base maps. These geometrically corrected images could come from a variety of sources, e.g. rectified aerial photographs, classified images from remote sensing data collection exercises.

In the case that the data provider does not supply the geo-referencing file, it will have to be created manually. The contents of the geo-referencing file for the OS map tile SS68 looks like this:

1. 5.00
2. 0.00
3. 0.00
4. −5.00
5. 260002.50
6. 199997.50

The first parameter shows the horizontal resolution of the pixel. The second and third parameters show the orientation variables. The fourth parameter shows the vertical resolution of the pixel. The fifth parameter shows the horizontal position of the centre of the top-left-most pixel. The sixth parameter shows the vertical position of the centre of the top-left-most pixel.

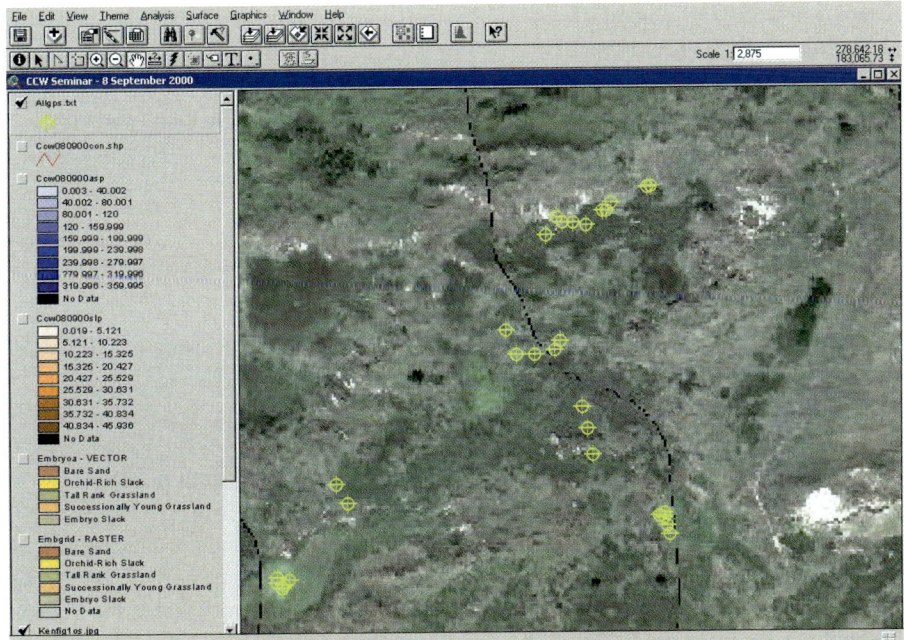

*Figure 15-1.* The GPS locations of ground-truth points at Kenfig: the backdrop is the CASI mosaic.

### 3.2.3      Step 3 – Importing GPS or other geographical data

Once the base maps are in place, we can go about adding the point-based GPS or other geographical data. First, we must assume that the geographical locations of the sample points have been captured using a modern positioning method such as GPS.

GPS has come a long way in the past decade, particularly in respect of size and accuracy. Some hand-held GPS can rapidly capture many hundreds of ground-truth points with positioning accuracy within 1 m. GPS manufacturers usually provide a range of software and hardware interfaces for downloading the positioning data into various well-established formats, e.g. the .CSV file format. These interchangeable files are essential for data exchange with GIS. Any commercial GIS will accept these CSV files. The key is the presence of well-defined geographical reference within the data file.

In the case of ArcView GIS, the .CSV file is renamed as a .TXT text file before being imported into the system. The "Add Event Theme" command will produce a geographical representation of all the data points held within the .TXT (.CSV) file. Fig. 15-1 shows the GPS locations of ground-truth points in Kenfig Dunes with a CASI mosaic as the backdrop.

### 3.2.4      Step 4 – Adding quadrat data to the GPS data points

The next step is to attach the quadrat survey data to the project. Conservation managers usually like to be able to access the full set of survey data while using the mapping system. The quadrat survey data are usually held in simple table-like files such as Microsoft EXCEL (.XLS) files.

This is one of the important functions that has long been lacking in traditional GIS systems. ArcView GIS provides a local programming (scripting) function, which allows the integration of two applications at the system level. First, you have to create a script containing the required commands, e.g.

*System Execute("c:\Program Files\Microsoft Office\Office\Excel.EXE c:\naw\gis\gcmi99\ccw-quad\bm.xls").*

This scripting function can then be called from within the GIS using a "Hot Link" function to bring up the quadrat data sheets.

### 3.2.5      Step 5 – Adding habitat photographs to the GPS data points

It is also an advantage to be able to associate the samples with photographic records of the habitat at the same locations. Again, this can be achieved by using the "Hot Link" function. First, the habitat photographs should be stored in a directory within the system. Secondly, a database field storing the full file path and filename of the photographic record is added to the GPS data file. Thirdly, a "Hot Link" is set up within the system. This can be done by selecting "Theme – Hot Link", and set the "Field" variable to the added database field, "Predefined Action" to "Link to Image File". The end-users can access the photographic records by pressing the "Hot Link" button.

### 3.2.6    Step 6 – Storing the completed monitoring project

Storing the entire monitoring project is a relatively straightforward task. ArcView GIS stores the definition of the project in an .APR file. It is important to note that the .APR file stores only the definition of the project and not the actual data. It is therefore essential to take note of the locations of all the necessary data files that are required by the project, for example, the OS base maps, their associated geo-referencing files, the GPS data file, the quadrat data files and the habitat photograph files.

It is good practice to store all the environmental monitoring data for a particular project under a single directory. In certain cases, it is possible to store all these data and the project in a single CD-ROM. This is a good way of both backing up the data and, if appropriate, distributing them to remote monitoring sites. The command to save the completed project is "File – Save Project".

### 3.2.7    Step 7 – Adding new data from repeat monitoring exercises

It is likely that new monitoring data will need to be added to the system in the future. We can do this by repeating Steps 3, 4 and 5 in the previous section. Steps 1 and 2 are not required unless the new survey extends to areas not covered previously by the base maps, or new geometrically corrected data have become available. The completed project can again be stored by issuing the "File – Save Project" command to the system as in Step 6.

## 4.    SECTION 2 – ADVANCED DATA HANDLING

This section is designed for conservationists with more experience of GIS. It discusses advanced data-handling of existing habitat maps and Digital Elevation Models (DEMs).

## 4.1    Existing habitat maps

As a project develops, we may want to convert historic records into GIS. During the research into an environmental monitoring project, it is not unusual to uncover historic records, particularly in the format of aerial photographs or habitat maps.

The key to a successful integration of historic records into GIS is the quality of the base map that these historic records were derived from. For example, it is extremely difficult, if not impossible, to transfer the habitat boundaries of a hand-drawn habitat map derived from a small-scale aerial photograph. However, it would be a different scenario if the habitat map were derived from an old base map. As long as the map projection parameters can be identified, there is no reason why the historic habitat map cannot be digitised and the digital boundaries converted to the GIS.   The procedure for converting historic habitat boundaries to GIS is as follows:

- Identify the historic habitat maps;
- Identify the base maps underlying the habitat maps;
- Identify the map projection parameters of the base maps;
- Digitise the historic habitat map; and
- Convert the digital boundaries of the historic habitat map to GIS.

In the case of historic habitat boundaries based on aerial photographs, the following procedure applies:

- Identify the historic habitat maps;
- Identify the aerial photographs underlying the habitat maps;
- Identify the flight parameters of the aerial photographs (e.g. flying height, photographic lenses characteristics, calibration parameters);
- Rectify the aerial photographs to remove distortions;
- Digitise the historic habitat map; and
- Convert the digital boundaries of the historic habitat map to GIS.

## 4.2     Importing Digital Elevation Models (DEMs)

Digital Elevation Models (DEMs) are a three dimensional representation of the real world. They are invaluable to environmental analysis that involves topography. DEM data can be acquired in a number of ways. Height data can be acquired by digitising contours from base maps, conducting a traditional aerial photographic survey, or through a relatively new airborne technology called Light Detection and Ranging (LiDAR). Once acquired, the DEM height data can be uploaded to the GIS as described in Step 3 previously.

To handle DEM data, ArcView GIS requires the installation of the "Spatial Analysis" extension, an additional software component. Once installed, the "Surface – Interpolate Grid" command helps to generate a raster data grid with interpolated heights throughout the pre-defined area. The size of the grid is under the control of the practitioners.

## 5.     ANALYSIS OF DATA TO SUPPORT ENVIRONMENTAL DECISION-MAKING

When the DEM has been imported it will be possible to analyse various aspects of the data associated with height, aspect and slope in relation to the vegetation data. Fig. 15-2 shows a 3D view of a vegetation classification draped on the top of the DEM for interactive visualisation.

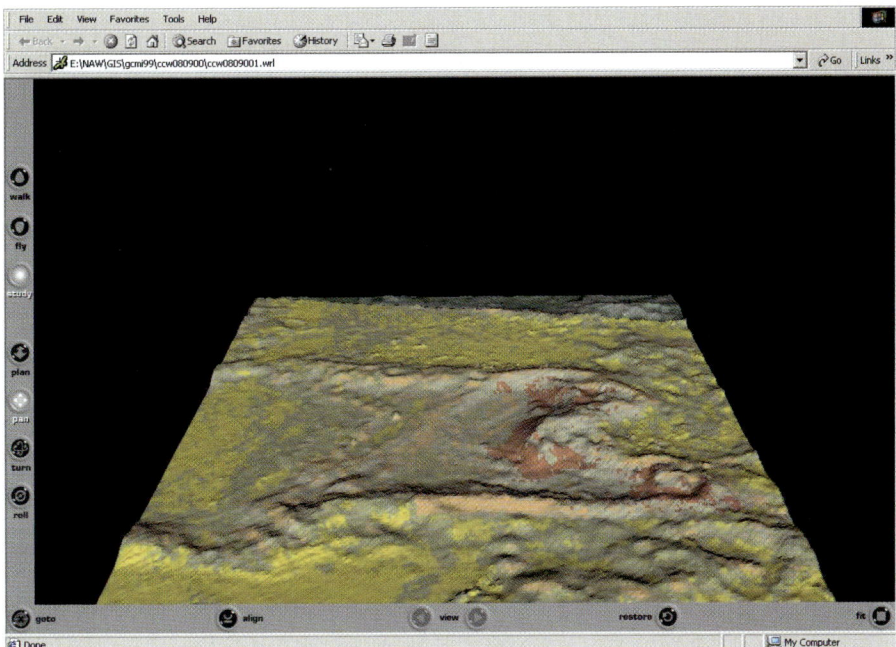

*Figure 15-2.* The 3D view of Kenfig Dunes with vegetation classification draped over the DEM.

# 6.    GIS TRAINING

To operate GIS to its full potential, you will need professional training in the use of a) the GIS toolkits and b) the optimal ways of organising the data. GIS training can be a complex and expensive exercise. The initial capital costs for acquiring a GIS and its associated data soon seem negligible compared with the long-term revenue costs of its maintenance. Only the larger conservation organisations can afford dedicated staff operating the GIS, and then efforts often focus on corporate requirements rather the requirements of individual conservation practitioners.

Over the past decade, I have personally witnessed professional ecologists attempting to double up as GIS operators, after only a few days of "training". This is an inefficient use of resources. A practical, and cost-effective solution is for GIS specialists to provide an easy-to-use environmental application tailored for the requirements of conservationists, where the operator of the system requires minimal training. With the aid of recent technological advances in the development of GIS, it is now possible to develop a user-friendly interface that allows conservation practitioners to collate the various strands of monitoring data into one project as easily as you can import an image into a Word document. This type of interface has been developed for many professional industries, and could easily be adapted for the purposes of conservation management and monitoring.

# PART IV

## THE CASE STUDIES

# CHAPTER 16

# THE MONITORING CASE STUDIES

## CLIVE HURFORD

*Countryside Council for Wales, Plas Penrhos, Ffordd Penrhos, Bangor, Gwynedd, LL57 2BQ*
*clive.hurford@serapias.net*

## MICHAEL SCHNEIDER

*County Administration, Västerbotten County, SE-901 86, Umeå, Sweden*
*Michael.Schneider@ac.lst.se*

## 1.  INTRODUCTION

The book so far has focused on the problems associated with habitat monitoring and ways of overcoming them: the remainder of the book primarily comprises case studies that demonstrate the practical application of monitoring. The case study sites are distributed in five countries: England, Finland, Germany, Sweden and Wales, though most originate from Wales (Fig.16-1) and Sweden.

In selecting the case studies, we have tried to cover as many broad habitat types and species groups as possible. The habitat case studies include arable land, neutral grassland, coastal heaths, broad-leaved woodland, coastal dunes, fens and coastal lagoons, while the species case studies cover large mammals, bats, birds, butterflies and snails. Most of the case studies are included in this part of the book, though the broad-leaved woodland case study is in Part V, and those for sand dunes, fens and coastal lagoons are in Part VI.

## 2.  KEY ISSUES FOR SPECIES MONITORING

The problems associated with habitat monitoring have been discussed extensively in earlier chapters, so this brief introduction focuses on some of the issues that we have to consider when we are developing a monitoring project for a species.

For conservation management purposes at least, species can be divided into two very broad groups: habitat specialists and habitat generalists. Many, though by no means all, of the rare and threatened species in Europe are habitat specialists.

*C. Hurford & M. Schneider (eds.), Monitoring Nature Conservation in Cultural Habitats, 149–156.*
© 2007 *Springer.*

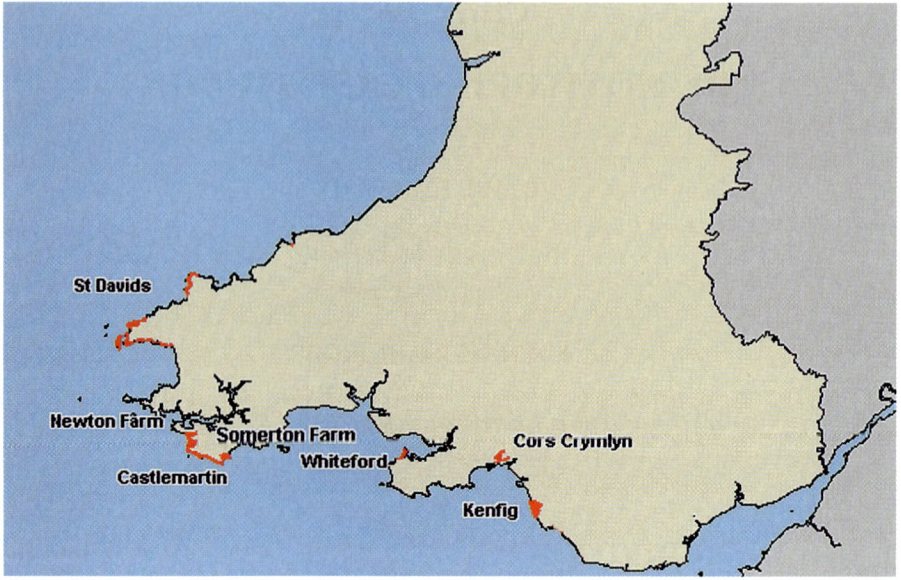

*Figure 16-1*. The distribution of the case study sites in Wales. Reproduced from Ordnance Survey mapping on behalf of Her Majesty's Stationery Office © Crown Copyright 100043571 2004-12-10.

These species are strongly associated with a particular habitat, and often with a specific habitat state. Habitat generalists, by contrast, tend to be far more catholic in their requirements. The case studies include examples from both groups.

Where the condition of the species is linked to the condition of its habitat, we recommend monitoring both the species and its habitat. There are several reasons for this, including:

- If the viability of a species is linked to the condition of its habitat, then detecting changes in habitat condition will allow us to make a management response before the species population is threatened.
- If the species is strongly associated with a cultural habitat, our most likely management response to a decline would involve habitat restoration;
- Many species can be difficult to find, let alone count. Therefore, we can be more confident that the species population is secure if we know that a) the species still is present at several locations and b) that its habitat is in optimal condition; and
- If we monitor only the species, when the population falls below its lower limit, we will have no record of the habitat condition when the population was at its peak: by monitoring both we will have a series of reference points to inform habitat restoration.

Although there can be many reasons for the decline of species, in many cases, either or both of habitat degradation and persecution will be major contributory factors. These are problems that can be addressed directly though habitat and species management. We should not commit valuable time and resources to monitoring the habitat, however, if the condition of the species is independent of habitat condition.

## 2.1    Selecting attributes for monitoring a species

The underlying principles of monitoring are the same for both habitats and species (Chapter 2). The monitoring differs only in our selection of attributes and the way we collect the data.

With respect to the most appropriate attributes to monitor, these will vary according to the species. In most cases, however, we will know enough about a threatened species to identify the critical attributes. For example, we will almost certainly know:

- How to identify it, at least in certain stages of its life cycle;
- Its breeding habits;
- Something about its habitat requirements;
- Its main prey or foodplants; and
- Something about its behaviour and life cycle.

For higher plants, mammals, birds and the larger, more obvious species of invertebrate, much of this information will be available, one way or another, from popular texts on the relevant species groups, e.g. Asher *et al.* (2002): Niethammer & Krapp (1978) *et seq.* Similar information is also likely to be available for rare species from the less popular groups, though it may take a thorough literature search or contact with species experts to find it.

## 2.2    Avoiding unnecessary monitoring effort

As with habitats, there is any amount of information that we might like to know about a species. For the purposes of monitoring, however, we should focus on the information that we need know. Therefore we must consider what assumptions we can make. For example, if we have a large population with a good age structure, can we make assumptions about available food resources and survivorship? Similarly, if survivorship is high, can we assume that predation levels are low? If we always have a healthy population of adults, can we make assumptions about breeding success? In the latter case, this will depend, to some degree, on the life span of the species. If we can make these types of assumption, then we can increase the efficiency of the monitoring project.

## 2.3    Monitoring indirect attributes

For many animal species, it will not be economically feasible to find and count the number of individuals in a population. In this case, monitoring tracks or signs of the species can be an alternative to direct observation. In Sweden, the tracking of large mammalian predators in the snow is a highly developed art used for both monitoring and surveillance (Aronson & Eriksson, 1990). Brown Bear *Ursus arctos* droppings are collected for DNA analysis and provide information on bear distribution, population size and sex structure, and individual home ranges (Taberlet *et al.*, 2001). We can also use signs of feeding to monitor some invertebrate species, such as wood-boring insects,

*Photograph by Clive Hurford.*

*Figure 16-2.* Animal tracks feature prominently in monitoring projects developed for large carnivores in Scandinavia. These tracks of American Mink *Mustela vison* in the mountains of north-west Sweden illustrate that the effects of human activities can be far reaching.

*Photograph by Clive Hurford.*

*Figure 16-3.* These tracks, in the sand dunes at Coto Doñana in south-west Spain, give an indication of the level of animal activity the previous night. Tracks are one source of information on movements of Iberian Lynx *Lynx pardinus* in this area.

instead of destructively sampling individuals from threatened populations in protected areas (Ehnström & Axelsson, 2002). The Bear and Wolverine case studies (Chapters 20 and 21) provide examples of how indirect attributes can be incorporated in a monitoring project.

## 3. TARGET SETTING

As is the case for habitats, when it comes to setting targets for a species, we should always set the lower limit at the point that we would become concerned, and not at the point that it is threatened. We should also consider whether we need an upper limit for the species: if the population is too strong, will it have a negative impact on another rare species? If we are protecting a predator, we will certainly be discriminating against its prey. We need to think about the point at which we would consider a recovering species to be restored. This is unlikely to be at the same point we became concerned about it, but there should still be a stage in its recovery when we would consider the species to be secure. Too often, we will set up a species recovery programme without defining when the species will be considered to have recovered. Consequently, we find ourselves committing resources to protection long after there is any justification for it, and sometimes with devastating consequences for other vulnerable species.

*Photograph by Clive Hurford*

*Figure 16-4.* If we are monitoring dragonflies, then the most reliable evidence of breeding is either freshly emerged adults, like this Common Darter *Sympetrum striolatum*, or the presence of exuviae on emergent vegetation.

## 4.      WHEN TO MONITOR

As with habitats, we must consider the best time of year to carry out the monitoring, and the most appropriate age class to monitor.  For example, if we want to know that a rare species of dragonfly (Odonata) is still breeding on our site, counting the number of adults will not answer that question: we will need to look for either emerging larvae or exuviae.  In this case, we would have to time our monitoring to coincide with the peak emergence period for the species in question, and get up early on mornings when the weather conditions are conducive to emergence, typically after clear calm nights.


## 5.      USING ASSOCIATED SPECIES AS INDICATORS OF HABITAT CONDITION

It is difficult to understand the concept of assessing whether a habitat is in optimal condition without taking into account the species that should be associated with it.  In the UK, for example, if an area of grassland supports an important, but declining, population of Marsh Fritillary *Eurodryas aurinia* butterflies, can we really claim that habitat is in optimal condition simply because the vegetation conforms to a particular plant community type?

The major flaw in the argument for monitoring against generic standards is that every site has different conservation potential, both for the extent of the habitat, and for the diversity of fauna and flora associated with it.  Therefore, the definition of optimal (or favourable) condition for a habitat must take into account the fauna and flora that we would expect to be associated with the habitat if it truly was in optimal condition; this definition will essentially be site-specific.  This is the only sensible interpretation of the term 'typical species' as used in the EC Habitats Directive.  The habitat case studies deal with this in two ways.  They either:

- Take account of the requirements of the key associated fauna and flora in the definition of optimal condition; or
- Include a requirement for the key associated species to be present.

Similarly, most species case studies include a definition of optimal habitat in the condition indicator table, on the basis that we cannot consider a species to be in optimal condition if its habitat is degrading.  If we have clearly identified the conservation priority on a site (Chapter 7), we will not find ourselves with conflicting definitions of optimal habitat.


## 6.      SPECIES AS ENVIRONMENTAL INDICATORS

One area that has not been exploited to its full potential in the case studies is the inclusion of known environmental indicators in the condition indicator tables.  The obvious examples are sensitive species of lichen (Luigi Nimis *et al.*, 2002: Richardson,

1992) and freshwater invertebrate, which are known to be good indicators of atmospheric pollution and water pollution respectively. Many of these indicators are relatively easy to identify and could be incorporated in habitat monitoring projects at very little additional cost. For example, if we were monitoring the condition of aquatic vegetation in a slow-flowing stretch of river, noting the presence and distribution of the Banded Demoiselle damselfly *Calopteryx splendens* would provide us with an indication of water quality at the same time. Similarly, if the more sensitive species of epiphytic lichen, e.g. species of *Lobaria* or *Usnea* (see Chapters 27 and 29), are well distributed on trees in our wood, we can be relatively confident that any problems have not resulted from atmospheric pollution. We could easily include a lower limit in our condition indicators for the occurrence and distribution of these types of environmental indicator without compromising the efficiency of the monitoring project.

## 7. ANTICIPATING CLIMATIC CHANGE

There may be no practical management response to a retraction of range resulting from climatic change. However, if we expect this to happen, we should at least assess whether our species of interest exists in the projected climatic conditions elsewhere and, if so, a) identify the habitats it occupies in those areas and b) find out how these habitats are managed. In the case of climatic warming, and in the absence of catastrophic change, we should expect species at the southern extremes of their range to retract to the north,

*Photograph by Clive Hurford*

*Figure 16-5.* The Banded Demoiselle *Calopteryx splendens* damselfly is sensitive to water pollution and therefore can be a good indicator of water quality that is easy to find and identify.

and those near their northern limit to expand from the south.  It may be difficult to unravel the reason for declines in some species, but if a species is maintaining its population levels in similar habitats further south, the decline is more likely to be related to habitat degradation than climatic factors.  There may be little that we can do to prevent some species retracting northwards, but we can at least ensure that suitable habitat is available for species that we would expect to persist in our region or expand into it from the south.  If we don't do this, then the conservation losses resulting from climate change will be far more pronounced than the conservation gains.

Agriculture will adapt to changes in climate and so must nature conservation.  Many changes can be predicted by looking at the distribution of our fauna and flora elsewhere in Europe, and by looking at which species can co-exist in habitats elsewhere.  It is debatable whether we should ever target 'edge of range' species for conservation management, because they are usually common elsewhere, and because their presence is determined by factors mostly beyond our control.  This is particularly true at present. 'Edge of range' species are likely to be the most reliable indicators of climate change, but the least reliable indicators of habitat condition.  As a general rule, the best indicators for conservation management are the species that we would expect to persist irrespective of oscillating climatic conditions.

## 8.      REFERENCES

Aronson, Å. and Eriksson P. 1990. *Djurens spår och konsten att spåra.* – Bonniers, Stockholm. (in Swedish).

Asher, J., Warren, M., Fox, R., Harding, P., Jeffcoate, G. and Jeffcoate, S. (2001). *The Millennium Atlas of Butterflies in Britain and Ireland.* Oxford University Press.  Oxford.

Ehnström, B. and Axelsson, R. 2002. Insektsgnag i bark och ved. – ArtDatabanken, Swedish University of Agricultural Sciences, Uppsala. (in Swedish)

Luigi Nimis, P., Scheidegger, C. & Wolseley, P. (eds.). (2002).  *Monitoring with Lichens – Monitoring Lichens.*  Kluwer Academic Press.  Dordrecht.

Niethammer J. and Krapp F. (eds.) 1978 – 2003. Handbuch der Säugetiere Europas (Handbook of European mammals), six volumes. Aula Verlag, Wiesbaden. (in German)

Richardson, D.H.S. (1992).  *Pollution monitoring with lichens.*  Naturalists' Handbook 19. Richmond Publishing Co. Ltd. Slough.

Taberlet, P., Luikart, G. and Geffen, E. 2001. New methods for obtaining and analyzing genetic data from free-ranging carnivores. - In: Gittleman, J. L., Funk, S. M., Macdonald, D. W. and Wayne, R. K. (eds.), *Carnivore conservation.* Cambridge University Press, Cambridge, pp. 313-335.

# CHAPTER 17

# MONITORING COASTAL HEATHS AT ST DAVID'S

CLIVE HURFORD

*Countryside Council for Wales, Plas Penrhos, Ffordd Penrhos, Bangor, Gwynedd, LL57 2BQ*
*clive.hurford@serapias.net*


STEPHEN EVANS

*Glanymor, Dinas Cross, Pembrokeshire, Wales, UK*
*glanymor.dinas@virgin.net*

## 1.    INTRODUCTION

The Natura 2000 site at St David's covers more than 50 km of coastline in southwest Wales, from Strumble Head in the north to Cwm Mawr in the south (see Fig.16.1).  The site also includes Ramsey Island, which is managed by the Royal Society for the Protection of Birds (RSPB) and lies 1-2 km offshore from Whitesands Bay near St David's.   Much of the site is under private ownership, with the rest mostly owned by the National Trust, though Pembroke Coast National Park leases the foreshore from the Crown Estate.

Every year, the coastal path at St David's attracts thousands of visitors, particularly in spring and early summer when Thrift *Armeria maritima*, Hairy Greenweed *Genista pilosa*, Spring Squill *Scilla verna* and Sea Campion *Silene uniflora* produce spectacular displays of colour.  The more enthusiastic botanists can also see several rare and locally scarce plant species, including Chives *Allium schoenoprasum* (Fig. 17-5),

*C. Hurford & M. Schneider (eds.), Monitoring Nature Conservation in Cultural Habitats, 157–168.*
© 2007 Springer.

Roseroot *Sedum rosea,* Yellow Centaury *Cicendia filiformis,* Spiked Speedwell *Veronica spicata,* Pale Dog-violet *Viola lactea,* Lanceolate Spleenwort *Asplenium obovatum* and endemic "species" of Sea Lavender *Limonium* spp.

A wide range of coastal habitats is present, ranging from rock-crevice communities on the most exposed cliff faces to maritime grassland, heath and scrub in the hinterland (Fig.17-3). These vegetation types present reflect the edaphic, physical, and geological factors, e.g. soil type, soil depth, degree of exposure to salt spray, aspect, and angle of slope.

The Natura 2000 site comprises three Sites of Special Scientific Interest (SSSI): St David's Peninsula Coast SSSI; Ramsey SSSI; and Strumble to Llechdafad Cliffs SSSI. For management purposes, each SSSI has been divided into 'managed' and 'minimum intervention' sections (22 in all), with conservation management aims decided for each managed section.

## 1.1    The minimum-intervention sections

The areas selected as minimum-intervention sections were mostly self-perpetuating or 'near natural' habitats. These were largely situated on the more exposed seaward parts of the site, e.g. the cliff ledges and the narrow, precipitous stretches of coast between the headlands. In contrast, other minimum-intervention sections were in some of the more sheltered cliffs, where ancient wind-pruned scrub of Blackthorn *Prunus spinosa* with patches of Privet *Ligustrum vulgare*, unmodified by man, still clings to steep, narrow cliff-faces.

*Photograph by Clive Hurford.*

*Figure 17-1.* The Pembrokeshire coast, seen here at Pwllderi, attracts many thousands of visitors every year.

The extent, distribution and condition of the vegetation types in these sections is largely dictated and maintained by natural forces, such as exposure to wind, salt deposition (and associated conditions of drought), and climatic extremes. Active management and, by association, monitoring of these habitats is neither practical nor necessary.

## 1.2    The managed sections

All of the main headlands and their adjoining sheltered inland areas were selected as managed sections, i.e.

- Areas where the extent and distribution of the habitats has been influenced by management activities, e.g. livestock grazing, cutting or burning over many centuries; and
- Accessible areas where succession to scrub or woodland is likely to occur in the absence of practical habitat management.

Over the period from the mid 1960s to the mid 1990s, pastoral practices had declined in these managed sections of the site, resulting in the spread of Bracken *Pteridium aquilinum* and scrub (mainly Gorse *Ulex* spp. and Bramble *Rubus fruticosus*). The managed sections include all of the larger headlands, because these invariably support vegetation types that require active habitat management and monitoring.

## 1.3    Key factors influencing the management decisions

The conservation manager at St David's (SE) was an experienced ecologist with a good understanding of coastal habitats. He had been studying sea-cliffs in Pembrokeshire since 1973 and before that date, had been a member of the former Nature Conservancy's UK Sea-cliff Vegetation study group (Mitchell & Malloch, 1991). He had a clear vision of what would represent optimal condition for the site, and recognised that the continuation of low-intensity grazing and neglect would result in a long-term decline in conservation value. The following factors influenced the management decisions at St David's.

- Many of the coastal headlands of the St David's peninsula had been enclosed by stone-walled fields during the Iron Age and, consequently, their vegetation had been grossly modified by man for many centuries. Grazing and burning and episodes of cultivation had created extensive areas of non-maritime heathland inland of the truly maritime heathland zone.
- Current management varied according to ownership, restrictions on 'common land', and accessibility to grazing stock.
- Several sections of coast had no conservation management agreement.
- Scrub control by patch burning had declined, resulting in many large areas of over-mature heath.
- Most sections of the coast suffered occasional uncontrolled winter and summer burns (independent of management agreements).

- Many sections were not grazed, because graziers feared that stock would lose condition on the cliffs and were worried about disturbance from walkers and their dogs.
- Where grazing did occur, stock levels were low.

## 1.4    The management priority at St David's

On sites of high conservation interest, it can be difficult to decide which species or habitat should drive the management of the site. However, we rarely want to manage a site for one feature at the expense of all others. At St David's, the low level management regime of the late 1990s represented a direct threat to both the local (Red-billed) Chough *Pyrrhocorax pyrrhocorax* population and to the coastal and maritime heath habitats, including several rare heathland plants. Many of the rare plant populations in the St David's Natura 2000 site were associated with the cliff crevices or thin soils, and were relatively secure, despite the retreat of pastoral practices in recent years.

The conservation manager decided that the priority at St David's should be the coastal and maritime heaths, because the restoration management for these habitats would also benefit the Chough population, and would be unlikely to have a detrimental effect on the maritime grassland and rare rock crevice plants.

*Photograph by Clive Hurford.*

*Figure 17-2.* An area with rabbit-grazed open, species-rich, heath vegetation on Ramsey Island.

## 2.     THE SURVEY PHASE

Although the heaths at St David's were surveyed in 1995/96 using the National Vegetation Classification (NVC), the data for the maritime grassland and ledge vegetation were incomplete.  Consequently, we carried out a survey of the grassland and ledges in the first two weeks of June 1997, a year before the monitoring was scheduled.  This involved:

- Walking the site, noting the typical vegetation types, the vegetation structure and species composition;
- Taking colour transparencies of the main vegetation types from fixed-points located with a GPS;
- Taking detailed samples in the key vegetation types (using 50 x 50 cm quadrats divided into sixteen 12.5 x 12.5 cm cells, and recording the cell presence of all vascular plant species); and
- Presenting annotated base maps of the site.

This information, together with the maps from the 1995/96 NVC survey, provided the conservation manager with a detailed account of the vegetation types and their distribution on the site.  The results from the survey data supported his opinion that the coastal heath habitats needed restoration management.

## 3.     THE SURVEILLANCE PHASE

The surveillance fieldwork was carried out in June 1998. The recording concentrated on the coastal heath habitat, as most of the maritime grassland occurred on steep slopes in minimum intervention sections.  The surveillance method involved:

- Locating ten permanently-marked 'L-shaped' linear plots in managed heathland sections;
- Placing five 50 x 50 cm quadrats, each divided into 16 cells, along the length of each plot;
- Recording the 'cell' presence of selected 'stress-tolerating' species within each quadrat;
- Estimating the percentage cover of selected 'competitive' species within each quadrat;
- Using quadrat photography to support the estimates of ericoid cover in each quadrat;
- Locating each plot with a differential Global Positioning System (DGPS) and using buried transponders to mark both ends of the plot and the angle of the 'L'; and
- Taking fixed-point photographs of the ledge vegetation at selected points within each SSSI.

# 4.      THE CONSERVATION AIMS

A conservation aim was set for each managed section of the site. This took account of:

- Information provided during the survey phase of the project;
- The management history of the site;
- The prevailing climatic conditions; and
- The local geology.

The conservation aims focused on the condition of the coastal and maritime heath habitats because:

- Few sizeable stands of maritime grassland occurred in the managed sections, and those that did would benefit if the adjacent heathland was managed by appropriate levels of grazing; and
- The extent of the heaths had declined as a result of past and present levels of grazing and burning.

Without management, much of the existing heathland would undergo succession to scrub or woodland habitats, initially by forming dense stands of *Ulex europaeus* and *Rubus fruticosus*. This has already happened - and continues to happen - where the maritime influence is least pronounced. This succession would not only reduce the extent of an internationally important habitat, it would also have a detrimental effect on the local Chough population, which uses the heath vegetation for feeding after dispersal from the nest sites. Occasional light burning, combined with heavy cattle grazing, should result in an increase in the younger growth phases of heath, and ensure that family parties of Chough can still access the habitat.

## 4.1    The condition indicators

Table 17-1 below outlines the evidence that the conservation manager would accept as an indication that the restoration management had been successful. We spent two days in the field with the conservation manager testing the definitions in the condition indicator table (and modifying them) until we were certain that they would deliver the required information. Monitoring was only carried out when we were confident that the criteria were effective.

*Table 17-1.* The condition indicator table for the coastal and maritime heath habitats in the managed sections at St David's.

| Condition indicator table for habitat restoration | The coastal and maritime heath habitat at St David's will be restored to optimal condition when, in each managed section: |
|---|---|
| **Recovery target** | the coastal and maritime heath comprises vegetation where<br><br>>70% of the sampling points are either open heath or open species-rich heath<br><br>and where<br><br>>30% of the sampling points are open, species-rich heath<br><br>and when<br><br>*Pteridium* and scrub cover is <20% of the section |
| **Definition of coastal and maritime heath** | Within a 1 m radius of each sampling point there is:<br>>10% ericoid cover |
| **Definition of open heath** | Within a 1 m radius of each sampling point there is:<br>>30% sub-shrub cover of *Calluna vulgaris, Erica cinerea* or *Genista pilosa*<br>AND<br>>10% bare soil or vegetation < 3 cm in height (excluding areas of exposed rock) or *Scilla verna* |
| **Definition of open, species-rich heath** | Within a 1 m radius of each sampling point there is:<br>> 30% sub-shrub cover of *Calluna vulgaris, Erica cinerea* or *Genista pilosa*<br>AND<br>>10% bare soil or vegetation < 3cm in height (excluding areas of exposed rock) or *Scilla verna*<br>AND<br>>2 of the following are present:<br>*Festuca rubra/Festuca ovina, Lotus corniculatus, Viola riviniana, Hypochaeris radicata, Anthoxanthum odoratum* |

## 4.2 Explanation of the condition indicators

The heath habitat is divided into three categories:
- Coastal and maritime heath;
- Open heath; and
- Open, species-rich heath.

The management aim is to ensure that more than 70% of the heathland is restored to a successionally-young growth phase, where patches of ericoids are separated by bare ground or closely grazed vegetation. Management based on repeated burning can create young heath with an open structure, but can also result in species-poor stands of heath over a hard topsoil crust. Because such shallow soiled species-poor heathland would be

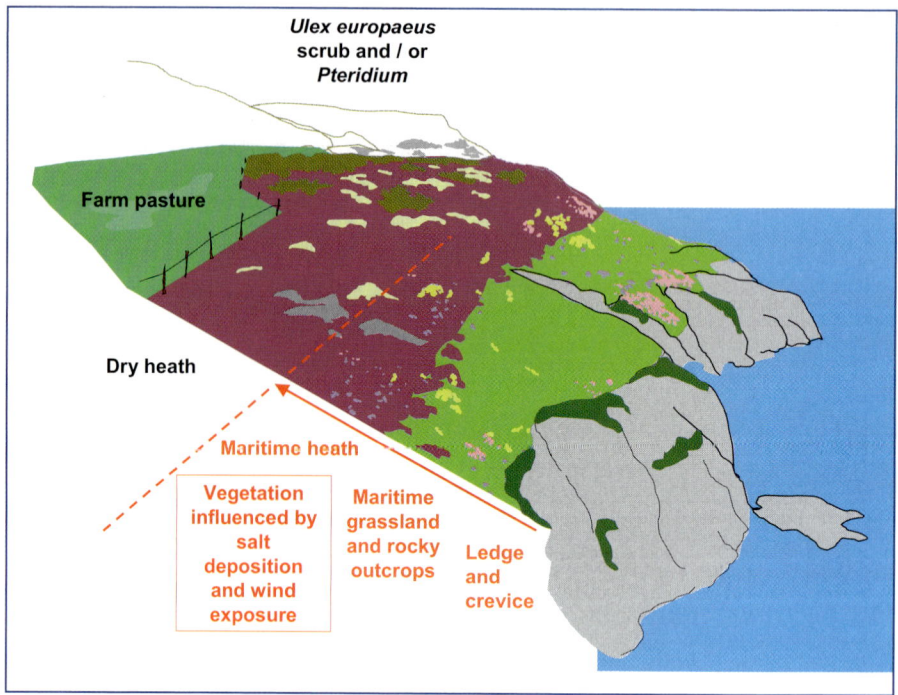

*Figure 17-3.* A diagram showing the typical habitat progression inland from the coast at St David's.

unlikely to provide good feeding areas for the Choughs, at least 30% of the heathland will have to be managed by appropriate levels of heavy-stock grazing. The logic behind the selection of attributes for the different states of heath is given below:

## 4.2.1    Coastal and maritime heath

The assumption here was that, if there was more than 10% ericoid cover, then the habitat either was, or had the potential to be, coastal or maritime heath. This level of ericoid cover suggested that the soil conditions were suitable for heath vegetation, and any vegetation with >10% ericoid cover was recorded as coastal and maritime heath. It is recognised that ericoids could persist at levels below 10% cover in transitional areas. However, by monitoring the areas with low levels of ericoid cover, over time we should detect whether the restoration management is increasing the extent of heath with >30% ericoid cover.

## 4.2.2    Open heath

This category of heath, with >30% sub-shrub cover, is already what we would consider to be 'true heath' vegetation, but the requirement for >10% cover of bare soil, vegetation <3 cm high or *Scilla verna* means that this habitat must in a successionally-young growth phase, either through burning, grazing or a combination of both.

### 4.2.3    Open, species-rich heath

Another 'true heath' category with an open structure and >30% sub-shrub cover. In addition, there is a requirement for three or more of the listed species to co-occur: these species were selected from data collected during the survey phase of the project. In management terms, this habitat state will be delivered through appropriate levels of heavy stock grazing.

## 5.    THE MONITORING PHASE

In the condition indicator table, it clearly states that the coastal and maritime heath in all managed sections of the site has to pass the criteria before the habitat will be restored. This allowed us to sample strategically.

Prior knowledge of the site, including experience gained during the survey phase, suggested that most of the coastal and maritime heath vegetation at St David's would fail to meet the criteria in the condition indicator table. Therefore, we started monitoring in those sections of the site that the conservation manager considered most likely to pass. If these sections failed to meet the criteria, then it would be safe to assume that most, if not all, of the other sections would also fail.

## 5.1    The monitoring at Point St John

The first managed section that we monitored was Point St John: a relatively small headland under grazing management.

Initially, we estimated the extent of *Pteridium* and scrub on the habitat map of Point St John. If the cover had been more than 20% of the section, the habitat would have failed the criteria in the condition indicator table and point-based monitoring would have been unnecessary. In the event, the combined cover of *Pteridium* and scrub was less than 20% of the section and we went ahead with monitoring the heath vegetation.

During the monitoring exercise, we recorded on a systematic grid (see Chapter 13), locating each sampling point with a DGPS accurate to <50 cm. The monitoring involved pacing out a grid and stopping at c.20 m intervals to record whether the vegetation within the 1 m area of search passed each of the criteria in the condition indicator table. This information was recorded on a monitoring form.

Points that landed in scrub, rock and grassland were excluded, but all points in vegetation with more than 10% ericoid cover were recorded.

*Figure 17-4.* An area of open, species-rich heath vegetation in the central area of Ramsey Island. This area would be accessible to feeding Choughs.

*Figure 17-5.* Chives *Allium schoenoprasum*, one of the rarer plants found in the Natura 2000 site at St David's.

# 6. RESULTS AND ANALYSIS

The results from the monitoring at St David's are shown in the table below.

*Table 17-2.* The monitoring results from Point St John, Strumble Head and Ramsey Island.

| Site name | Total number of points in coastal and maritime heath | Number passing as open heathland | Number passing as open, species-rich heathland | % passes |
|---|---|---|---|---|
| Point St John | 35 | 3 | 11 | 40 |
| Strumble B | 72 | 10 | 8 | 25 |
| Strumble C | 73 | 23 | 6 | 40 |
| Ramsey | 186 | 10 | 19 | 16 |

The percentage of monitoring points passing the criteria for open heathland at Point St John fell well short of the restoration target of 70%. Therefore the condition of the coastal and maritime heath habitat was sub-optimal.

We also monitored the three other managed sections of the site thought most likely to pass the criteria in the condition indicator table: Sections B and C in the Strumble to Llechdafad Cliffs SSSI, and Ramsey Island SSSI. The results from these sections are also shown in Table 17-2.

These results mean that the best examples of coastal and maritime heath in each SSSI failed to meet the criteria in the condition indicator table. On this evidence, there can be no doubt that the overall condition of the heathland habitat at St David's is currently sub-optimal and in need of restoration management.

It took us two staff days to monitor each section, but only two staff days in total to establish that the condition of the coastal and maritime heath habitat within the Natura 2000 site at St David's was sub-optimal: we knew that after monitoring Point St John.

# 7. DISCUSSION

There are three main areas of discussion raised by this case study. Firstly, the implications for site management from the monitoring results, then the implications for future monitoring, and finally the implications for reporting.

## 7.1 The implications for site management

Although the monitoring was carried out in only four sections of the site, these were the sections where the coastal and maritime heath was most likely to meet the criteria in the condition indicator table. As such, these were also the sections most conducive to restoration management.

In conservation terms, it would be sensible to prioritise the sections that we monitored for restoration management as, in the short term, these have the best chance of being restored. This will ensure that at least four sections of the coast will support an open heathland structure of benefit to plants and invertebrates, and optimise feeding opportunities for Choughs in the immediate future. The alternative would be to prioritise the sections of coast that are most overgrown for restoration. In these sections, it would take longer to deliver the management aims and demand more resources. There is also a risk that the few remaining areas of reasonably well-structured heath would close up while we were putting our efforts and resources into restoring these overgrown sections.

Only when the appropriate restoration management is in place in the better sections, should we continue the process in the remaining sections of the site.

## 7.2    The implications for repeat monitoring at St David's

The methods used for monitoring the heath vegetation at St David's were user friendly, i.e. they could be carried out by non-specialists, and have been proved to deliver consistent results between observers. Recording a small number of key attributes at each monitoring point, and using a DGPS to log the location of the points, means that we can obtain information on changes in habitat extent and habitat quality over a short period of time. Two recorders would spend only four days repeating the monitoring at St David's.

This monitoring was targeted at only four sections of the site, however, and the other managed sections will also need monitoring. In these situations, we recommend that the remaining sections are monitored immediately before they come under restoration management, spreading the monitoring effort over time.

The attributes in the condition indicators have been defined in terms that will allow observers to look down and assess whether the vegetation where they are standing will pass the criteria. This makes it possible to gain a reasonable impression of the condition of the habitat in each managed section without recording the monitoring points until a) we are no longer certain of what the result will be, or b) we need to confirm that the habitat has been restored before switching to a programme of maintenance management.

## 7.3    The implications for reporting

Because all of the managed sections were believed to be in a sub-optimal state before monitoring, it was possible to monitor the three sections most likely to pass the criteria in the condition indicators and to make the assumption that if these failed so would the others. This information puts the conservation manager in the position of being able to plan the restoration strategy without having monitored the whole of the site. The monitoring result also provides reliable site-based evidence for statutory reporting obligations.

## 8.    REFERENCES

Mitchley, J. & Malloch, A.J.C., (1991). Sea Cliff Management Handbook for Great Britain. IEBS, University of Lancaster, Lancaster.

# CHAPTER 18

# MONITORING ARABLE WEEDS AT NEWTON FARM

CLIVE HURFORD

*Countryside Council for Wales, Plas Penrhos, Ffordd Penrhos, Bangor, Gwynedd, LL57 2BQ*
*clive.hurford@serapias.net*

## 1.    INTRODUCTION

In contrast to most of the case studies in the book, this chapter describes a project designed to monitor the arable weed flora on a farm in the wider countryside, as opposed to within a protected area. The primary aim was to develop an efficient method for monitoring the effects of the management prescriptions on the conservation value of arable farms in agri-environment schemes, instead of simply monitoring compliance with management prescriptions. Although very few arable farms are designated for nature conservation, arable habitats cover large areas of land in Europe and are important for a wide range of animals and plants, many of which are in steep decline. Therefore, the need for efficient and effective monitoring in these cultural habitats is just as great as for semi-natural habitats protected in nature reserves.

## 1.1    General background information

After thousands of years of struggling to eradicate weeds from their arable fields, major advances in technology during the 20th century gave farmers the ability to produce 'clean' crops. Wilson & King (2004) suggest that the most significant developments leading to the declining conservation value of arable land in the UK were:
* Efficient seed-cleaning techniques;
* The widespread application of herbicides;
* The development of highly nitrogen-responsive crops;
* An increase in nitrogen applications;
* The near-complete mechanisation of farming;
* Changes in crop rotations;
* Hedgerow removal (reducing the area of field edge refuges); and
* Efficient field drainage.

*C. Hurford & M. Schneider (eds.), Monitoring Nature Conservation in Cultural Habitats, 169–184.*
© 2007 *Springer.*

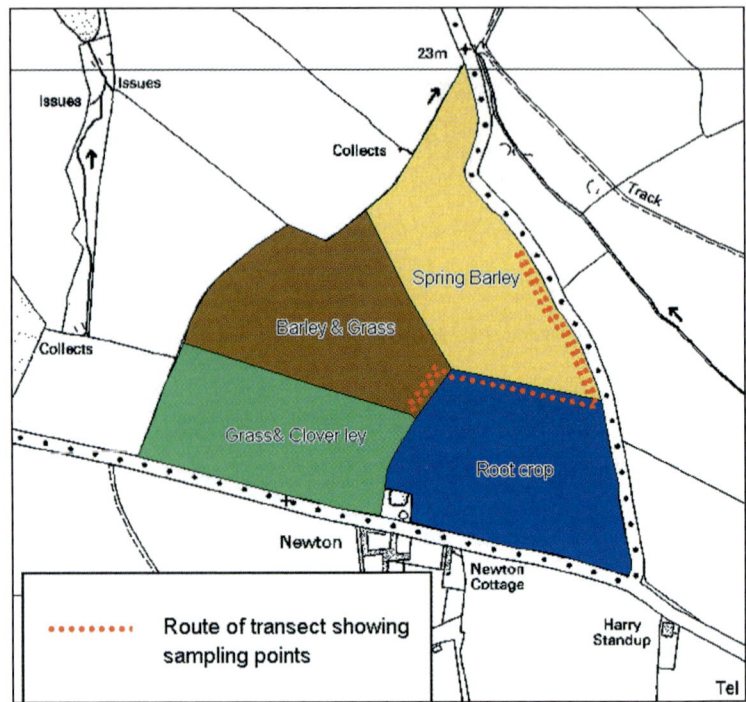

*Figure 18-1.* The arable fields at Newton Farm, the red dots are the monitoring points. Reproduced from Ordnance Survey mapping on behalf of Her Majesty's Stationery Office © Crown Copyright 100043571 2004-12-10

*Photographs by Clive Hurford*

*Figure 18-2.* The clean Barley crop on the left has been regularly treated with herbicides; the crop on the right, with abundant Corn Marigold *Chrysanthemum segetum*, has rarely been treated.

These changes have had a profound effect on both the arable weed flora and on the birds and invertebrates associated with it. Arable weeds represent c.20% of the most rapidly declining species of higher plant in the UK (Preston *et al.*, 2002).

However, unlike the large cereal-producing farms in southeast Britain, many farms in Wales have traditionally produced an arable crop to provide winter feed for animal stock (rather than human consumption). In the west of Wales, at least, arable weeds have suffered more from the trend towards cleaner and tidier farms, than from any desperate need to eradicate them from the crops.

## 1.2 Newton Farm

Newton Farm is a medium-sized farm of c. 80 hectares situated near Rhoscrowther in South Pembrokeshire. The farm business centres on a small herd of suckler cows and sheep, with four fields in a rotation to produce arable crops for winter-feed.

The farm is exceptional because the landowner has maintained the traditional crop rotation (Table 18-1) and has rarely applied herbicides. As a consequence, the arable fields are renowned for occasional, but spectacular, displays of Common Poppy *Papaver rhoeas* and Corn Marigold *Chrysanthemum segetum*.

*Table 18-1.* The sequence of crop rotation at Newton Farm.

| Sequence of crop rotation at Newton Farm |
| --- |
| Grass and clover ley |
| Unsprayed root crop |
| Unsprayed spring barley |
| Cereal stubble with light grazing |
| Plough – then spring barley under-sown with legumes |
| Grass and clover ley |

## 2. KEY FACTORS INFLUENCING THE MANAGEMENT DECISIONS AT NEWTON FARM

Although the owners of Newton Farm are sensitive to the nature conservation interest of the arable fields, the farm is a family business that must remain financially viable. To achieve this, the farm has had to diversify and now uses one of its fields as a small caravan and camping park for summer tourists. Additional resources are generated through farming subsidies, which include a management agreement with the local agri-environment scheme (Tir Gofal).

## 2.1 The management priority

The priority of the landowner is to run an economically viable farm and to accommodate the needs of the local fauna and flora wherever possible. From a nature conservation perspective, the management priority on this site is to maintain the diverse and abundant arable weed flora.   There is no reason why these different priorities should be mutually exclusive.

## 3.      RESEARCH AND SURVEY

The research phase of the project involved a basic literature search on the arable weed flora of southwest Wales.  We also contacted local botanists to ask if they knew of farms where the arable weed flora was well represented, and were given directions to farms on the Gower peninsula and in the St Ishmaels area of Pembrokeshire.  After contacting the landowners for access permission, we visited these farms in order to gain field experience of the scarcer arable weed species, and also to develop a feel for the local arable weed flora.

Subsequently, we carried out a crude survey of the crops on the Castlemartin peninsula in south Pembrokeshire, by checking fields from the roadside and public footpaths for indications of a diverse arable weed flora.  Initially, we used Corn Marigold *Chrysanthemum segetum* as the key indicator for sites worthy of a more detailed search: this species proved to be a good indicator species because it is tall, brightly coloured, and easy to identify from a distance.   The presence of *Chrysanthemum segetum* did not guarantee an interesting weed flora, but it was relatively abundant at all but one of the more interesting farms that we visited, and only once gave a false impression of conservation interest.

Having looked, albeit briefly, at several of the arable fields in the area, we then contacted the landowner at Newton Farm to ask for permission to develop and trial a monitoring method on the farm.  We selected this site because:

- The farm has a local reputation for supporting a good range of arable weed species;
- Arable weeds are unusually abundant in the field margins; and
- The landowner is genuinely interested in the wildlife on the farm.

The landowner not only granted us permission to develop a monitoring project at Newton Farm, he also took an active interest in our work and provided valuable information on past and current management practices.

The survey at Newton Farm involved recording all species of arable weed present in the field margins, i.e. within 5 m of the field edges.  We checked each of the fields in the arable rotation, including the grass and clover ley, though we excluded this field from the subsequent monitoring project. The survey revealed 39 species (Table 18-2) of arable weed, including several that are locally scarce, e.g. *Chrysanthemum segetum*, Sharp-leaved Fluellen *Kickxia elatine*, Henbit Dead-nettle *Lamium amplexicaule*, Cut-leaved Dead-nettle *Lamium hybridum* and Weasel's Snout *Misopates orontium*.

*Table 18-2.* The species of arable weed recorded at Newton Farm in August 2004.

| Species | Under-sown Barley | Barley | Roots |
|---|---|---|---|
| *Anagallis arvensis* | + | + | + |
| *Aphanes arvensis* | + | | |
| *Atriplex prostrata* | | + | |
| *Capsella bursa-pastoris* | | | + |
| *Cerastium fontanum* | | + | + |
| *Cerastium glomeratum* | | + | |
| *Chenopodium album* | | + | + |
| *Chrysanthemum segetum* | + | + | |
| *Convolvulus arvensis* | + | | |
| *Erodium cicutarium* | + | | |
| *Euphorbia helioscopa* | + | | + |
| *Fallopia convolvulus* | + | | |
| *Fumaria muralis* | | | + |
| *Fumaria bastardii* | + | | |
| *Geranium dissectum* | | + | + |
| *Gnaphalium uliginosum* | | | + |
| *Kickxia elatine* | + | + | + |
| *Lamium amplexicaule* | | | + |
| *Lamium hybridum* | | | + |
| *Lamium purpureum* | + | + | + |
| *Misopates orontium* | | + | + |
| *Polygonum aviculare* | + | + | + |
| *Polygonum lapathifolium* | + | | |
| *Polygonum persicaria* | + | + | + |
| *Raphanus raphanistrum* | + | | + |
| *Senecio vulgaris* | | + | |
| *Sinapis arvensis* | | + | |
| *Solanum dulcamara* | + | + | |
| *Solanum nigrum* | | | + |
| *Sonchus arvensis* | | + | + |
| *Sonchus asper* | + | + | |
| *Spergula arvensis* | + | + | + |
| *Stachys arvensis* | + | + | + |
| *Stellaria media* | + | + | + |
| *Tripleurospermum inodorum* | | + | + |
| *Urtica urens* | | | + |
| *Viola arvensis* | + | + | + |
| **Additional species recorded in the grass and clover ley** | | | |
| *Papaver rhoeas* | | | |
| *Silene vulgaris* | | | |

## 4.        THE CONSERVATION AIM

A reasonable conservation aim for Newton Farm is to maintain a diverse arable weed flora. We considered including additional aims for farmland birds and butterflies, but decided against this on the basis that, if the management is conducive to the persistence of the arable weed flora, then the associated fauna will almost certainly benefit too.

### 4.1     The condition indicators

*Table 18-3.* The condition indicator table for the arable weed flora at Newton Farm.

| Condition indicators | To maintain the arable weed flora at Newton Farm in optimal condition where: | |
|---|---|---|
| Extent and distribution | Lower limit | In one margin of at least three different fields planted with cereal or root crops: <br><br> >50% of the vegetation has two or more of *Kickxia elatine, Spergula arvensis* and *Stachys arvensis* within a 50 cm radius of each sample point, <br><br> And when <br><br> Each of the three species is present in >30% of the sample points <br><br> And when <br><br> *Chrysanthemum segetum, Fallopia convolvulus, Lamium amplexicaule, Lamium hybridum* and *Misopates orontium* are present in at least one of the cereal or root crop field margins. |
| Site-specific habitat definitions | | |
| Field margin | Vegetation within 4 m of any field boundary | |
| Root crop | Fields planted with Swedes or Turnips | |
| Cereal crop | Fields planted with Spring Barley, Wheat or Oats | |

### 4.2     Explanation of the condition indicators

The eight species in the condition indicator table were chosen for one or more of the following reasons, being:
- Sensitive to herbicides and/or nitrogen applications; or
- Listed among the 100 most rapidly declining species in the UK; or
- Locally scarce.

*Photographs by Clive Hurford*

*Figure 18-3.* Arable weeds at Newton Farm. Clockwise from top left: *Kickxia elatine, Lamium amplexicaule, Tripleurospermum inodorum, Misopates orontium* and *Chrysanthemum segetum.*

*Table 18-4.* Reasons for the selection of the condition indicator species at Newton Farm

| Indicator species | Sensitive to herbicides and nitrogen input | Listed in 100 most rapidly declining species in UK | Locally scarce or declining in UK |
|---|---|---|---|
| *Kickxia elatine* | + | | + |
| *Spergula arvensis* | + | + | + |
| *Stachys arvensis* | + | + | + |
| | | | |
| *Chrysanthemum segetum* | + | + | + |
| *Fallopia convolvulus* | + | + | + |
| *Lamium amplexicaule* | + | | + |
| *Lamium hybridum* | | | + |
| *Misopates orontium* | + | | + |
| | | | |

The condition indicator table has a lower limit for three discrete attributes: 1) the co-occurrence of three key species at monitoring points, 2) the frequency of three key species in the monitoring areas, and 3) the continued presence of six locally scarce or nationally declining species within the farm boundary.

All of the species named in the condition indicator table are Archeophytes that were introduced to Wales before 1500 A.D. They all prefer moist, rather than damp, fertile soils and cannot tolerate dense shade (Hill *et al.*, 1999). Corn Spurrey *Spergula arvensis* and Field Woundwort *Stachys arvensis* also prefer arable land with a soil pH of 6.0 or less, which is becoming increasingly rare in the UK (Grime, 1990).

In this instance, we used the monitoring results to inform the lower limit for the co-occurrence of *Kickxia elatine, Spergula arvensis* and *Stachys arvensis* (see Discussion). These species were selected for the following reasons:

- All three species are notably abundant at Newton Farm by comparison with the other farms we visited;
- *Kickxia elatine* is locally scarce in Wales, while both *Spergula arvensis* and *Stachys arvensis* are among the fastest declining species in the UK (Perrin *et al.*, 2002);
- Neither *Spergula arvensis* or *Stachys arvensis* can tolerate rises in soil pH;
- All three species are vulnerable to shading and herbicide applications; and
- These species regularly co-exist within small habitat patches at Newton Farm.

## 5.     THE MONITORING PHASE

During the preparation of the monitoring project, we had to make a number of decisions, including:

- Where to monitor;
- How frequently to record monitoring points;
- The most appropriate area of search at each monitoring point; and
- Which attributes to record at each monitoring point.

## 5.1    Where to monitor

We decided to restrict the monitoring samples to the field margins, to minimise any damage that the sampling might cause to the crops.  In our opinion, this would not compromise the validity of the monitoring result as, typically, the highest density of arable weeds was found within 5 m of the field boundary.  The only exception to this was the root crop, where high densities of arable weeds also occurred within the main crop.  However, most, if not all, of the weedy species in the heart of the crop were also represented in the field margin, and we decided that they could be monitored effectively from there.

For efficiency, we also decided to sample only one edge of each field planted with cereals or roots.  In each case we chose the edge with the most diverse and abundant arable weed flora.  We recommend that the same approach is taken during repeat monitoring exercises.

## 5.2    How frequently to record monitoring points

Our initial instinct was to record monitoring points at 5 m intervals; but we retained the option of adjusting the frequency of points as we were carrying out the monitoring, for example, by recording at 10 m intervals.  We also decided to record points in pairs: one at 1.5 m and the second at 3.5 m from the field edge.  This would draw attention to any narrowing of the weedy margins in the future (Fig.18-1).

## 5.3    The most appropriate area of search at monitoring points

We tested for the most appropriate area of search by checking five points for the presence of weedy species, initially within a 50 cm radius and then within a 1 m radius of the point.  The results of this exercise revealed that we gained very little additional information for the extra effort of recording a 1 m area of search, as most species had already been recorded within a 50 cm radius of the point.  This meant that we would detect a decline in the co-occurrence of the selected species earlier using a 50 cm radius area of search, and so we chose this radius.

## 5.4    The attributes to record at each monitoring point

Although we had surveyed the arable weed species present on the farm, and we were aware that the abundance of *Kickxia elatine*, *Spergula arvensis* and *Stachys arvensis* was unusually high, we did not have the confidence to select an indicator species assemblage before carrying out the baseline monitoring.  Therefore, we decided that, for the baseline monitoring exercise at least, we would record the presence of all arable weed species within a 50 cm radius of each monitoring point, and use the baseline results to inform the selection of the indicator species assemblage.  We did not feel that this would be too time-consuming, as we expected only a small number of species to be present at each monitoring point.

*Figure 18-4.* Field Woundwort *Stachys arvensis*, Scarlet Pimpernel *Anagallis arvensis* and Field Pansy *Viola arvensis* in the arable crop margin at Newton Farm.

## 5.5 The monitoring method

We monitored the arable weed flora along three stretches of field margin:
- One in a spring barley crop that was under-sown with grass;
- One in a spring barley crop; and
- One in a root crop (Swedes).

We started recording at 5 m intervals along the edge of each field, and every 5 m recorded two points, one at 1.5 m and another at 3.5 m in from the field boundary. At each monitoring point we recorded the presence of every arable weed species within a 50 cm radius.

After recording four pairs of points, we realised that the vegetation was relatively homogeneous, with a similar suite of species recorded at each point. Thereafter, we adjusted the frequency of the points to 10 m intervals, which allowed us to cover a greater length of field margin in the time allocated for the monitoring.

Each monitoring point was located with a differential Global Positioning System accurate to <2 m. In retrospect, this was probably unnecessary, as we would not necessarily recommend repeating the same points in the future (see Discussion).

## 6. RESULTS AND ANALYSIS

In total, 60 monitoring points were recorded in the field margins at Newton farm: 28 in the spring barley under-sown with grass, 18 in the root crop, and 14 in the spring barley crop. In all, 31 species of arable weed were recorded in the samples, including *Chrysanthemum segetum*, *Kickxia elatine*, *Lamium amplexicaule* and *Lamium hybridum*. The species that occurred most frequently were *Stachys arvensis* (81%), Scarlet Pimpernel *Anagallis arvensis* (62%), *Spergula arvensis* (57%), Common Mouse-ear *Cerastium fontanum* (54%), Field Pansy *Viola arvensis* (54%), *Kickxia elatine* (45%), Fat-hen *Chenopodium album* (34%), Scentless Mayweed *Tripleurospermum inodorum* (28%), Common Chickweed *Stellaria media* (24%), Cut-leaved Crane's-bill *Geranium dissectum* (21%) and Knotgrass *Polygonum aviculare* (15%). The mean number of species per monitoring point (Table 18-5) ranged from 5 to 7 in the different crop margins and peaked at just over 7 in the spring barley.

*Table 18-5*. The mean number of arable weed species recorded at monitoring points.

| Crop type | Mean number of species per 50 cm radius |
|---|---|
| | |
| **Spring barley and grass** | 5.71 |
| **Roots** | 5.16 |
| **Spring barley** | 7.29 |
| | |
| **Overall mean** | 5.9 |

## 6.1    The co-occurrence of *Kickxia elatine, Spergula arvensis* and *Stachys arvensis*

Two of *Kickxia elatine, Spergula arvensis* and *Stachys arvensis* co-occurred at 22 points (37%), and all three species co-occurred in a further 15 points (25%). This means that at least two of these species co-occurred at >50% of the monitoring points, which was the lower limit set for co-occurrence in the condition indicator table. All three species were absent from only three samples (see Table 18-6).

*Table 18-6.* The co-occurrence of the three key indicator species at the monitoring points

| Crop type | Number of key indicator species found together at monitoring points | | | |
|---|---|---|---|---|
| | **0** | **1** | **2** | **3** |
| | | | | |
| **Spring barley and grass** | 0 | 1 | 13 | 14 |
| **Roots** | 1 | 11 | 6 | 0 |
| **Spring barley** | 2 | 8 | 3 | 1 |
| | | | | |
| **Overall** | **3 (5%)** | **20 (33%)** | **22 (37%)** | **15 (25%)** |

Critically, for monitoring purposes at least, the co-occurrence of two or more of these species in 62% of the samples means that it will be possible to detect both increases and declines in co-occurrence from repeat monitoring results.

## 6.2    The frequencies of *Kickxia elatine, Spergula arvensis* and *Stachys arvensis*

All three key indicator species were present in each of the field margins. Both *Stachys arvensis* and *Spergula arvensis* achieved their highest frequencies in the margin of the spring barley crop that was under-sown with grass, and maintained relatively high frequencies in the root crop margin (Table 18-7). *Kickxia elatine* was also relatively frequent in the spring barley under-sown with grass, but achieved its highest frequency in the spring barley crop margin.

*Table 18-7.* The frequency of the key species at monitoring points in the field margins.

| Crop type | Frequency of occurrence (no. of samples / %) | | |
|---|---|---|---|
| | *Kickxia elatine* | *Spergula arvensis* | *Stachys arvensis* |
| | | | |
| **Spring barley and grass** | 15 (54%) | 26 (93%) | 28 (100%) |
| **Roots** | 1 (6%) | 7 (39%) | 16 (89%) |
| **Spring barley** | 11 (79%) | 1 (6%) | 5 (28%) |
| | | | |
| **Overall** | 27 (45%) | 34 (57%) | 49 (81%) |

## 6.3    The presence of the scarce and declining species

As the selection of the five scarce and declining species listed in the condition indicator table was informed by the survey phase of the project, it was no great surprise to find that they were all still present during the monitoring phase.    *Chrysanthemum segetum* was locally abundant on the farm, particularly in the grass and clover ley (Fig.18-5).    Black Bindweed *Fallopia convolvulus* was generally scarce but seen occasionally, while *Lamium amplexicaule*, *Lamium hybridum* and *Misopates orontium* were all restricted to one or two locations on the farm.

## 7.    DISCUSSION

This case study raises a number of interesting monitoring issues, particularly in relation to target setting and the scope for repeating the monitoring.

## 7.1    Target setting

One of the reasons for testing the monitoring method at Newton Farm was to get a feel for the standard that we should be aspiring to on farms where the arable flora has been suppressed by herbicides.    The data that we collected during the baseline monitoring exercise went some way towards achieving that goal.

This was also one of the reasons that we delayed finalising the details in the condition indicator table until after collecting the baseline data (as opposed to our recommended approach of setting the targets first and then monitoring against them).    In this respect, the baseline recording was not a true monitoring exercise: it was a stepping-stone to developing a well-informed set of site-specific condition indicators to monitor against in the future.

In many ways, this is a sensible way of developing site-specific condition indicators, particularly on sites that are clearly in optimal condition.    Unfortunately, more often than not, our conservation areas are in sub-optimal condition and in need of restoration, which means that we are in the position of having to describe what we are trying to recreate, rather than what we would like to maintain.    That information is far more likely to come from historic survey data, or what we know about the habitat from other sites.

## 7.2    Recommendations for repeat monitoring

In terms of repeatability, the arable weed flora at Newton Farm presented us with two difficult and related problems:
- All of our indicator species were annual plants; and
- Each field was in a temporary state, and guaranteed to change as part of the annual crop rotation.

*Photograph by Clive Hurford*

*Figure 18-5.* Corn Marigold *Chrysanthemum segetum* growing in a clover and rye grass ley at Newton Farm, from a distance these plants could easily have been overlooked as buttercups *Ranunculus* spp.

In more stable habitats, we could have identified the areas with an unusually high abundance of *Spergula arvensis* and *Stachys arvensis*, for example, and used this information to set a lower limit for these species at those locations. This would not make sense at Newton Farm, however, because the fields are in a rotation cycle, and best field margins in 2004 could be under a grass and clover during the repeat monitoring year. This is not a problem for the plants, but it is for the monitoring.

In some ways, the situation was analogous to setting up a monitoring project for spring annual plants on a dynamic dune system: we had an interesting assemblage of annual plants occupying habitats that were guaranteed to change, probably both in extent and location, on a regular basis. In this situation, we would not be concerned about where the species were, as long as they were present above a given abundance level at several locations on the site.

At Newton, we had only two sensible options, which were to:

- Set a lower limit for the key species in the field margins that we monitored in August 2004, and specify that the monitoring should be repeated only during years with an identical crop distribution on the farm; or
- Set a lower limit for the abundance and distribution of the key species on the farm, but not identify specific locations.

We chose the second option, because the species that we selected have long-lived seeds that will persist in the seed bank for many years. These species are likely to appear in whichever field margins have been disturbed that year.

For repeat monitoring exercises, we would monitor one margin in each of the three fields not under a grass and clover ley, and use the abundance of *Spergula arvensis*, *Stachys arvensis* and *Kickxia elatine* to decide which margins to monitor. The field margins with the greatest abundance of the three key indicator species would be monitored. In effect, this means that we are monitoring the sections of field margin that are most likely to pass the criteria in the condition indicator table. We would note the presence or absence of the other indicator species during the process of selecting the monitoring locations.

*Photographs by Clive Hurford*

*Figure 18-6.* The populations of Yellowhammers *Emberiza citrinella* and Corn Buntings *Miliaria calandra* have both suffered steep declines in the UK as a result of agricultural improvements to arable land. Where appropriate, i.e. within their natural range, these species should be considered for inclusion in the condition indicator table as typical species of arable habitat.

We also considered recording the frequencies of weeds that are believed to be important for farmland birds, such *Anagallis arvensis*, *Chenopodium album* and *Stellaria media*. However, these species are widely distributed and abundant on the farm, and far less susceptible to herbicides than the species that we have selected as condition indicators.

## 7.3    The advantages of adopting a similar approach for monitoring agri-environment schemes

All of the field recording at Newton Farm took place over two days and cost less than €600 in terms of surveyor time. This is a very small sum by comparison with that committed to the management agreement. During the course of that two days we:
* Compiled a list of the arable weed species at Newton Farm;

- Assessed the co-occurrence of the arable weed species at 60 sample points in three field margins;
- Set site-specific targets for the condition of the arable weed flora; and
- Identified a subset of condition indicators for the purposes of repeat monitoring.

All of this information could be used to assess the success of the management prescriptions used in an agri-environment scheme. Using the same methods, it would take even less time to monitor whether the prescriptions were being effective on a site where the arable weed flora was impoverished after years of being suppressed by herbicides. It would even be possible to assess progress towards optimal condition, perhaps after five years of participation in a scheme, by simply looking at the mean number of arable weed species at the monitoring points. This would not be a time-consuming exercise.

## 8.    REFERENCES

Hill, M.O., Mountford, J.O., Roy, D.B. & Bunce, R.G.H. (1999). *Ellenberg's indicator values for British plants.* ECOFACT Volume 2. London Department of the Environment, Transport and the Regions.

Grime, J.P., Hodgson, J.G. & Hunt, R. (1990). *The abridged comparative plant ecology.* Chapman & Hall. London.

Preston, C.D., Pearson. D.A. & Dines, T.D. (2002). *New Atlas of the British and Irish Flora.* Oxford University Press. Oxford.

Wilson, P. & King, M. (2003). Arable Plants – a field guide. WildGuides Ltd. Hampshire.

# CHAPTER 19

# MONITORING NEUTRAL GRASSLAND AT SOMERTON FARM

DAN GUEST[1] & CLIVE HURFORD[2]

*Countryside Council for Wales, Plas Penrhos, Ffordd Penrhos, Bangor, Gwynedd, LL57 2BQ*
*D.Guest@ccw.gov.uk*[1]
*clive.hurford@serapias.net*[2]

## 1. BACKGROUND INFORMATION

Somerton Farm is a 22 ha holding situated roughly 4 km from the coast in southwest Pembrokeshire. Historically, the farm was part of the much larger Orielton Estate, but was sold, along with most of the other estate properties, in the 1920s. Up until the late 1960's, Somerton was managed as a mixed farm with a resident sheep flock (lowland breeds including Suffolk and Border Leicester) with store cattle bought in and fattened during the summer. Arable crops were also grown; including early potatoes, oats, barley and kale, and an annual hay crop was taken for winter fodder.

Exceptionally, during the war years (1939-1945) some of the flatter pastures were ploughed to grow flax. Arable cropping ceased in the late 1960s and the cultivated fields reverted to improved pasture. Until the mid 1970s, year-round sheep grazing was augmented with summer cattle grazing. Thereafter until 1996, the only grazing was by cattle in the summer months.

Since 1996 the farm has comprised mostly semi-improved permanent pasture, with smaller areas of marshy grassland and a broad-leaved plantation. The pasture is neutral grassland (overlaying old red sandstone) that is managed through year-round grazing with pedigree Dexter cattle (Fig.19-2), a small hardy breed capable of wintering outside and tackling rough grass and advancing scrub. This herd, of c.35 animals, is managed

*C. Hurford & M. Schneider (eds.), Monitoring Nature Conservation in Cultural Habitats, 185–194.*

organically and comprises six to eight suckler cows, their offspring, and additional stock
that are bought in for finishing.

## 1.1    The conservation priority

The conservation priority at Somerton Farm is the assemblage of grassland fungi
associated with the neutral grassland (see Table 19-1).  The farm is notable for supporting
a high diversity of Waxcaps *Hygrocybe* spp., Earthtongues (*Geoglossaceae*) and Fairy
clubs (*Clavariaceae*) within a relatively small area of land.  The presence of 18 species of
Waxcap, and 12 species of Earthtongue and Fairy club, including two BAP priority
species *Hygrocybe calyptriformis* and *Microglossum olivaceum*, suggests that the farm is
of national importance for grassland fungi (Anon, 1999: McHugh, *et al.*, 2001).

Grassland fungi are found in all of the drier fields at Somerton, but by far the greatest
variety of species is confined to two enclosures (labelled A and B in Fig. 19-1).  The
steep, west-facing bank that runs along the north-western edge of these two enclosures
remains ungrazed during the summer and supports stands of relatively unimproved neutral
grassland.  These slopes attract an interesting invertebrate fauna, including the nationally
rare Shrill Carder Bee *Bombus sylvarum*.

Several species of farmland bird breed at Somerton, including some species that have
suffered steep declines in the UK during the latter part of the 20th century, e.g. Bullfinch
*Pyrrhula pyrrhula*, Yellowhammer *Emberiza citrinella*, Linnet *Carduelis cannabina* and
Skylark *Alauda arvensis*  (Crick, *et al.*, 2004).  In addition, Barn Owls *Tyto alba* use the
farm for hunting and Hen Harriers *Circus cyaneus* have visited occasionally in winter.

Mammal interest focuses on regular visits by two other Natura 2000 species: Otters
*Lutra lutra*, that use a small man-made lake for feeding, and Greater Horseshoe Bats
*Rhinolophus ferrumequinum*, that feed along the hedgerows and over the areas of neutral
grassland: a testimony to the rich insect life these fields support.

## 1.2    The management priority

From the perspective of conservationists, the management priority at Somerton is the
neutral grassland habitat, notably the three fields along the north-west edge of the farm.
The west and north-facing slopes of these fields are unlikely to have received any
cultivation or chemical treatment in the past, as they are too steep for horse or tractor-
drawn implements.  The more level 'plateau' areas have not been ploughed for at least 60
years (if ever), but were treated periodically with basic slag and chemical fertilizers until
the late 1960s.

Until the late 1960s, cattle and sheep grazed these fields throughout the year.
Thereafter, until 1996, this management was replaced by cattle grazing in summer.  Since
then, cattle have grazed the level field plateaux in spring and summer, and the west-facing
slopes and small associated areas of marshy ground in autumn and winter.  This grazing is
controlled by temporary electric fences, which are moved regularly to help maintain a
reasonably tight sward.

## 2.   FUNGAL SURVEYS

During the autumns of 2003 and 2004, considerable effort was committed to surveying the fungi associated with the neutral grassland at Somerton. These surveys focused mostly on compiling a species list for the site, noting the distribution of the various species, and measuring vegetation height. This work was carried out primarily by the landowners (Holly and David Harries), with the support of the British Mycological Society conservation officer and a colleague from the mycological section of the Royal Botanic Gardens, Kew, who surveyed the site as part of a Wales-wide survey of the mycoflora of lowland grassland sites commissioned by the Countryside Council for Wales (Griffiths, *et al.*, in prep.). Table 19-1 lists the species of Waxcap, Earthtongue and Fairy Club fungi recorded in the grassland at Somerton Farm. We supplemented this information by carrying out a baseline survey on a systematic grid in Field A to inform the condition indicators for the site (Plot 1 in Fig. 19-1).

*Table 19-1.* Species of Waxcap, Earthtongue and Fairy Club fungi recorded at Somerton Farm in 2004. The species in red text are indicators of high quality grassland; the species in blue text indicate medium quality grassland.

| Waxcaps | Earthtongues |
|---|---|
| *Hygrocybe calyptriformis* | *Geoglossum fallax* |
| *Hygrocybe citrinovirens* | *Geoglossum glutinosum* |
| *Hygrocybe ceracea* | *Microglossum olivaceum* |
| *Hygrocybe chlorophana* | *Trichoglossum walteri* |
| *Hygrocybe coccinea* | |
| *Hygrocybe conica* | **Fairy clubs** |
| *Hygrocybe pratensis* | *Clavaria acuta* |
| *Hygrocybe fornicata* | *Clavaria fumosa* |
| *Hygrocybe pratensis* var. *pallida* | *Clavaria vermicularis* |
| *Hygrocybe psitticlna* | *Clavulinopsis corniculata* |
| *Hygrocybe punicea* | *Clavulinopsis helvola* |
| *Hygrocybe quieta* | *Clavulinopsis umbrinella* |
| *Hygrocybe russocoriacea* | *Clavaria incarnata* |
| *Hygrocybe glutinipes* | *Ramariopsis kunzei* |
| *Hygrocybe irrigata* | |
| *Hygrocybe insipida* | |
| *Hygrocybe intermedia* | |
| *Hygrocybe virginea* | |

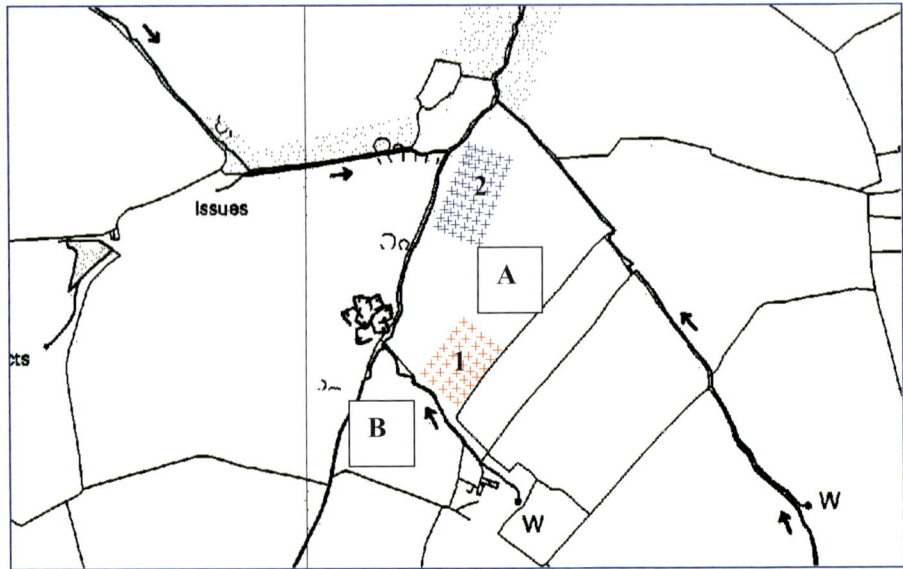

*Figure 19-1.* A map of Somerton Farm showing the locations of Fields A and B, and Plots 1 and 2. Reproduced from Ordnance Survey mapping on behalf of Her Majesty's Stationery Office © Crown Copyright 100043571 2004-12-10.

*Photograph by David Harries*

*Figure 19-2.* Part of the Dexter herd grazing in Field A at Somerton Farm. The grazing levels are tightly controlled by temporary electric fencing, which is moved on a daily basis at critical times of the year.

Our survey in Plot 1 showed that 60% of the vegetation had at least one Waxcap, Earthtongue or Fairy club fruiting body within a 1 m radius, and that the mean vegetation height (where fruiting bodies were present) was 7.9 cm for Waxcap fungi, and 7.3 cm for Fairy Clubs. No Earthtongues were noted, maybe because these appear to have a preference for open soil in very short vegetation, which was rarely encountered in the survey plot.

In general, fungal fruiting bodies were strongly associated with the shorter vegetation in the survey area: if the vegetation was above 11 cm high, it was not worth searching for fruiting bodies, as the larger species were obviously not there and the smaller species, i.e. the Fairy Clubs and Earthtongues, would reliably prove to be absent after diligent searching.

## 2.1    Vegetation survey

To augment the fungi survey data, and as preparation for selecting the indicator assemblages at Somerton Farm, we carried out an additional survey in July 2004. On this occasion, we surveyed the higher plant flora on the west-facing slopes of the most northerly field. This descriptive exercise highlighted the main differences in diversity between the herb-rich vegetation on the lower slopes, and the vegetation dominated by *Dactylis glomerata* and *Avenula pubescens* on the middle and upper slopes. The vegetation survey revealed that:

- *Lotus corniculatus* and *Centaurea nigra,* two species closely associated with relatively unimproved neutral grasslands (Rodwell, 1992), were well-distributed on the bank, and remained frequent even in patches dominated by rank grasses; and
- Several other species indicative of unimproved conditions were also present on the bank, notably *Succisa pratensis, Leontodon hispidus, Carex caryophyllea* and *Stachys officinalis,* but these have a much more local distribution and are in the main restricted to the steeper mid-lower slopes.

## 3.    THE CONDITION INDICATORS

Table 19-2 shows the condition indicators for the neutral grassland at Somerton Farm. These focus on the most northerly field on the farm, and incorporate aims for both the waxcap population and the flora on the west-facing slopes.

## 3.1    Explanation of the condition indicators

The condition indicators in Table 19-2 provide lower limits for the overall extent of neutral grassland in the two key parts of the site for the habitat, Fields A and B; for the proportion of the neutral grassland on the plateaux areas of Fields A and B that should be optimal condition for Waxcap fungi; and for the proportion of the vegetation on the west facing slopes of Fields A and B that should be in optimal condition for attracting invertebrates.

Although the conservation priority at Somerton is undoubtedly the diversity and abundance of fungi associated with the neutral grassland, the management should also be able to accommodate the requirements of the flora and associated invertebrate fauna on the slopes. This can be achieved through the continued use of temporary electric fencing to control grazing management.

*Table 19-2.* The condition indicators for the neutral grassland habitat at Somerton Farm.

| Condition indicators | To restore the neutral grassland at Somerton Farm to optimal condition where: | |
|---|---|---|
| **Habitat Extent** | **Lower limit** | Extent of neutral grassland in Fields A and B as mapped in 2004 (see Fig. 19-1) |
| **Habitat quality** | **Lower limit** | **In Fields A and B:** <br><br> >90% of the plateau vegetation is in optimal condition for Waxcap fungi. <br><br> **And when** <br><br> >70% of the vegetation on the west-facing slopes is in optimal condition |
| **Site-specific habitat definitions** | | |
| **Neutral grassland** | Vegetation dominated by grasses and short to medium height herbaceous plants. Ericoids and marshy grassland species notably *Molinia, Juncii* and *Filipendula* are scarce or at only low covers. | |
| **Plateau vegetation in optimal condition for Waxcap fungi** | Vegetation with a sward height of <7-cm (in October and November) with one or more fruiting body of Waxcap, Earthtongue and Fairy Club fungi present within a 1-m radius (in a good fruiting season); <br><br> **And** <br><br> The continued presence at Somerton of *Hygrocybe punicea, H. calyptriformis, H. citrinovirens, H. fornicata, H. pratensis* var. *pallida, H. quieta, H. irrigata, Clavaria fumosa, Trichoglossum walteri* and *Microglossum olivaceum.* | |
| **Slope vegetation in optimal condition** | **Within a 25-cm radius** <br> Vegetation with >50% cover of herbs and sedges and three or more of *Lotus corniculatus, Centaurea nigra, Stachys officinalis, Leontodon hispidus, Succisa pratensis,* and *Carex caryophyllea* present; and <br><br> **Within a 1-m radius** <br> < 10% cover of *Dactylis glomerata*; and trees, saplings and scrub, and *Filipendula ulmaria* absent. | |

*Photographs by David Harries*

*Figure 19-3.* Four of the Waxcap species found at Somerton Farm. Clockwise from top left these are *Hygrocybe punicea*, *Hygrocybe calyptriformis*, *Hygrocybe coccinea* and *Hygrocybe conica*.

## 3.2 The condition indicators for the Waxcap fungi

The results from our survey plot, and from casual observations elsewhere on the farm, indicated that the optimum sward height for fruiting bodies of the key fungi groups (Waxcaps, Earthtongues and Fairy Clubs) was 7 cm or less: there was no indication that the vegetation could be too short. Furthermore, in areas where the vegetation height was appropriate you would expect to see fruiting bodies of at least one of the key groups within a 1 m radius of any point. The continued presence of the indicators of high and medium quality grassland should allow us to assume that the vegetation has the capacity to support the less demanding species associated with them.

## 3.3    The condition indicators for the vegetation on the west-facing slopes

The condition indicators for the neutral grassland focus primarily on a suite of species characteristic of relatively unimproved neutral grasslands.   All of these species are primarily stress-tolerators (see Chapter 8) in neutral grassland and are relatively intolerant of nutrient enrichment (Grime, *et al.*, 1988).   In addition five out of the six species selected flower profusely in mid to late summer providing an important nectar source for several species of invertebrate.   The survey phase revealed that on the steeper parts of the bank where the least modified grassland occurs the sward is very fine-grained, with at least three of the six indicators occurring at most 50 x 50 cm recording points.

*Photograph by Clive Hurford*

*Figure 19-4.* This herb-rich vegetation, with abundant Common Knapweed *Centaurea nigra* and Bird's-foot Trefoil *Lotus corniculatus*, was typical of the lower west-facing slopes at Somerton Farm in July 2004.

This intimate texturing of the better quality vegetation allowed us to concentrate our search for positive indicators on a very small area, which carried with it several benefits:
- Search time at each point was reduced;
- Observer error was reduced, as we could be confident that even the more inconspicuous plants were unlikely to be overlooked; and

- We could rely on presence / absence as a good indicator of quality rather than having to include some measure of abundance or within sample frequency, both of which are associated with much higher levels of measurement error.

The main threat to the conservation interest of the neutral grassland on the banks was seen to be undermanagement and the associated increase in cover of rank grasses (particularly *Dactylis glomerata*) and spread of scrub and *Filipendula ulmaria*. The latter only represents a significant threat very locally, along the bottom of the field where the neutral grassland grades into damper pasture. These species have a relatively coarse distribution on the bank and in order to detect them sensitively, we employed a larger 1 m radius search area.

# 4.    THE MONITORING METHOD

The monitoring at Somerton will need two visits: one in July to monitor the vegetation on the west-facing slopes and another in October or November to monitor the plateau vegetation and associated fungi. The fungi monitoring should take place only in 'good years' for Waxcap fungi.

As the site is relatively small, the whole of Fields A and B should be monitored, recording systematically at 10 m intervals. At each monitoring point, the vegetation should be assessed against the appropriate criteria in the condition indicator table. We estimate that two days will be required to complete the monitoring at Somerton: one day in July and another in October or November.

Because both Fields A and B had to pass the criteria in the condition indicator table, we focused the monitoring on the field that was most likely to fail: this was Field A.

# 5.    RESULTS AND ANALYSIS

Plot 1, which we used to examine the relationship between the key fungi groups and vegetation height, was situated in the area of Field A most likely to pass the criteria for plateau vegetation in the condition indicator table. The survey results showed that only 26% of the survey points met the criteria for optimal Waxcap habitat, against a target of >70%. As this was the area of the field most likely to meet the criteria, we can assume that the current condition of Field A is sub-optimal. As both Fields A and B must pass the targets in the condition indicator table, the overall condition of the neutral grassland habitat at Somerton must also be considered to be sub-optimal. Vegetation height was the main reason for the neutral grassland failing to meet the criteria for optimal Waxcap habitat: 63% was >7cm high.

Plot 2 was monitored in July 2004, and only 16% of the monitoring points met the criteria for optimal neutral grassland vegetation, with most points failing for three reasons: >10% *Dactylis* cover; <50% herb and sedge cover; and less than three species of the indicator assemblage present. On this basis the neutral grassland vegetation at Somerton must also be considered to be in sub-optimal condition.

## 6.  REFERENCES

Anon. (1999). UK Biodiversity Group: Tranche 2 Action Plans, Volume III - plants and fungi. English Nature, Peterborough.

Boertmann, D. (1996). The genus Hygrocybe. Fungi of Northern Europe, Volume 1. Danish Mycological Society, Greve, Denmark.

Crick, H.Q.P., Marchant, J.H., Noble, D.G., Baillie, S.R., Balmer, D.E., Beaven, L.P., Coombes, R.H., Downie, I.S., Freeman, S.N., Joys, A.C., Leech, D.I., Raven, M.J., Robinson, R.A. & Thewlis, R.M. (2004) *Breeding Birds in the Wider Countryside: their conservation status 2003.* BTO Research Report No. 353. BTO, Thetford.

Crofts, A. & Jefferson, R.G. (Eds). (1999). *The Lowland Grassland Management Handbook* (2$^{nd}$ Edition). English Nature / The Wildlife Trusts.

Grime, J.P., Hodgson, J.G. & Hunt, R. (1988). Comparative Plant Ecology: a functional approach to common British species. Unwin Hyman. London.

Griffiths, G.W., Evans, S., Holden, L., Evans, D., Aron, C., Mitchel, D & Jones, P. (in prep.). Mycological survey of selected semi-natural grassland in Wales. CCW contract report no: FC 73-01-403.

Marren, P. (1998). *Fungal flowers: the waxcaps and their world.* British Wildlife **9**:3, 164-172.

McHugh R., Mitchel D., Wright M. & Anderson R. (2001). *The fungi of Irish grasslands and their value for nature conservation.* Biology and Environment: Proceedings of the Royal Irish Academy Vol 101B, 3 225-242.

Newton, A.C. (1997). *Waxcap grassland survey: interim report on results.* British Mycological Society.

Rodwell, J.S. (Ed). (1992). *British Plant Communities: Volume 3, Grasslands and montane communities.* Cambridge University Press.

Rotheroe, M., Newton, A., Evans, S. & Feehan, J. (1996). *Waxcap-grassland survey.* Mycologist, 10:1, 23-25.

Rotheroe, M. (1997). *A comparative survey of waxcap-grassland fungi of Ireland and Britain.* JNCC Contract report No. F76-01-71.

# CHAPTER 20

# MONITORING THE BROWN BEAR *URSUS ARCTOS* IN VÄSTERBOTTEN COUNTY

## MICHAEL SCHNEIDER

*County Administration, Västerbotten County, SE-901 86, Umeå, Sweden*
*Michael.Schneider@ac.lst.se*

## 1.    BACKGROUND INFORMATION

### 1.1    Västerbotten County

Västerbotten is the second largest and second most northerly of Sweden's 21 counties, covering about 14% (55 000 km$^2$) of the country's area. The County stretches from the coast of the Bothnian Bay in the East up to almost 1800m above sea level in the mountain range near the Norwegian border in the West. The climate varies considerably between different parts of Västerbotten, but generally is characterized by cold winters with heavy snow. More than
50% of the County is covered by forest, which is intensively used by large-scale forestry. Mountains, mires and water bodies cover another 45%, while agricultural land is rare and mostly confined to the coastal plains and river valleys. Most of Västerbotten's 255 000 inhabitants live along the coast in the east, where the biggest cities are. Human population density decreases steadily from the East to the West, with few inhabitants in the forested inland areas and especially the mountain range.

There are about 160 nature reserves and 230 Natura 2000 sites in the County, protecting c.13% of its area. Most of the sites are relatively small, but some sites along rivers and some reserves in the mountains are huge; Vindelfjällen nature reserve alone covers 550 000 hectares of mountain habitats. Predators are not restricted to the protected areas, but occur in most of the County.

*C. Hurford & M. Schneider (eds.), Monitoring Nature Conservation in Cultural Habitats, 195–214.*
© 2007 *Springer.*

Five big predatory species co-occur in Västerbotten and are of conservation concern. These are Lynx *Lynx lynx*, Wolf *Canis lupus*, Wolverine *Gulo gulo*, Golden Eagle *Aquila chrysaetos* and Brown Bear *Ursus arctos*. For Wolverine, Golden Eagle and Brown Bear, Västerbotten holds an important proportion of the total Swedish population. Sweden, in turn, plays an important role in the conservation of big predators in Europe. This is why the management of these predators in Västerbotten is of national and international importance.

## 1.2    Ursus arctos

The Brown Bear is the biggest and most numerous of the big mammalian predators in the County. The Swedish population is currently increasing and was thought to number between 1635 and 2840 in 2004 (Kindberg *et al.*, 2004). We know a lot about bear biology and ecology because of the Scandinavian Brown Bear Project, which has researched the species in Sweden and Norway since 1984 (Sandegren & Swenson, 1997).

Scandinavian bears are among the least aggressive brown bears in the world. The last time an unarmed person was killed by a bear in Sweden was at the beginning of the 19th century, when a girl surprised a female bear with cubs at their den. Wounded bears killed hunters in 1902 and 2004, in the latter case when the bear was near its den. People who have been attacked and injured by bears during recent years were mostly hunters with guns (six out of seven cases between 1976 and 1995 in Norway and Sweden) (Skandinaviska Björnprojektet, 2000).

In autumn, the diet of Scandinavian bears consists mostly of plant material, especially berries (*Vaccinium myrtillus, V. idaeus, Empetrum hermaphroditum*), while in spring ants (*Formica* spp. *Camponotus* spp.) are favoured. Bears also consume Elk *Alces alces*, but mostly carcasses, leftovers from people's Elk hunting, and calves. Female Elks that lose their calf partially compensate for it with increased fecundity during the following year (Andrén *et al.*, 1999). Elk hunting is of economical, cultural and recreational importance for the human population in Västerbotten. About 13 000 Elks are killed every year by several thousand hunters in the County. Elk hunting is an important activity for many people, supplying the rural population with most of the meat needed during the year, and supplying parts of the urban population with much of their recreational needs.

Depredation on and damage to livestock by bears can be considerable in certain places and at certain times, but is generally not a big problem in the County. Victims are mostly sheep and Reindeer *Rangifer tarandus* calves. Beehives are attacked occasionally, too. A well-developed compensation system reimburses farmers and Reindeer owners for their losses (cf. Naturvårdsverket, 1991). While farmers have to prove their losses, Reindeer owners are paid if bears occur in the area, whether killed Reindeer are found or not. Many Reindeer herders maintain, though, that the sums paid for compensation are much too low.

In spring, bears are often seen close to or within human settlements. Typically, these are young bears, especially males. Presumably, most often this is related to a lack of food in the forest at this time of year, and the ongoing mating season, where the big and dominant males are moving around in the woods, pushing away young males. Usually, those bears disappear when resource levels in the forest increase as spring proceeds.

A limited number of individuals can be shot legally under strict regulations each year (the quota for Sweden in 2004 was 101 bears, of which 14 were in Västerbotten). Poaching of bears and other predators is a fact and seems to be considerable. However, the number of bears killed per year is unknown. The Brown Bear Project has estimated that as many bears are killed illegally as are shot legally (Swenson & Sandegren, 1999).

## 1.3 Key factors influencing management decision for the Brown Bear

The Brown Bear is listed as near threatened in the latest version of the Red List of Swedish Species (Gärdenfors, 2005) and included in Appendix 4 of the EU Habitats Directive. The County Administration has the responsibility to protect the species in Västerbotten in accordance with existing laws and other regulations.

There are four key factors influencing management decisions for the Brown Bear in Västerbotten:

- The national, coherent predator policy in Sweden;
- Reindeer husbandry;
- The feelings and attitudes of the human population in the County; and
- Bear dispersal biology.

### 1.3.1 A coherent predator policy at the national level

After two years of surveying and objective setting, involving hundreds of people in Sweden and elsewhere (SOU, 1999: 146), the Swedish Parliament has decided that viable populations of the predator species Brown Bear, Lynx, Wolf, Wolverine and Golden Eagle shall exist in the country. The species shall be allowed to recolonise areas where they formerly occurred. This national policy provides the framework for predator management in the country (Regeringen, 2000).

The national target for the Brown Bear is a minimum of 100 reproducing females per year, corresponding to about 1000 animals in Sweden. Also, the species will be allowed to spread into the areas in between the current breeding areas, and continue its southward expansion.

The national minimum target of 100 reproducing females per year is modest compared with the 405–1215 reproductions that could be supported in the country annually with respect to the available habitat (Støbet Lande *et al.*, 2003).

At a regional scale, county administrations have to produce management plans for the five large predator species, taking into account specific regional circumstances such as the level of economic losses inflicted by carnivores, the amount of natural habitat available for the animals, human attitudes, and other factors. Each county administration in Sweden has to decide on how big the predator populations should be in the county, how predators should be distributed in the county, and which methods should be used to reach these goals.

### 1.3.2    Reindeer husbandry

Wild Reindeer are extinct in Sweden, but domestic Reindeer occur in the northern half of the country. Reindeer husbandry is important from an economical, cultural and conservation perspective. In Västerbotten County, about 60 000 free-ranging Reindeer migrate annually between their wintering grounds along the coast and their summer grazing grounds in the mountains, and are thoroughly surveyed, and sometimes even transported by van, by their owners, the Sami people. Reindeer owners are organised in so-called Sami villages, which are areas of land and administrative units rather than settlements. In Västerbotten, most Sami villages stretch from the coast up to the mountains.

Reindeer are among the favourite food for all five predator species in the County. The conflict between Reindeer owners and predators has previously resulted in heavily diminished predator populations.

### 1.3.3    Human feelings and attitudes

Many people have strong feelings towards large carnivores. These feelings may be positive or negative, encompassing everything from hatred to deep fear to enthusiasm. Often, it is not the predators *per se* that are the problem, but a conflict between the central administrative power and the countryside community (Skogen *et al.*, 2003). All of these feelings and attitudes have to be taken seriously when managing big predators.

Human attitudes depend on:

- Levels of damage to livestock, Reindeer and game animals;
- Actual or experienced threat to humans; and
- Levels of local involvement during decision making.

Most of the County is suitable for Västerbotten's five predator species with respect to natural factors (Støbet Lande *et al.*, 2003). The human factor is of uttermost importance and determines predator numbers and distribution (Persson *et al.*, 2004). Poaching may be a problem for the Brown Bear in certain parts of the County.

### 1.3.4    Bear dispersal biology

In Västerbotten County, the annual rate of population increase of bears has been estimated to be ca. 7% (according to a regression model in Kindberg *et al.* 2004). Female bears first reproduce at the age of 4–5 years. There are on average 2.4 cubs in each litter, and the mean interval between litters is 2.5 years. The females' probability of survival to reproductive age is 80%, and the annual survival rate of adult females is 90%. Bears reach a maximum age of 30 years (Sandegren & Swenson, 1997).

Most female Brown Bears in Sweden occur in four so-called core areas. These are the areas where the species survived the population low (about 130 individuals in Sweden) at the beginning of the 20th century (Ekman, 1910: Lönnberg, 1929: Bjärvall, 1978). Västerbotten touches two of these core areas. One core area is shared with Jämtland County to the South, and the other one with Norrbotten County to the North.

Bear density in the core areas is 2–3 individuals per 100 km$^2$ (Sandegren & Swenson, 1997, Swenson pers. com.). The dispersal of young bears from inside core areas seems to be limited, while the edges of core areas appear to be the most important sources for dispersers (Swenson, pers. comm.). Young bears spend their first 16 or 28 months with their mother. Young females tend to stay close to their mother's home range even later, and only 40% disperse. The maximum dispersal distance for young females is 80 km (average 40 km), while young males can move more than 500 km (average 140 km) (Swenson & Sandegren, 2000).

Home range size is different for different categories of bears, during different seasons, and in different parts of Sweden. In northern Sweden, the smallest home ranges (61 km$^2$, post-mating 169 km$^2$) are occupied by females with cubs during the mating season, when they try to avoid meeting adult males, which could kill their young. The largest home ranges are used by adult males during the mating season (736 km$^2$, post-mating 424 km$^2$), when they search for oestrous females. The home range sizes of other categories of bears lie in between these extremes (adult females 278 and 123 km$^2$, mating and post-mating, respectively, females with yearlings 226 and 261 km$^2$, non-dispersing 2-year-old bears 69 and 76 km$^2$) (Dahle & Swenson, 2003).

## 1.4    The management priority

Management objectives for the Brown Bear are:
- A more even distribution of females;
- An overall bigger population; and
- Differentiated densities in the County.

The level of damage inflicted by bears is currently acceptable and should not be allowed to increase. To allow hunting is one means of reaching a positive attitude of the people towards bears. The Brown Bear population has adapted to low adult mortality rates. Therefore, the killing of adult males and females, by poaching as well as legal hunting, has a big effect on the social organisation and the population dynamics of the species. In consequence, hunting is an effective management tool, which has to be used carefully and wisely. The current bear hunting system in Sweden is very much influenced by international regulations (EU Habitats Directive, Bern Convention) where the Brown Bear is listed as a species of conservation concern. Bears can be shot to protect humans, mostly to remove problem individuals and reduce the size of populations that are too big. Many people feel that international regulations hamper effective bear management by hunting.

Young females usually do not disperse very far from their mother. Therefore, the colonisation of a new area by a female bear is a rare event. If she survives, every such female will cause the creation of a new miniature core area, because most of her daughters will stay close to her. In consequence, every dispersing female is of great value for management. However, it is not possible to distinguish between males and females at a distance in the field (unless females have cubs with them). Therefore, hunting counteracts the establishment of newly dispersed females. Hunting restrictions in areas that become colonised by bears, accompanied by an information campaign, are a possible solution to this problem.

Managing for the Brown Bear means dealing with people's attitudes. Increasing public understanding of bear ecology and conservation should also improve people's understanding of other predators and could indirectly benefit those species. By contrast, increasing bear numbers could lead to a situation where people feel that the total predator load in a given area is becoming too big, which could result in demands to reduce numbers of other predator species, in this case especially Lynx and Golden Eagle. However, as Brown Bear and Golden Eagle generally do not inflict much damage on livestock, this should not be a problem. A carefully developed management strategy with local participation, an intense information campaign and repeated surveys to keep track of attitudes should suffice to guarantee the co-existence of these three predator species.

## 2.     RESEARCH AND SURVEY

Initially, we did not have enough information available about the Brown Bear in Västerbotten to make reasonable and responsible management decisions. The main questions that we had to answer were:

- What is suitable habitat for the Brown Bear in Västerbotten, and where do we find it in the County?
- How many bears are there in the County, and where?
- Which problems exist, both for bears and for humans?
- What does Västerbotten's human population think about Brown Bears?
- Which international, national and regional restrictions exist and have to be taken into account when making decisions regarding Brown Bear management in Västerbotten?

What we did to answer these questions was:

- Investigate the history of the Brown Bear in the County;
- Collate all existing recent data on bears in Västerbotten, including road and railway casualties, hunting statistics and reports on sightings of bears (all these data are collected in a national database for easy access);
- Start up the Regional Predator Council;
- Intensify our contacts with media and the public;
- Increase our contacts with researchers dealing with different aspects of predator management;
- Initiate a questionnaire survey on human attitudes towards bears and other predators;
- Start a thorough survey of bear numbers and distribution using DNA analysis of faecal samples; and
- Intensify our co-operation with national authorities and administrations in adjacent counties, both in Sweden and Norway.

The question regarding suitable habitat was answered by an international group of researchers, who conducted a GIS-based analysis, which resulted in maps covering all

Sweden, Norway and northern Finland (Støbet Lande *et al.*, 2003). We used these results rather than producing maps of our own.

We tried to take a holistic approach during the objective setting phase, which was prolonged because of this. However, secondary problems will decrease because the affected groups participated in the decision-making process. In the Regional Predator Council, several organisations are involved in setting objectives (numbers and distribution) for the five predator species. Reindeer owners, hunters, foresters, farmers, ornithologists, conservationists, local politicians, police, the state attorney and County Administration gather several times each year to discuss predator issues. The Predator Council is an advisory board supporting the County Administration, and its main duty is the flow of information within organisations, between organisations, and between organisations and administration. In consequence, decision-making takes a relatively long time. The reason for the tight involvement of non-governmental organisations is that local involvement is a prerequisite for long-term success when working with big predators. Without local participation, people in the countryside who are negatively affected by predators will not necessarily obey decisions that are made centrally (in Stockholm or Umeå, the capital of Västerbotten). Locals who are involved from the beginning will be more eager to support the whole project and to participate in any necessary fieldwork, which will help to reduce the costs of surveillance and monitoring.

*Photograph by Michael Schneider*

*Figure 20-1.* The heart of the southern core breeding area of Brown Bears near Svanabyn, Dorotea municipality in Västerbotten (July 2003). Forest stands of different ages, clear-cuts, mires and lakes comprise the landscape. Agricultural areas are rare.

A questionnaire survey was conducted in co-operation with researchers from the Swedish University of Agricultural Sciences in Umeå. Although initially planned only for Västerbotten, this survey was subsequently conducted in most counties in northern Sweden.

A bear survey using DNA analysis of faecal samples was initiated in co-operation with the hunters in the County. Sampling was conducted between 21 August and 31 October 2004. Results of the analysis are due in spring 2006.

Variables determining the quality of bear habitat include:

- Productivity of the Elk population;
- Productivity of berry-producing dwarf shrubs (especially *Vaccinium* spp. and *Empetrum hermaphroditum*);
- Density of ants (*Formica* spp. and *Camponotus* spp.);
- Tree size (big trees are refuges for bear cubs from adult males intent on infanticide):
- Traffic intensity (bears actively avoid roads but, nevertheless, are occasionally killed by cars and trains);
- Density of hunters (most bears are shot during Elk hunting, not by specialised bear hunters); and
- Attitudes of the local human population.

Optimal bear habitat in Västerbotten thus consists of large areas of continuous forest, where the landscape includes both mature forest (supplying big trees, berries, and ants) and young-growth forest and clear-cuts with a productive Elk population (Fig. 20-1). Optimal habitat has few roads and railways and a human population with a positive attitude towards bears. The human population is likely to have a positive attitude if levels of damage are low, which is the case outside Reindeer calving grounds and in areas without free-ranging livestock.

Optimal bear habitat is to be found in the central parts of the County, where modern forestry has created a mosaic landscape, with different age classes of coniferous forest and where the densities of humans, roads, railways, Reindeer calves and livestock are low.

Marginal bear habitat is to be found in the coastal areas, where human population and livestock density as well as traffic intensity are high. Also, the mountains are marginal bear habitat, because huge areas are used as Reindeer calving grounds, because overall productivity is low, because the remaining snow in spring enables poachers to find the bears, and because big trees are absent from large parts of the area.


## 3.        BACKGROUND SURVEILLANCE

When the monitoring protocol is fully working, background surveillance for the Brown Bear will consist of regular attitude surveys and faecal sampling for DNA analysis, conducted every fifth year.

Predator surveillance is regulated by law in Sweden, as it is crucial for the compensation system for killed Reindeer and the overall management of predators in the country. Each year the County Administration has to determine, for each Sami village, the number of reproductions of Wolf, Lynx and Wolverine. If reproduction does not occur, it has to be determined whether single animals occur regularly or only temporarily in a Sami village. Currently, for Brown Bear and Golden Eagle, it merely has to be determined whether the species occurs: presumably this system will be changed when we start to have better data on these two species. The results are sent to the Sami Parliament (Sametinget), where decisions are made regarding the sums paid for compensation.

Several methods are used when surveying predators. Passive methods mostly comprise the collecting of reports from the public on predator sightings or predator damage. Active methods differ between species. Wolf, Lynx and Wolverine are surveyed by snow tracking, which is done by professional trackers (the County Administration's field staff) in co-operation with Sami villages and hunters. For the Golden Eagle, we conduct nest surveys in co-operation with ornithologists.

Existing methods for mammals based on tracks in the snow do not work for the Brown Bear, because bears hibernate. Currently, there are three methods used to collect information on bear numbers and distribution.

1. Bear observations by hunters: the hunters report all sightings of bears and other predators during the first seven days of Elk hunting. Because Elk hunting starts more or less simultaneously in all parts of the County and several thousand hunters are involved, the area is covered reasonably well. The result of this activity is an index of bear (and bear cub) occurrence in different parts of the County. Changes in the index between years reflect changes in bear numbers. However, this index does not tell us anything about bear density unless it has been calibrated against actual bear density (see below) (Tab. 20-1, no. 1).

2. Statistics on dead bears from hunting and road casualties give very detailed information about individual bears, as samples are taken from each killed animal for a variety of analyses. When combining data from different years, hunting statistics return valuable information about the distribution of male and female bears in the County and about the age structure of the population, but not about densities (Tab. 20-1, no. 2).

3. Reports from the public are collected concerning the observation of female bears with cubs. These reports return information about the distribution of reproducing females, but are of variable quality, depending on the competence of the observer. Often, it is not possible to decide if cubs where born during the current year or the year before, which makes it difficult to determine the actual number of reproductions per year. This method provides a minimum number of reproducing females in the County (Tab. 20-1, no. 3).

In 2004 we tested for the first time a non-invasive method, based on DNA analysis of droppings collected by volunteers, mostly hunters during Elk and small-game hunting. This faecal analysis is the most detailed and most expensive method, returning scientifically based information about number, sex, distribution and relatedness of bears in the County (Bellemain *et al.*, 2005). Shortcomings are the current astronomic costs for the DNA analyses and the time lag between sampling and the return of the results (Tab. 20-1, no. 4).

When surveying the human population for their attitudes towards bears in the County, the relevant organisations in Västerbotten are screened for existing data on predators (within the Regional Predator Council). Also, we collect information on organisation-specific requirements and wishes regarding large predators (within the Regional Predator Council and otherwise). Furthermore, we conducted a questionnaire survey in a research project in co-operation with the Swedish University of Agricultural Sciences to assess human attitudes towards predators in the County with satisfactory spatial resolution (Tab. 20-1, no. 5).

*Table 20-1.* Time schedule for different surveillance and monitoring activities regarding the Brown Bear in Västerbotten. The years 2004 and 2009 are survey years; 2005–2008 are monitoring years. When the monitoring programme is running properly, public reports will not be needed.

|   |                      | 2004 | 2005 | 2006 | 2007 | 2008 | 2009 | 2010 | Etc. |
|---|----------------------|:----:|:----:|:----:|:----:|:----:|:----:|:----:|------|
| 1 | Bear observations    | ● | ● | ● | ● | ● | ● | ● | |
| 2 | Hunting statistics   | ● | ● | ● | ● | ● | ● | ● | |
| 3 | Reports              | ● | ● | ○ | ○ | ○ | ○ | ○ | |
| 4 | DNA analysis         | ● |   |   |   |   | ● |   | |
| 5 | Questionnaire survey | ● |   |   |   |   | ● |   | |

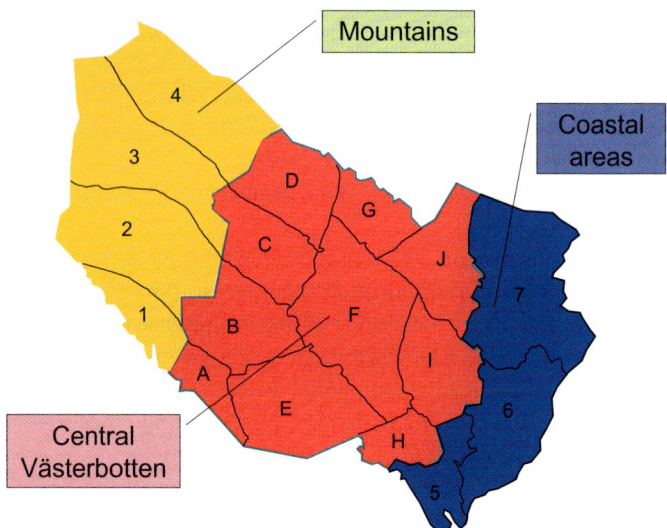

*Figure 20-2.* Västerbotten County is subdivided into three management zones. The Mountains consist of four management areas (1–4). Here, bear density should be low. In Central Västerbotten, consisting of areas A – J, bear density should be high. In the Coastal Areas (areas 5–7), only a few bears should occur.

## 4.    THE CONSERVATION AIMS

The bear population in Västerbotten is currently in a restoration phase. The population needs to increase to between 250 and 600 individuals to reach favourable condition. General conservation management aims for the species and the County are:

- The bear population should hold between 25 and 60 reproducing females each year;
- Bear densities should be different in the coastal region (low), central Västerbotten (high) and the mountain areas (low density); and

- Females should colonize the areas in between existing core breeding areas.

According to a GIS analysis conducted by Støbet Lande and co-workers (2003), the area suitable for Brown Bears in Sweden is 405 000 km$^2$, when taking into account prey density, human population density, infrastructure, habitat type and elevation. About 23% of the most suitable Swedish Brown Bear habitat is found in Västerbotten.

Although bear populations in the existing core areas are not believed to be saturated, the density seems satisfactory. We find this density in only four areas in Sweden, but the bear population is well above the minimum national level today, and many people think bears are already too numerous. The existing core areas are also the prime source of dispersal to unpopulated areas.

Bear density in core areas is 2–3 individuals/100 km$^2$. According to Støbet Lande *et al.* (2003) 98.5% of the County is suitable bear habitat. That means that there would be space for about 1 100 bears in Västerbotten, or 110 annual reproductions. However, to minimise conflicts between bears and Reindeer husbandry in the mountains and between bears and livestock in the coastal areas, only the central c.55% of the County should have a dense bear population. In consequence, the bear population in Västerbotten could consist of about 600 bears, or 60 reproductions per year, with few in the mountains or near the coast.

## 4.1   The condition indicators

It is not practical to survey the habitat of the species in the field, primarily because the County is much too big and too heterogeneous for meaningful and cost-effective sampling, but also because we know that bears do well in the managed forest landscape of Västerbotten. It is the human dimension we have to focus on if we want to gain an insight into differences and changes in bear habitat quality.

We do not think that Elk productivity, berry supply, ant density and occurrence of mature trees will deteriorate in the future. Human population density and traffic intensity, however, are likely to decrease in the future, because of an overall tendency of young people to leave the inner parts of the County and to move to the coastal areas or to southern Sweden. However, even a few humans can jeopardize our attempts to achieve favourable condition of the bear population if they have a negative attitude towards bears.

The variables that best describe human attitudes towards bears in the County are still to be determined. Researchers are working on this question, and the results of the attitude survey 2004 are due in summer 2005.

### 4.1.1   Poaching

Signs of poaching are:
1. Confirmed poaching;
2. Snow mobile tracks following bear tracks in the snow; and
3. Illegal dead bears being marketed.

*Table 20-2.* Condition indicator table during survey years.

| | | The Brown Bear *Ursus arctos* population in Västerbotten County will be restored to favourable condition when |
|---|---|---|
| Distribution | Lower limit | Reproducing females are present in all Areas A, B, C, D, E, F, G, H, I and J (see Fig. 20-2) |
| | Upper limit | Reproducing females are present in all Areas A-J (Central Västerbotten) and in >2 of Areas 1–4 (Mountains) and in >2 of Areas 5–7 (Coastal areas) |
| Population size | Lower limit | In Areas A-J 25 reproducing females are present in any year. The number of females per area is determined by area size and productivity, see Fig. 20-3 |
| | Upper limit | In Areas A-J 53 reproducing females are present in any year (the number of females per area is determined by area size and productivity, see Fig. 20-4) and in each of Areas 1–7 1 reproducing female is present in any year |
| Habitat quality | Lower limit | In each of Areas A-J human attitudes towards bears are as positive as, or more positive than in 2004 |

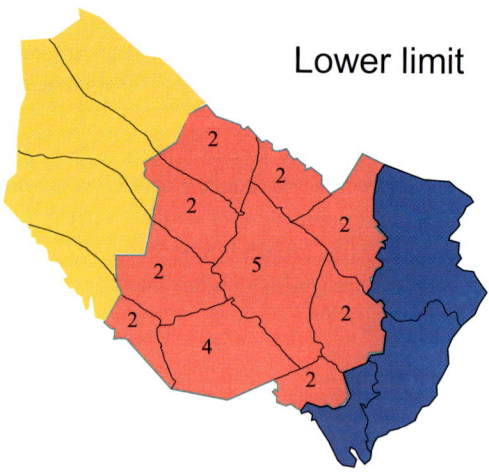

*Figure 20-3.* At the lower limit of population size, the numbers of reproducing females per management area should approach the distribution as shown in this map. The figures refer to actual numbers as found with the help of DNA analysis of bear droppings. The number of reproductions can be estimated from the number of females present via demographic models (still to be developed).

*Table20-3.* Condition indicator table during monitoring years.

| | | The Brown Bear *Ursus arctos* population in Västerbotten County will be restored to favourable condition when |
|---|---|---|
| Distribution | Lower limit | Reproducing females are reported in all Areas A-J (see Fig. 20-2) |
| | Upper limit | Reproducing females are reported in all Areas A-J and in >3 of Areas 1–7 |
| Population size | Lower limit | In each of Areas A–J (Central Västerbotten) ≥ 1 reproducing female is reported in any year. |
| | Upper limit | In each of Areas A-J >4 reproducing females are reported in any year and in each of Areas 1–7 >1 reproducing females are reported in any year |
| Habitat quality | Lower limit | In each of Areas A-J signs of poaching are at the same level or fewer than in 2004 |

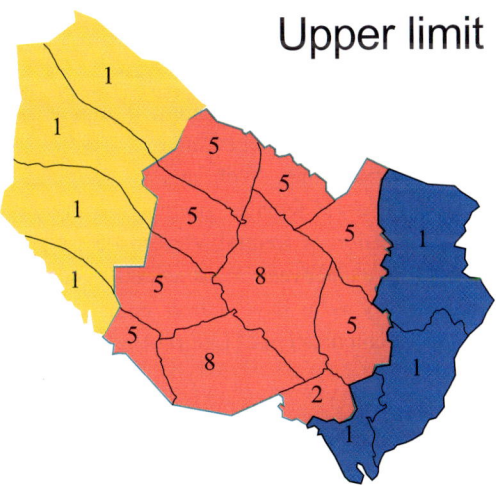

*Figure 20-4.* At the upper limit of population size, a few reproducing females can occur even in the Mountains and the Coastal areas. The figures refer to the actual numbers of females found per management area by DNA analysis of bear droppings.

### 4.1.2    Numbers of bears

In order to minimize sampling effort, it would be sufficient to have females reported from areas F, G, I and J to know if females occur in the whole of Central Västerbotten, because these areas traditionally have the lowest densities of bears in this part of the County. However, as hunters report sightings from the whole County anyway, it is not necessary to make any such geographical restrictions.

## 4.2    Explanation of the indicators

### 4.2.1    Human attitudes

At the beginning of the $21^{st}$ century, the bear population in Västerbotten was increasing, judging from the fact that bears were recolonising parts of the County and from the increase in bear numbers observed by hunters (see Fig. 20-4). Therefore, as long as attitudes to the bears do not deteriorate, the bear population will naturally increase, which means that the population will soon reach favourable condition.

### 4.2.2    Numbers of bears

In a previous study in central Sweden, there was a significant correlation between the number of bear observations made by hunters and the number of bear individuals found by DNA analysis of droppings. One bear seen per 1000 hrs of observation corresponded to c.2 individuals present (Kindberg *et al.*, 2004). According to Kindberg (pers. comm.) however, it cannot be assumed that the correlation will hold in Västerbotten. The relationship between observational index and bear numbers will be revealed with the help of the results from the faecal sampling. With this information, we will be able to extrapolate from numbers of "observed females" during monitoring years.

During the phase of colonisation of Central Västerbotten, we will have to accept a more uneven distribution of bears between areas within management zones, but also between zones. The first priority is to have a bear population in the County that is sufficiently big to guarantee the colonization of Central Västerbotten by females.

## 5.    THE MONITORING PHASE

Monitoring of the Brown Bear population will be conducted in Västerbotten County every year (see Tab. 20-1). Currently we use four different methods to obtain information on a yearly basis:

- Bear observations by Elk hunters;
- Statistics on dead bears from hunting and road casualties;
- Reports from the public; and
- Signs of poaching.

## 5.1    Description of monitoring method

Bear observations by hunters result in an index of bear and bear cub occurrence in different parts of the County. Changes in this index between years reflect changes in bear numbers. This method is relatively cheap, covers most of the County, and is relatively reliable. The method was introduced and first used in Sweden in 1998. No special equipment is needed. A questionnaire is sent to all hunting parties prior to the start of the Elk hunting season. During the first seven days of Elk hunting, hunters note all sightings of predators on their hunting grounds. Participation is not compulsory, but most hunters fill in the questionnaire and send it back to their regional office of the Swedish Association for Hunting and Wildlife Management.

The relationship between observations made by the hunters and the real number of bears will be established every fifth year by DNA analysis of bear droppings. During years without faecal sampling, extrapolation will be used to deduce yearly bear numbers from the indices derived from hunters' observations (see Section 4).

Statistics on dead bears from hunting and road casualties give information about individual bears and about the distribution of male and female bears in the County. County Administration staff collect tissue samples from bears shot during the hunting season, while the Police deal with bears killed on roads and railways. Both samples and dead bears are sent to the Swedish National Veterinary Institute for analysis.

Reported observations of female bears with cubs by the public tell us about the distribution and minimum number of reproducing females in the County. We should consider abandoning the collection of their reports when the other methods are working properly, as it is seldom possible to check details such as species identity and age of cubs. Furthermore, uneven sampling is a fact, because people are more willing to report in areas where bears are rare or have recently arrived. This means that relatively few reproductions will be reported in areas where people are used to seeing bears.

Signs of poaching are collected by County Administration staff all year round for all predator species during ordinary fieldwork and during special activities in cooperation with Police, Customs Authority and Coast Guard.

## 6.    RESULTS AND ANALYSIS

As we have only recently started the Brown Bear monitoring programme in Västerbotten, there are no final results to present yet. The figures below show the current status of some of the variables used during monitoring and survey. Final results from the bear work during 2004 will be presented in 2005.

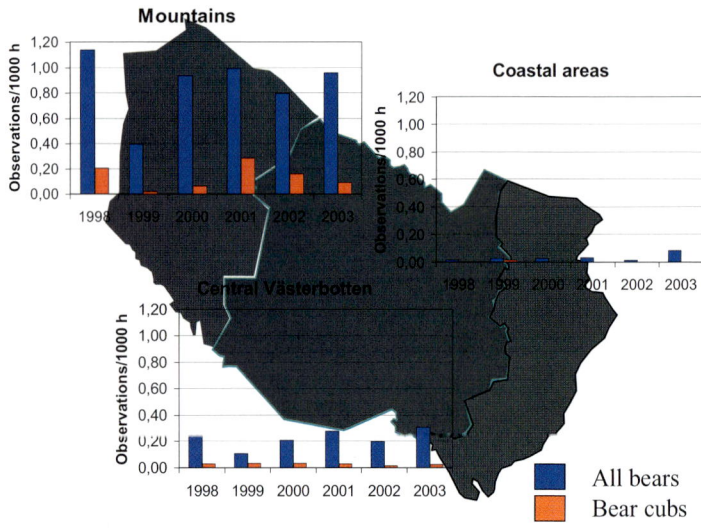

*Figure 20-5.* The number of bears observed during 1000 hours of observation in the Mountain region, Central Västerbotten and the Coastal areas, respectively, between 1998, when this survey started, and 2003. According to these data, there are more bears in the mountains than in Central Västerbotten, which is not in accordance with the plans for bear distribution in the County.

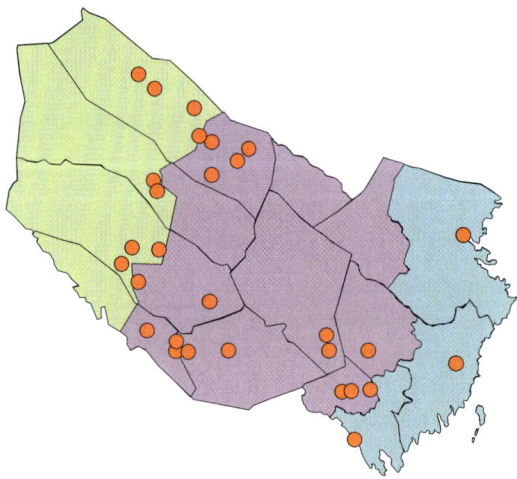

*Figure 20-6.* Reports by the public regarding observations of bear cubs in Västerbotten County 2004. In many cases it is unclear if the young were born in 2004 or the year before. No effort has been made to find out if different observations in a given area belong to the same family group.

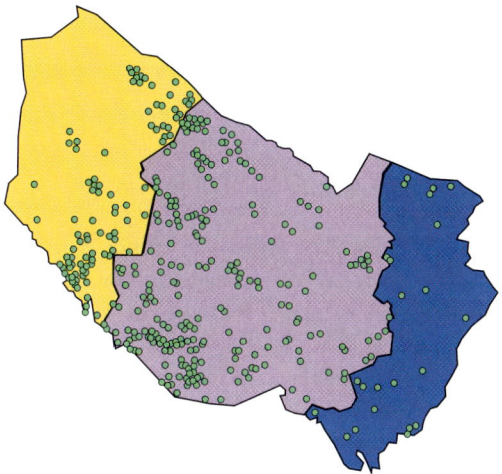

*Figure 20-7.* Faecal samples were collected in most parts of the County in autumn 2004. In this figure, 350 of 940 samples are depicted. A pattern of highly uneven distribution of bear droppings in the County emerges, and this pattern does not change when all samples are plotted on the map.

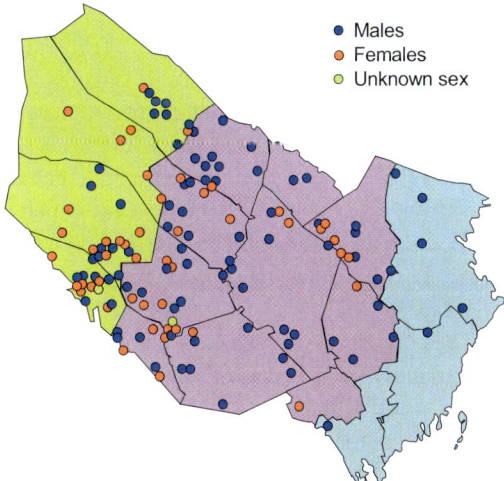

*Figure 20-8.* The localities in Västerbotten where bears died between 1981 and 2003. Most of the bears were killed during hunting; a few were killed by cars and trains. While the distribution of males is relatively uniform, female distribution shows distinct clusters.

# 7.    DISCUSSION

In this chapter I have presented ideas about the monitoring of Brown Bears. Details about the reasoning behind bear numbers, bear distribution, and management techniques to be applied, are presented in the Management Plan for the Brown Bear in Västerbotten County (Schneider, unpubl.).

After a couple of years of data collection, brainstorming and objective setting, the monitoring scheme is effective as from the year 2004. Adjustments of this scheme will eventually be made, when better knowledge of bear distribution and population dynamics in the County make it necessary.

Västerbotten County is huge. It would be difficult and costly for County Administration staff alone to survey the whole area each summer. It would be very helpful if we could look at smaller areas or places that indicate the condition of larger areas. However, currently there are no realistic survey methods for bears that could be used by trained field personnel in small geographical areas in short periods of time, which would render reliable information on the status of bears that could be used for management decisions in the entire County.

Realistically, surveys of bears are best carried out either by hunters or the public (relatively cheap, with some uncertainty in the data), or based on DNA analysis and capture-mark-recapture models (expensive, but relatively precise data) (cf. Taberlet *et al.*, 2001). For this DNA-based method to work with Brown Bears, the investigated areas have to be huge. Also, faecal sampling and DNA analysis are too costly to be conducted every year. However, as more laboratories become interested in these methods, the costs for DNA analyses will presumably decrease.

Bears receive very much public attention. Therefore, it is relatively easy to engage hunters and the public in the surveillance and monitoring of the species. Surveys by hunters usually develop an internal dynamic that should be supported, and not hampered by geographically and temporally structured sampling schemes. Either everyone samples, everywhere, or sampling is not conducted at all. Involving hunters and the public makes them co-owners of the results, which means that figures for bear numbers and distribution will be more widely accepted than if the County Administration alone had produced the results. This is very positive for the management of bears (Zachrisson, 2004).

There are, however, some uncertainties regarding the future:

- The minimum size of single management areas within management zones has still to be determined and may change from those shown on the maps above.
- The willingness of hunters to participate will depend on the performance of the manager when dealing with bear questions. Management decisions against hunters may result in a decreased willingness to co-operate.
- The age distribution of hunters is of some concern, as the recruitment of young people into hunting organisations is low. Eventually, hunters may become too few for effective sampling of the entire County.
- The cost of bear monitoring may increase when we leave the stage of experiment and development and enter the phase of routine attitude surveys. Presumably, the present interest of researchers will decrease, and managers will ultimately have to pay for most data collection and analysis.

At the national scale, the Swedish Environmental Protection Agency has to decide whether Västerbotten's monitoring scheme can be applied to the rest of Sweden. If so, a programme should be established whereby different counties conduct faecal analysis in different years. A national programme would have relatively constant financial requirements from year to year, which would simplify funding. Such a programme would also support the establishment of laboratories able to do the analysis, supplying them with a regular flow of samples, which in turn should result in more cost-effective analyses.

Monitoring is one of many tools of predator management. Successful management of predators is labour- and cost-intensive (cf. Macdonald, 2001), and requires support from administrative bodies at different levels. This support depends directly on the level of funding available, and this depends on political decisions. If politicians change direction, funding priorities may change, which could force administrators to change management priorities.

# 8. ACKNOWLEDGEMENTS

My sincerest thanks to Clive Hurford for helping to structure my thoughts about bear monitoring, and to Västerbotten County Administration and the Countryside Council for Wales for making our joint monitoring project possible. Jon Swenson kindly discussed many aspects of bear ecology and management and commented on the manuscript. Gunnar Ledström was of great help during the planning of the DNA project. Sven Brunberg gave advice on sampling of bear droppings. The staff of the regional office of the Swedish Association for Hunting and Wildlife Management was very helpful while starting the DNA project. Mats Nilson took care of incoming faecal samples and gave comments. Magnus Kristoffersson shared his knowledge of bear ecology and his experience with practical bear management in Jämtland County. Jonas Kindberg made available the results of the hunters' yearly predator survey. Sonja Almroth assisted while inspecting optimum bear habitat. I am especially grateful to Björn Jonsson for giving me the freedom and the opportunity to work on this case study.

# 9. REFERENCES

Most of the literature I used has been published in Swedish. A list of publications on Scandinavian bears, many of them in English, can be found at http://www.bearproject.info

Andrén, H., Liberg, O. and Sand, H. (1999). De stora rovdjurens inverkan på de vilda bytestammarna i Sverige. - Bilagor till Sammanhållen rovdjurspolitik; Slutbetänkande av Rovdjursutredningen. - Statens offentliga utredningar 1999: 146. Stockholm, pp. 119-182. (in Swedish)

Bellemain, E., Swenson, J., Tallmon, D., Brunberg, S. and Taberlet, P. (2005). Estimating population size of elusive animals using DNA from hunter-collected feces: comparing four methods for brown bears. – *Cons. Biol.* **19**: 150-161.

Bjärvall, A. (1978). Björnen i Sverige. Statens naturvårdsverk, rapport - LiberFörlag/Allmänna Förlaget, Vällingby. (in Swedish)

Dahle, B. and Swenson, J. (2003). Seasonal range in relation to reproductive strategies in brown bears *Ursus arctos*. – *J. Anim. Ecol.* **72**: 660-667.

Ekman, S. (1910). *Norrlands jakt och fiske.* Facsimile of the original edition - Två Förläggares Bokförlag, Umeå. (in Swedish)

Gärdenfors, U. (2005). *Rödlistade arter i Sverige 2005.* The 2005 Red List of Swedish Species. - ArtDatabanken, Uppsala. (bilingual)

Kindberg, J., Swenson, J., Brunberg, S. and Ericsson, G. (2004). Preliminär rapport om populationsutveckling och -storlek av brunbjörn i Sverige, 2004. – Unpublished report to the Swedish Environmental Protection Agency. (in Swedish)

Lönnberg, E. (1929). Björnen i Sverige 1856-1928 - Almqvist & Wiksells Boktryckeri AB, Uppsala and Stockholm. (in Swedish)

Macdonald, D. W. (2001). Postscript - carnivore conservation: science, compromise and tough choices. - In: Gittleman, J. L., Funk, S. M., Macdonald, D. W. and Wayne, R. K. (eds.), *Carnivore conservation.* Cambridge University Press, Cambridge. pp. 524-538.

Naturvårdsverket (1991). Förslag till nytt ersättningssystem för rovdjursdödade renar. Naturvårdsverkets rapport 3899 - Naturvårdsverket, Solna. (in Swedish)

Persson, J., Zachrisson, A., Sandström, C. and Ericsson, G. (2004). Lokal förvaltning av stora rovdjur. En kunskapssammanställning. FjällMistrarapport nr 3. - FjällMistra, Umeå. (in Swedish)

Regeringen 2000. Regeringens proposition 2000/01:57 Sammanhållen rovdjurspolitik. Stockholm. (in Swedish)

Sandegren, F. and Swenson, J. (1997). Björnen. Viltet, ekologi och människan. - Svenska Jägareförbundet, Spånga. (in Swedish)

Schneider, M. unpubl. Björnen *Ursus arctos* i Västerbottens län. Förvaltningsplan för åren 2005-2006. Länsstyrelsen i Västerbottens län, Umeå. (in Swedish with English summary)

Skandinaviska Björnprojektet 2000. Är björnen farlig? - Björnprojektet i Orsa, Orsa. (in Swedish)

Skogen, K., Haaland, H., Krange, O., Brainerd, S. M. and Hustad, H. (2003). Utredninger i forbindelse med ny rovviltmelding. Lokale syn på rovvilt og roviltforvaltning. En undersøkelse i fire kommuner: Aurskog-Høland, Lesja, Lierne og Porsanger. Nina Fagrapport 070 - Norsk Institutt for Naturforskning, Trondheim. (in Norwegian)

SOU 1999: 146, Sammanhållen rovdjurspolitik, Slutbetänkande av rovdjursutredningen. - Fakta Info Direkt, Stockholm. (in Swedish)

Støbet Lande, U., Linnell, J. D. C., Herfindal, I., Salvatori, V., Broseth, H., Andersen, R., Odden, J., Andrén, H., Karlsson, J., Willebrand, T., Persson, J., Landa, A., May, R., Dahle, B. and Swenson, J. (2003). Utredninger i forbindelse med ny rovviltmelding. Potensielle leveområder for store rovdyr i Skandinavia: GIS-analyser på et økoregionalt nivå. NINA Fagrapport 64 - Norsk institutt for naturforskning, Trondheim. (in Norwegian)

Swenson, J. and Sandegren, F. (1999). Misstänkt illegal björnjakt i Sverige. - In: Bilagor till Sammanhållen rovdjurspolitik; *Slutbetänkande av Rovdjursutredningen. - Statens offentliga utredningar 1999*: **146**. Stockholm, pp. 201-206 (in Swedish)

Swenson, J. and Sandegren, F. (2000). Ekologi och förvaltning av brunbjörnen i Skandinavien. Summarising report from the Scandinavian Brown Bear Research Project (in Swedish). - Published on the Internet at http://www.jagareforbundet.se/files/ursslutrappF2C.pdf

Taberlet, P., Luikart, G. and Geffen, E. (2001). New methods for obtaining and analyzing genetic data from free-ranging carnivores. - In: Gittleman, J. L., Funk, S. M., Macdonald, D. W. and Wayne, R. K. (eds.), *Carnivore conservation.* Cambridge University Press, Cambridge, pp. 313-335.

Zachrisson, A. (2004). Co-management of Natural Resources. Paradigm Shifts, Key Concepts and Cases - FjällMistra, Umeå.

# CHAPTER 21

# MONITORING THE WOLVERINE *GULO GULO* IN VÄSTERBOTTEN COUNTY

MICHAEL SCHNEIDER

*County Administration, Västerbotten County, SE-901 86, Umeå, Sweden*
*Michael.Schneider@ac.lst.se*

## 1.  BACKGROUND INFORMATION

Västerbotten County in northern Sweden is home to five big predatory species that are threatened in Europe and subject to regulations within the Natura 2000 framework. In a previous chapter, I presented our work with the Brown Bear (*Ursus arctos*) in Västerbotten. Here, I present ideas and activities related to the monitoring and management of a very different species. For general information about Västerbotten and the predator work done there, please refer to Chapter 20.

## 1.1  Gulo gulo

The Wolverine is the largest terrestrial member of the weasel family (Mustelidae). Relatively little is known about Wolverine biology and ecology. Two web sites (www.wolverinefoundation.org, www.lcie.org) present compilations of much of the information available and extensive bibliographies. From a Swedish point of view, key references are Haglund (1966), Pulliainen (1993), Pasitschniak-Arts & Larivière (1995), Landa (1997), Landa *et al.* (2000a), Naturvårdsverket (2003) and Persson (2003). The Swedish Wolverine Research Project is currently acquiring new information in the Sarek area in Norrbotten County (cf. Persson, 2003).

*C. Hurford & M. Schneider (eds.), Monitoring Nature Conservation in Cultural Habitats*, 215–230.
© 2007 *Springer.*

# 1.2   The Wolverine in Västerbotten County

Historically, Wolverines were found in mountainous and forested areas in central and northern Sweden (Ekman, 1910: Krott, 1960: Persson, 2003 and references therein). Due to human persecution, the population declined dramatically from the middle of the 19th century, until the species was protected in 1969. Having been restricted to a small population in the mountain range along the Swedish-Norwegian border, the Wolverine population increased again during the first decades after protection. However, population size has been relatively stable but small during recent years and was estimated to be ca. 360 individuals in 2002 in Sweden (Naturvårdsverket, 2003). Most (about 70%) of the Swedish Wolverines live in Norrbotten County. The distribution in Sweden is very patchy, with large areas of vacant, but presumably suitable, habitat in between occupied areas.

In Västerbotten, as in most counties, Wolverines are restricted to the mountain areas and associated forests. Only in two counties, Västernorrland and Gävleborg, do Wolverines regularly occur and even breed in the forest landscape closer to the coast (Kilström, 2004). The forest Wolverines in Västernorrland live about 50 km south of Västerbotten in an area with rugged terrain and seem to mainly feed on leftovers from people's Elk hunting. These Wolverines and their habitat are models for the future expansion of Wolverine distribution in Västerbotten County.

# 1.3   Key factors influencing management decision for the Wolverine

The Wolverine is listed as endangered in the latest version of the Red List of Swedish Species (Gärdenfors, 2005) and included in Annex II of the EU Habitats Directive. In accordance with existing laws and regulations, the County Administration has the responsibility to protect the species in Västerbotten.

There are four key factors influencing management decisions for the Wolverine in Västerbotten:

- The national, coherent predator policy in Sweden;
- Reindeer (*Rangifer tarandus*) husbandry;
- Wolverine denning and dispersal biology, and
- Other predators.

### 1.3.1   A coherent predator policy at the national level

According to a decision by the Swedish Parliament, a viable Wolverine population shall exist in the country. The species shall be allowed to recolonise areas where it formerly occurred. This national policy sets the framework for Wolverine management in the country (Regeringen, 2001).

The national target for the Wolverine is a minimum of 90 reproducing females per year, corresponding to about 575 animals in Sweden, as a first step. When this level has been reached, new discussions shall define the minimum level for the population to be viable in the long term. A more even distribution of individuals within the species´ natural range is the goal for Wolverine distribution.

According to the national predator policy, the County Administration in Västerbotten has to come up with a management plan for the Wolverine, taking into account specific regional circumstances such as the level of economic losses inflicted, the amount of suitable habitat available and human attitudes. We have to decide how big the Wolverine population should be in the County, how Wolverines should be distributed, and which methods should be used to reach these goals.

## 1.3.2    Reindeer husbandry

Wolverines and some 60 000 semi-domestic Reindeer co-exist in Västerbotten. Reindeer herding is thoroughly associated with the Sami culture in northern Scandinavia and, as such, is both a way of earning a livelihood and a part of the identity of an ethnic minority (Landa *et al.*, 2000b, cited after Persson, 2003). In the summer months, the Reindeer are left unattended for long periods, as they roam the mountains. Conversely, during migration and on the winter grounds in coastal Västerbotten, the Sami pay close attention to their herds. Wolverines prey upon Reindeer, and losses may be heavy in certain places and at certain times (Bjärvall *et al.*, 1990). Wolverine predation on Reindeer has been one of the main reasons for their persecution and historical decline in Sweden in the 19th and 20th century. Wolverines in Västerbotten do not disturb hunting and do not prey upon other species of livestock. Therefore, no groups in Swedish society, other than Reindeer owners and conservationists, are particularly interested in Wolverines. The existing conflict between Wolverines and Reindeer herders demands a compromise between predator conservation on one side and the conservation of the indigenous Sami culture on the other (Persson, 2003).

## 1.3.3    Wolverine denning and dispersal biology

Females start to occupy reproductive dens (extensive tunnels in the snow) in February or March, and the young stay in and around this natal den during the first two months of their life. In the current breeding area of the Wolverine in the Scandinavian mountain chain, most dens are situated on mountain slopes near or above timberline, often in deep snow near cliff areas. The same denning site can be used year after year (see Fig. 21-5). Several denning sites comprise a den area, and sites can shift within this area between years, depending on snow conditions. Wolverines recolonising an area often use the traditional denning sites, indicating that optimal sites may be a limited resource (Andersen *et al.*, 2002).

It is unclear if Wolverines are sensitive to disturbance by humans during the denning period. Protection of natal denning habitat from human disturbance may be critical for the persistence of the Wolverine. It has been found in some studies that the association between Wolverine presence and refugia may be linked to a lack of available denning habitat outside protected areas. It is hypothesized that an increased use of snowmobiles and increased recreational activities in winter may displace Wolverines from potential denning habitat. Also, direct persecution and killing of Wolverines in natal dens may be a problem in certain places and at certain times (see discussions in Landa *et al.*, 2000a).

Juveniles stay within their mother's home range until late August, female young sometimes longer. Dispersing Wolverines have been observed to move very long distances (males $\leq$ 500 km, Flagstad *et al.* 2004) and therefore should be able to recolonise vacant Wolverine habitat. The large proportion of unoccupied habitat could be explained by a high turnover rate in the population, rather than low dispersal capacity. If the turnover rate is high, because of legal harvesting or illegal killing, a high proportion of females should be sedentary (occupying the territories of killed females), thus reducing recolonisation rates (Vangen *et al.*, 2001).

### 1.3.4     Interactions between carnivores

Although Wolverines are looked upon as not being very efficient hunters, they are capable of taking large live prey such as domestic sheep, Reindeer and, exceptionally, Elk (*Alces alces*). Predation on such big game is believed to occur mainly under specific snow conditions, when Wolverines due to their big feet float on deep snow while the movement of ungulates is hindered. Other predator species, such as Wolf (*Canis lupus*) and Lynx (*Lynx lynx*), may be important for the Wolverine as suppliers of ungulate carrion (Haglund, 1966: Pulliainen, 1993: Landa, 1997).

## 1.4     The management priority

Management objectives for the Wolverine are a more even distribution of individuals, an overall larger population, and the recolonisation of vacant habitat in the County. The level of damage inflicted by Wolverines is currently relatively high and should be thoroughly surveyed. Until today, a lack of basic information on Wolverine biology and habitat requirements has resulted in little management beyond administrative protection.

The occurrence of other, more efficient predators supports Wolverines by supplying them with remains of large prey. This directly positive effect can be counteracted by negative ones. Wolves can kill Wolverines, and especially so in mountain areas without trees that could be used for retreat. In Västerbotten, only single, dispersing wolves occur in the mountains, and an establishment of wolf packs in this area is not desirable with respect to Reindeer husbandry.  A high density of Lynx can indirectly affect Wolverines by decreasing the willingness of Reindeer herders to tolerate a high density of Wolverines as well. This problem has been accounted for in the management plan for Lynx in Västerbotten, which states that Lynx numbers should be reduced in the mountain areas when Wolverines have reached a sufficiently high density there. However, total predation pressure per Wolverine will increase in areas where the Lynx population is decreasing, as Wolverines use the leftovers of Lynx kills when such are available instead of killing Reindeer themselves.

High levels of predation could jeopardize the survival of the Sami Reindeer herding tradition. Here, conservation of our natural heritage has to be traded against the conservation of a cultural heritage. In this case, decisions have to be made at the political

*Figure 21-1.* Björnlandet national park and surroundings are supposed to be good Wolverine habitat in the forested landscape of central Västerbotten. The area is similar to that part of Västernorrland County where stationary forest Wolverines are to be found. The terrain is rough, lynx are relatively numerous and the Elk population is productive. Today, Wolverines do not occur in this area.

level. According to political decisions that already have been made, predators must not make Reindeer husbandry impossible.

## 2.      RESEARCH AND SURVEY

There was not enough information available about the Wolverine in Västerbotten to make reasonable management decisions. The questions we had to answer were:

- What are the basic biological parameters in the life of Wolverines, and what are their ecological requirements?
- What is suitable habitat for the Wolverine in Västerbotten, and where do we find it in the County?
- Which problems are inflicted by Wolverines?
- What do people in Västerbotten think about Wolverines? And
- Which international, national and regional restrictions will affect Västerbotten's decisions regarding Wolverine management?

What we did to answer these questions was:

- Survey the existing literature with regard to Wolverine history in Västerbotten;
- Analyse our results from the ongoing Wolverine surveillance;
- Start up the Regional Predator Council;
- Intensify our co-operation with the Swedish Wolverine Research Project;
- Increase our contacts with researchers dealing with different aspects of predator management;
- Initiate a questionnaire survey on human attitudes towards Wolverines and other predators;
- Collect faecal samples for DNA analysis; and
- Intensify our co-operation with national authorities and regional administrations in adjacent counties, both in Sweden and Norway.

An international group of researchers conducted a GIS-based analysis of habitat suitability for big predators, which resulted in maps covering all Sweden, Norway and northern Finland (Støbet Lande *et al.*, 2003). However, the analysis made with respect to Wolverines seems rather conservative, bearing in mind the success of the forest Wolverines in Västernorrland County and the historic distribution of Wolverines in Västerbotten. We believe that a greater area of Västerbotten is suitable for Wolverines than is shown in the map of Støbet Lande *et al.* (2003).

*Photograph by Michael Schneider*

*Figure 21-2.* Vindelfjällen nature reserve is the largest of all mountain reserves in Västerbotten (ca. 550 000 ha), and in fact one of the largest protected areas in Europe. Every year, several reproductions of Wolverine occur in Vindelfjällen.

Wolverines were discussed in the Regional Predator Council, where different groups of society are represented. We examined the knowledge about Wolverines present in different organisations, sampled attitudes, compiled data on damage levels, and discussed regional targets concerning the numbers and distribution of Wolverines in Västerbotten.

We conducted a questionnaire survey in co-operation with researchers from the Swedish University of Agricultural Sciences in Umeå. This survey will give us information about the attitudes of the people in Västerbotten to Wolverines.

Faecal samples are taken to get insight into Wolverine population structure and dynamics. DNA analysis can determine the identity, sex, and the genetic relationship of individuals. This sampling is conducted by County Administration staff only but will give valuable results because of yearly replication.

To answer some of the questions in Wolverine ecology and management, we are considering starting a LIFE-Nature project, we will participate in the First International Wolverine Symposium, and we plan to initiate new research on Wolverine foraging behaviour and habitat choice.

Variables determining the quality of Wolverine habitat include the:

- Occurrence of undisturbed denning sites;
- Ruggedness of the terrain;
- Snow conditions;
- Number and condition of Reindeer;
- Dynamics of small mammals;
- Extent of Elk hunting;
- Occurrence of Lynx; and
- Attitudes of local Reindeer herders.

Optimal Wolverine habitat in Västerbotten thus consists of areas where the terrain *per se* or regulations prevent human disturbance at denning sites, where enough snow accumulates for denning and the storage of food, where Reindeer or Elk are abundant and where Lynx or people's hunting activities guarantee a sufficient food supply. Optimal habitat also requires a human population with a positive attitude towards wolverines, or a rugged terrain that prevents Wolverines from being harassed by snowmobile drivers. Reindeer herders are the only people who are actually negatively affected by Wolverine. Presumably, their attitude will be more positive if the total predator pressure is acceptable and if levels of financial compensation for losses are sufficiently high.

Good Wolverine habitat is to be found in the huge mountain reserves, where strict regulations restrict the use of snowmobiles and other recreational activities, where Reindeer are numerous at least in summer, where Lynx occur, where snow conditions are benign, and where occasional lemming outbreaks may be positive for the Wolverine population during summer. Furthermore, good habitat is to be found in many parts of central Västerbotten, where human population density is low, where the physiognomy of the terrain ensures undisturbed sites, and where leftovers from people's intensive hunting of a productive Elk population supply Wolverines with the food needed for winter survival and successful reproduction.

# 3.      BACKGROUND SURVEILLANCE

Background surveillance of the Wolverine population consists of a yearly survey of all known denning sites, snow tracking and faecal sampling. The attitude of the human population is also studied.

Several methods are used when surveying Wolverines. Passive methods mostly include the collection of reports from the public on sightings of individuals or family groups, or tracks in the snow. Active methods include snow tracking, which is done by professional trackers (the County Administration's field staff) in co-operation with Sami villages. During January and February each year, when the Lynx population is surveyed in the whole County, the snow tracking also renders information about the distribution of Wolverines, hot spots of activity, and the possible location of new denning sites.

During March to May each spring, all known denning sites in the County are visited and the presence or absence of Wolverines is stated. If Wolverine activity is found, the site will be visited several times to locate a natal den and, if present, to determine if young have been born, and, if so, how many. In uncertain cases, denning sites will be visited after the snow has melted. At this time, the contents of the tunnel system will be exposed and will reveal if young have been born (NFS, 2004:17). The number of active dens is used as an indicator of Wolverine population size (Landa *et al.*, 1998). During site visits, faecal samples are taken for DNA analysis.

When surveying the human population in the County, the relevant organisations in Västerbotten are screened for existing data on Wolverines (within the Regional Predator Council). Also, we collect information on organisation-specific requirements and wishes regarding the species (within the Regional Predator Council and otherwise). Furthermore, we conduct regular surveys of human attitudes towards Wolverines in the County with a questionnaire survey every fifth year.

We are currently considering the use of remote camera surveys to determine the reproductive status of dens and to prevent the illegal killing of family groups at natal dens.

*Table 21-1.* The time schedule for different surveillance and monitoring activities regarding the Wolverine in Västerbotten. Den surveys, snow tracking in the whole County and faecal sampling for DNA analysis are conducted every year, while an attitude survey is done every fifth year only.

|   |                     | 2004 | 2005 | 2006 | 2007 | 2008 | 2009 | 2010 | Etc. |
|---|---------------------|------|------|------|------|------|------|------|------|
| 1 | Den survey          | ●    | ●    | ●    | ●    | ●    | ●    | ●    |      |
| 2 | Snow tracking       | ●    | ●    | ●    | ●    | ●    | ●    | ●    |      |
| 3 | DNA analysis        | ●    | ●    | ●    | ●    | ●    | ●    | ●    |      |
| 4 | Questionnaire survey | ●    |      |      |      |      | ●    |      |      |

# 4.      THE CONSERVATION AIMS

The Wolverine population in Västerbotten is currently in a restoration phase. The population has to increase to reach favourable condition. A Wolverine population of >170 individuals in the County would represent favourable condition, if there were regular genetic exchange with the rest of the Scandinavian Wolverine population. General management aims for the species and the County are:

- The Wolverine population should increase from today's level (ca. 15 reproductions, see Fig. 21-4) and hold a minimum of 26 reproducing females each year;
- Wolverine densities should be similar in different areas of Reindeer husbandry (Sami villages); and
- Wolverines should not only occur in the mountains, but also recolonise their historical range in the forested inland of Västerbotten.

Reindeer are the most important source of food for Wolverines in Sweden. Reindeer husbandry occurs in five counties only: Norrbotten, Västerbotten, Jämtland, Västernorrland and Dalarna. As a first step, the minimum number of yearly reproductions of Wolverines in Sweden has been set to 90. If these 90 reproductions are distributed between counties according to the number of Reindeer available for Wolverines in each of them, Västerbotten ends up with 26 Wolverine reproductions per year. Therefore, the minimum number of reproductions per year should be 26 in Västerbotten. The maximum number should be set with regard to the levels of damage inflicted by Wolverines to Reindeer husbandry, and with regard to minimum and maximum national levels that guarantee a viable population (this still has to be defined). As we have not reached the minimum level in the County yet, this will be subject to discussions in the future.

## 4.1   The condition indicators

It is not practical to survey the habitat of the species in the field, because the County is much too big and too heterogeneous for meaningful and cost-effective sampling. Information on long-term changes in snow conditions can be attained from the Swedish Meteorological and Hydrological Institute. Reindeer numbers can be extracted from national statistics from the Swedish Board of Agriculture. Information about Elk hunting extent and intensity in different parts of the County is very detailed, because of strict regulations regarding hunting areas and bag limits. In particular, however, it is the human dimension that we have to look at if we want to get an insight into differences and long-term changes in Wolverine habitat quality.

We do not think that Reindeer density, Elk productivity and Elk hunting intensity will deteriorate in the near future. According to our management plan, Lynx will increase in Västerbotten. Rodent population dynamics are hard to predict and may be negatively influenced by large-scale forestry in the inland and coastal areas. Human density and traffic intensity are likely to decrease in the future, because of an overall tendency of young people to leave the inner parts of the County and to move to the coastal areas or to southern Sweden. However, even a few humans can jeopardize the condition of the Wolverine population if they have a negative attitude towards the animals.

Presumably, in the future, there will be fewer Reindeer herders in Västerbotten, and each of them will have larger herds than today. Fewer people and more animals mean an even higher workload per person, which will result in less time available to attend the animals. Guarding the Reindeer will therefore be even more difficult than today.

*Table 21-2.* Condition indicator table.

| Condition indicator table for population restoration | | The Wolverine *Gulo gulo* population in Västerbotten County will be restored to favourable condition when |
|---|---|---|
| Distribution | Lower limit | Natal dens are present in all Areas A, B, C, D, E, F and G (i.e. in all Sami villages outside the coastal area, see Fig. 21-3) |
| | Upper limit | None set |
| Population size | Lower limit | In Areas A1–F1 (the mountain breeding area) >15 natal dens in total are present in any year and in Area A2–G (the forested inland) >10 natal dens in total are present in any year and in each of Areas A–G (Sami villages) >2 natal dens are present in any year |
| | Upper limit | Will be determined starting from the levels of damage inflicted by Wolverines when lower limit has been passed |
| Habitat quality | Lower limit | In Västerbotten County attitudes of Reindeer herders towards Wolverines are more positive than in 2004 and indications of poaching are at the same level or fewer than in 2004 |

## 4.2    Explanation of the indicators

The lower limit for distribution draws attention to the need for the Wolverine population to expand out from the mountain area (Areas A1–F1) into the surrounding forests (Areas A2–G). The lower limit for population size in the mountains is based on the maximum number (15, 15 and 16) of dens found there since surveillance started in 1996. The lower limit for the total number of dens in the County is based on the minimum national level for Wolverine reproductions and the relative abundance of Reindeer in Västerbotten.

The number of natal dens is a crude indicator of the number of Wolverines present in an area (Landa *et al.*, 1998). Not all females reproduce every year, and the number of females reproducing may differ significantly between years. Therefore, the results of den surveys of several consecutive years should be used to find trends in the population, rather than drawing conclusions from the results of one single year. Multiplying the number of dens by 6.4 renders an approximate number of individuals in a given area.

Human attitudes are the most important habitat variable for Wolverines (and other predators) in the County. The Wolverine population has not increased greatly since the species was protected in 1969. It is assumed that illegal poaching is the main reason for this. If this is the case, and attitudes remain the same, the Wolverine population should stay at the same level, whereas if attitudes become more positive, mortality should decrease and the population should increase, and approach favourable condition.

For obvious reasons, the illegal killing of Wolverines is difficult to measure. However, there are indicators that we can use to get an idea about the extent of poaching:

- Confirmed poaching (if Police investigations show that illegal killing has occurred);
- Snow mobile tracks following Wolverine tracks in the snow;
- Killed Wolverines are found;
- Injured animals are observed or tracked;
- The numbers of dead young in natal dens are unusually high. Multiple dead kits in a den indicate that the mother has died or the young have been killed inside the den.

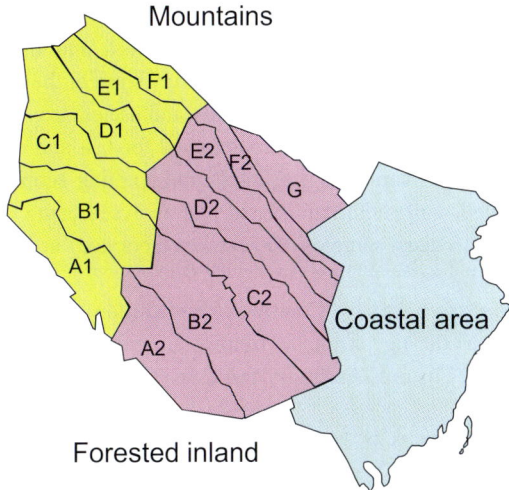

*Figure 21-3.* Västerbotten County is subdivided into three zones, the current breeding area of Wolverines in the mountains, an expansion zone in the forested inland, and the coastal area, where Wolverines are welcome but where no active management towards Wolverine establishment is conducted. Mountains and inland are subdivided according to the Sami villages (A–G) having their grounds there. This division is not made for the wintering grounds along the coast, where in total 15 Sami villages are represented in the County.

## 5.  THE MONITORING PHASE

We monitor the Wolverine population in Västerbotten County every year, except for the attitude survey, which we conduct every fifth year (see Tab. 21-1). Currently we use four different methods to obtain information on a yearly basis:

- Survey of natal dens.
- Snow tracking.
- DNA analysis.

- Questionnaire survey.

See Section 3 for more information on the different methods.

## 5.1    Description of monitoring method

The results of the annual Wolverine survey contribute to the monitoring of the species. Combining the results from several years enables us to see trends in the population. Results from DNA analysis render information about individual home range size and location, multi-annual den utilization, and genetic relationships between individuals.

The winter survey of Wolverines is a demanding undertaking, particularly with respect to training, the knowledge and experience of field staff, and the equipment needed. Temperatures can be as low as $-35°C$ and days are very short at the beginning of the year. We purchase snowmobiles with four-stroke engines and use environmentally adapted fuel. During summer controls of dens, helicopters are used for fast and easy access to remote sites. GPS-receivers, digital cameras, binoculars and spotting scopes are all part of the basic equipment needed, with field data stored and analysed using GIS and an internet-based national database. Faecal samples are sent to a laboratory at Uppsala University for analysis.

Signs of poaching are noted by County Administration staff all year round for all predator species, both during ordinary fieldwork and during special activities in cooperation with Police, Customs Authority and Coast Guard.

## 6.    RESULTS AND ANALYSIS

Surveillance results show that neither distribution nor population size meet the levels set for the Wolverine population to be in favourable condition (Figs. 21-4 and 21-5). The number of natal dens found (Fig. 21-4) is far below the minimum level of 26 reproductions in Västerbotten County. Most of the dens are situated in the mountains, while very few have been found in the expansion area in the forested inland. The distribution of Wolverine denning sites, and their use during different years, has been very uneven in Västerbotten between 1996 and 2003 (Fig. 21-5). Some sites are obviously preferred and used every year, while others are used only occasionally. At the time of writing, Wolverine reproduction does not occur across large parts of the County (Fig. 21-6). An attitude survey was conducted in 2004. Results are due in the beginning of 2005, which means that, at the time of writing, we are not very well informed about attitudes towards Wolverines in different parts of society.

Because we have not reached the minimum targets for the Wolverine in Västerbotten, we will have to work intensively on the management of the species. The regional management plan for the Wolverine in Västerbotten County provides a framework for this project (Schneider, unpubl.).

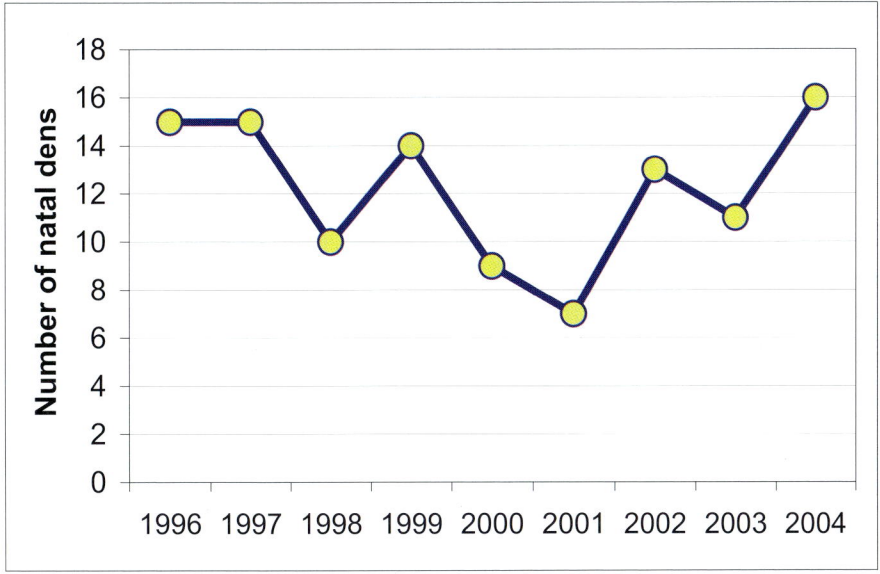

*Figure 21-4.* The number of natal dens found in Västerbotten County differed pronouncedly between years. No clear trend can be detected.


# 7.    DISCUSSION

In this study, I have presented ideas on the monitoring of Wolverines. Wolverine management is discussed elsewhere (Landa, 1997: Landa *et al.*, 2000a, 2000b: Schneider, unpubl.).

Regional targets for the distribution and numbers of big predators in Västerbotten have been developed during discussions in the Regional Predator Council. In contrast to the Brown Bear, there is no broad public interest in the Wolverine, neither in Västerbotten County nor elsewhere in Sweden. The only group in society directly affected by the Wolverine are Reindeer owners among the Sami people. In neighbouring Norway, the situation is very different: there, sheep move freely and unattended in the mountains during the summer, and many of them are killed by Wolverines. The Norwegian answer is legal harvesting of parts of the Wolverine population (see Miljøverndepartementet, 2003).

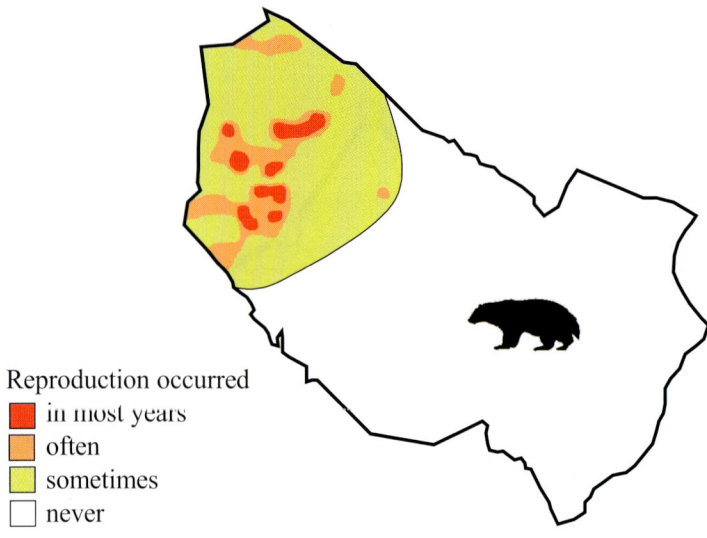

Reproduction occurred
- ⬛ in most years
- 🟧 often
- 🟨 sometimes
- ⬜ never

*Figure 21-5.* The distribution of Wolverine denning sites, and their use during different years, has been very uneven in Västerbotten between 1996 and 2003.

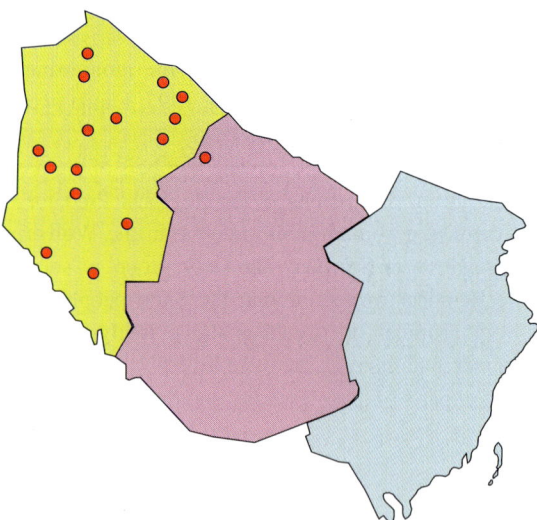

*Figure 21-6.* The approximate location of natal dens found in 2004. Only one out of 16 dens was situated in the forested inland, while the remaining 15 were found in the traditional breeding grounds in the mountains.

The monitoring of the species is straightforward, as the well-developed survey system for big predators in Sweden renders most of the information needed. The management and conservation of the species, on the other hand, are more complicated, due to the huge geographical areas that have to be dealt with, illegal killing that has to be stopped, and political conservation priorities that favour the Reindeer herding culture.

Surveying the whole County with respect to Wolverine occurrence is an expensive undertaking. In part, the work is co-ordinated with the Lynx survey, which helps to save some money and renders information about the Lynx population that is important for Wolverine management as well. Also, Sami Reindeer herders are involved in Wolverine survey, which has both negative and positive implications. The involvement of the Sami gives them an insight into the work of the County Administration and hopefully increases their understanding of the system of predator management in Västerbotten. It also makes them co-owners of the data, which should increase their trust in the figures. Furthermore, the Sami survey parts of the County which otherwise would have to be covered by County Administration staff, and they contribute their valuable knowledge about predator occurrence and distribution. On the other hand, this involvement also makes the system susceptible, as people with negative attitudes might use the information to the disadvantage of Wolverines.

We are currently considering some changes in the survey and monitoring system. Surveying the County every second or third year, or subdividing the County into areas that are surveyed during different years, would make the system cheaper, but result in a lower resolution of predator data. Using remote camera surveys instead of visits by snow mobile would decrease costs and increase data acquisition as well as aid the struggle against the illegal killing of Wolverines.

More detailed information about Wolverine history, ecology, management and conservation in Västerbotten County is to be found in the regional management plan for the species (Schneider, unpubl.). However, the chronic lack of basic data on Wolverine biology and ecology severely hampers the development of functioning conservation strategies. There is a great need for intensified research, as managers need facts on which to build their management and conservation strategies.

# 8.    ACKNOWLEDGEMENTS

I am grateful to Lars Danielsson and Mats Jonsson for fruitful discussions and for assistance when inspecting optimal Wolverine habitat. Jens Persson shared his knowledge of Wolverine biology and ecology, discussed Wolverine management, and commented on the manuscript. Gunnar Ledström and his colleagues in Västernorrland County freely shared their experience with forest Wolverines.

# 9.    REFERENCES

Andersen, R., Landa, A., Brøseth, H. and Linnell, J. D. C. (2002). Instruks för yngleregistrering av jerv. A - bakgrunnsinformasjon og overvåkningsmetoddik - NINA, Trondheim. (in Norwegian)

Bjärvall, A., Franzén, R., Nordkvist, M. and Åhman, G. (1990). Renar och rovdjur - Naturvårdsverkets Förlag, Solna. (in Swedish)

Ekman, S. (1910). *Norrlands jakt och fiske*. Facsimile of the original edition - Två Förläggares Bokförlag, Umeå. (in Swedish)

Flagstad, O., Hedmark, E., Landa, A., Broseth, H., Persson, J., Andersen, R., Segerström, P. and Ellegren, H. (2004). Colonization history and noninvasive monitoring of a reestablished wolverine population. – *Cons. Biol.* 18: 676-688.

Gärdenfors, U. (2005). *Rödlistade arter i Sverige 2005*. The 2005 Red List of Swedish Species. - ArtDatabanken, Uppsala. (bilingual)

Haglund, B. (1966). De stora rovdjurens vintervanor. I (Winter habits of the Lynx (*Lynx lynx* L.) and Wolverine (*Gulo gulo* L.) as revealed by tracking in the snow). - *Viltrevy 4*: 80-310. (in Swedish with English summary)

Kilström, Å. (2004). *The wolverine population in the boreal forest area*. – Degree project in Biology, Uppsala University, Uppsala.

Krott, P. (1960). Der Vielfrass oder Järv. Die neue Brehm-Bücherei 271 - A. Ziemsen Verlag, Wittenberg Lutherstadt. (in German)

Landa, A. (1997). *Wolverines in Scandinavia: ecology, sheep depredation and conservation*. - PhD-thesis, Norwegian University of Science and Technology, Trondheim.

Landa, A., Tufto, J., Franzén, R., Bo, T., Lindén, M. and Swenson, J. E. (1998). Active wolverine *Gulo gulo* dens as a minimum population estimator in Scandinavia. - *Wildl. Biol.* 4: 159-168.

Landa, A., Lindén, M. and Kojola, I. (2000a). Action plan for the conservation of wolverines (*Gulo gulo*) in Europe. – Council of Europe Publishing, Nature and Environment 115.

Landa, A., Linell, J.D.C., Lindén, M., Swenson, J.E., Røskaft, E. and Moksnes, A. (2000b). Conservation of Scandinavian wolverines in ecological and political landscapes. - In: Griffiths, H.I. (ed.). *Mustelids in a modern world*. Blackhuys Publishers, Leiden, pp 1-20.

Miljøverndepartementet (2003). Rovvilt i norsk natur. Stortingsmelding nr. 15 (2003-2004), Oslo (in Norwegian)

Naturvårdsverket (2003). Åtgärdsprogram för bevarande av järv (*Gulo gulo*). Åtgärdsprogram nr. 21 - Naturvårdsverket, Stockholm. (in Swedish)

NFS (2004:17) Naturvårdsverkets föreskrifter och allmänna råd om inventering samt bidrag och ersättning för rovdjursförekomst i samebyar. (in Swedish)

Pasitschniak-Arts, M. and Larivière, S. 1995. *Gulo gulo*. - Mamm. Species 499: 1-10.

Persson, J. (2003). Population Ecology of Scandinavian Wolverines. *Acta Universitatis Agriculturae Sueciae*. Silvestria 262 - Swedish University of Agricultural Sciences, Umeå.

Pulliainen, E. (1993). *Gulo gulo* (Linnaeus, 1758) - Vielfrass. - In: Stubbe, M. and Krapp, F. (eds.), *Handbuch der Säugetiere Europas. Band 5: Raubsäuger, Teil 1*. Aula Verlag, Wiesbaden. pp. 481-502. (in German)

Regeringen (2001). Regeringens proposition 2000/01:57 Sammanhållen rovdjurspolitik , Stockholm. (in Swedish)

Schneider, M. unpubl. Järven *Gulo gulo* i Västerbottens län. Förvaltningsplan för åren 2005-2006 - Länsstyrelsen Västerbotten, Umeå. (in Swedish with English summary)

Støbet Lande, U., Linnell, J. D. C., Herfindal, I., Salvatori, V., Brøseth, H., Andersen, R., Odden, J., Andrén, H., Karlsson, J., Willebrand, T., Persson, J., Landa, A., May, R., Dahle, B. and Swenson, J. (2003). Utredninger i forbindelse med ny rovviltmelding. Potensielle leveområder for store rovdyr i Skandinavia: GIS-analyser på et økoregionalt nivå. NINA Fagrapport 64 - Norsk institutt for naturforskning, Trondheim. (in Norwegian with English summary)

Vangen, K. M., Persson, J., Landa, A., Andersen, R. and Segerström, P. (2001). Characteristics of dispersal in wolverines. - *Can. J. Zool.* **79**: 1641-1649.

# CHAPTER 22

# MONITORING THE GREATER MOUSE-EARED BAT *MYOTIS MYOTIS* ON A LANDSCAPE SCALE

MICHAEL SCHNEIDER

*County Administration, Västerbotten County, SE-901 86, Umeå, Sweden*
*Michael.Schneider@ac.lst.se*

MATTHIAS HAMMER

*Coordination Officer for Bat Conservation in Northern Bavaria, Erlangen University,*
*DE-910 58, Erlangen, Germany*
*mhammer@biologie.uni-erlangen.de*

## 1.   BACKGROUND INFORMATION

In Bavaria in southern Germany, protected sites for the Greater Mouse-eared Bat *Myotis myotis*, a species listed in Annex II of the Habitats Directive, are usually small and consist of a single building containing a nursery colony (e.g. Fig. 22-1). In this chapter, we wonder to what degree the long-term conservation status of single colonies can be deduced from results of the ongoing surveillance, i.e. yearly counts of the number of individuals present in the colony. This question is founded on two aspects of Greater Mouse-eared Bat ecology:

1. Nursery colonies are not independent of each other. Females can use different maternity roosts in different years.
2. Nursery colonies need foraging grounds near by, and most often these hunting habitats are not protected.

A question that followed from these was, how large should monitoring units be to render reliable information on the status of the species?

In the following text, we present our ideas regarding a monitoring system that takes into account the metapopulation-system of the species and also renders information on the status of hunting grounds.

*C. Hurford & M. Schneider (eds.), Monitoring Nature Conservation in Cultural Habitats*, 231–246.
© 2007 *Springer*.

*Photograph by Michael Schneider*

*Figure 22-1*. A typical Natura 2000 site for the Greater Mouse-eared Bat in Bavaria, Laudenbach Castle harbors one of the largest nursery colonies in Main-Spessart district. The species likes attics of large buildings.

## 1.1    The Greater Mouse-eared Bat in Northern Bavaria

The Greater Mouse-eared Bat is a species for which we have reasonably detailed data over a relatively long period of time in northern Bavaria, regarding both population development and the basic ecology of the species. Female Greater Mouse-eared Bats form nursery colonies in buildings, often in large attics of churches and castles, during the summer. These nursery colonies can be very large and consist of >1000 individuals. Males are solitary during summer and occupy cavities in trees, nest boxes or buildings, where they wait for visiting females to mate (Schober & Grimmberger, 1998). Migrations between summer roosts and hibernacula are relatively short (< 200 km) and depend on the occurrence of suitable hibernation sites, which may be caves, mines, tunnels and cellars (Kulzer, 2003, Rudolph *et al.*, 2004).

Deciduous forests with a sparse understorey are the main foraging areas for the species in Bavaria (Meschede & Heller, 2002), where the main prey items are large ground-living insects such as carabid beetles. Females in nursery colonies utilise forests up to 15 km from the roost (Rudolph *et al.*, 2004 and references therein). The area of deciduous forest and its productivity determine the occurrence and size of nursery colonies (Rudolph *et al.*, 2004).

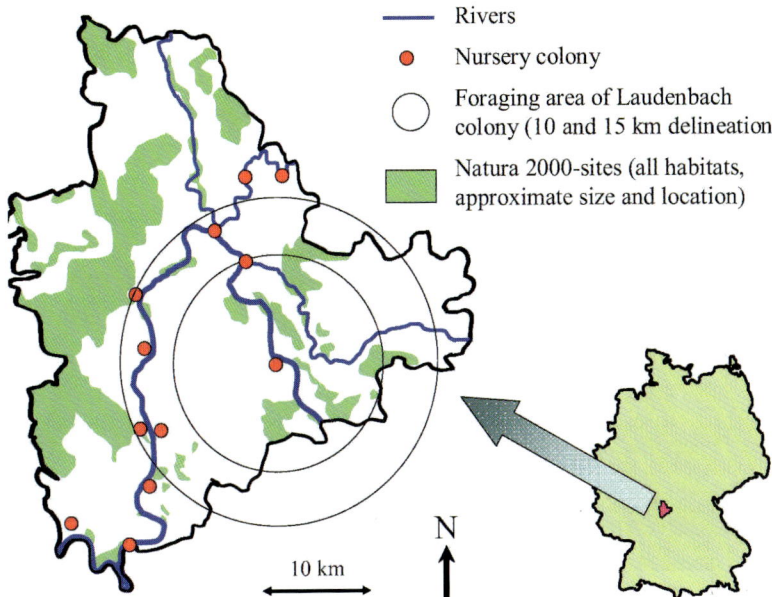

*Figure 22-2.* In Main-Spessart district, 12 nursery colonies of Greater Mouse-eared Bats are known. These are surveyed each summer and the animals are counted. In total, the colonies harbor about 6 000 animals. Each colony has the potential to use hunting grounds up to at least 10 km from the roost. Map after information on nursery colonies in Schönmann *et al.* (2001) and information on Natura 2000 sites from the Bavarian Environmental Protection Agency (http://gisportal-umwelt.bayern.de/ffh/finweb/karte_start.htm).

Below, we use Main-Spessart district (Landkreis Main-Spessart) to test which geographical scale should be used during monitoring. Districts are units with an administration (Landratsamt) that is responsible for issues relating to nature conservation, among many other things. Main-Spessart district has an area of 1 323 km² and a human population density of about 100 inhabitants/km². Two conservation officers work at the district administration, one of them specifically with bats. Three of Bavaria's 15 largest nursery colonies of the Greater Mouse-eared Bat are found among the 12 colonies in Main-Spessart (Boye, 2003).

## 1.2    Key factors influencing management decisions for the Greater Mouse-eared Bat

In a global perspective, the Greater Mouse-eared Bat is at lower risk but near threatened (Chiroptera Specialist Group, 1996) and has its centre of distribution in northern Bavaria and the surrounding areas (Biedermann *et al.*, 2003). Therefore, Germany and Bavaria have a special responsibility for the conservation of this species. The Greater Mouse-eared Bat is listed as vulnerable in the latest version of the red list of

*Photograph by Matthias Hammer*

*Figure 22-3.* Female Greater Mouse-eared Bats gather in large colonies to give birth to their young in summer. Part of the Wolfsmünster colony, July 1997.

German mammals (Boye *et al.*, 1998) and is included in both Annex II and Annex IV of the EU Habitats Directive. In Bavaria, the species is not threatened, but included in the list of species that are near threatened (Liegl *et al.*, 2003).

According to figures presented by Geiger (2003), Bavaria holds more than 50% of the total German population of Greater Mouse-eared Bats, with 290 known nursery colonies containing about 80 000 animals (Rudolph *et al.*, 2004). The densities of Greater Mouse-eared Bats in northwest Bavaria are much higher than in most remaining areas of the state and reach levels of up to 15 individuals per km$^2$ of suitable forest foraging habitat, the highest figures reported for any part of Europe (Rudolph & Liegl, 1990, Rudolph *et al.*, 2004).

Since 1985, when structured surveillance of Greater Mouse-eared Bats started, the number of known nursery colonies increased from 70 to 290 in Bavaria. Mean colony size also increased, from 267 to 447 animals between 1985 and 1995. Since 1998, mean colony size has been relatively stable or started to decrease, which is supposed to reflect saturation or degradation of the hunting habitats surrounding the colonies (Meschede & Rudolph, 2004).

There are three key factors influencing management decisions for the Greater Mouse-eared Bat in Main-Spessart district: metapopulation dynamics, suitable foraging grounds in deciduous woodland, and human attitudes regarding maternity roosts.

## 1.2.1 Metapopulation dynamics

Females can switch between maternity roosts between years. Therefore, the numbers of females in one single colony may differ between years although the total number within an area has not changed. The main problem is deciding the best geographical scale for monitoring and surveillance of the species.

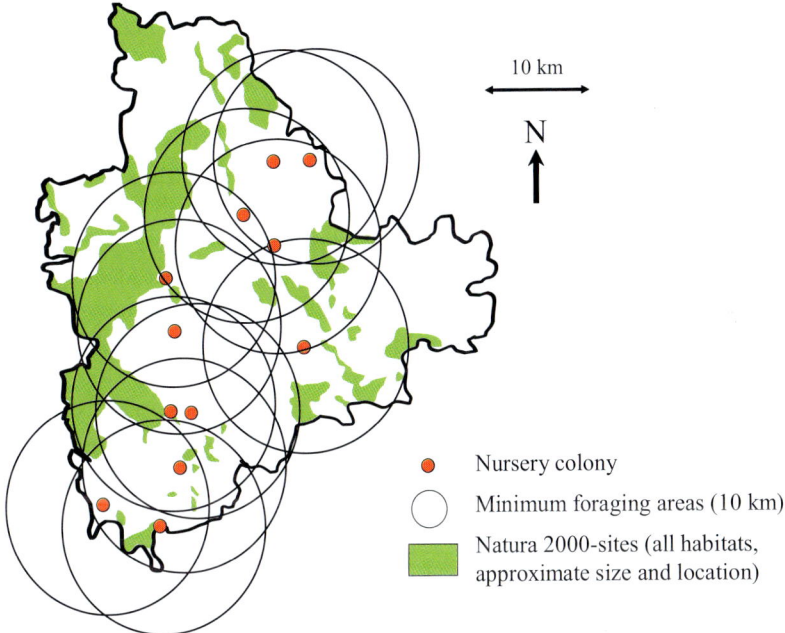

*Figure 22-4.* Theoretically, the hunting grounds of the colonies in Main-Spessart overlap considerably. Obviously, most of the district is used by bats, and the Natura 2000 woodland sites alone are not big enough to satisfy the foraging needs of all colonies in the district.

## 1.2.2 Foraging requirements of Greater Mouse-eared Bats

Female bats use hunting grounds in the deciduous woodlands surrounding the colony, travelling up to 15 km from the roost every night. Individuals normally forage as close to the roost as possible (Krebs & Kacelnik, 1991), and it has been proposed that the area of woodland within a 10 km radius of the colonies is the most important factor in the success of a colony. Each female uses an average hunting area of 30-35 ha, and the areas used by different females, even from different colonies, can overlap (Meschede & Heller, 2002 and references therein).

Bat managers are not very well informed about the status of these foraging areas, and it is hoped that the protection of potential foraging sites in forest habitats listed in Annex 1 of the Habitats Directive will suffice to keep the Greater Mouse-eared Bat in favourable conservation condition (Rudolph, 2004). However, protected forest sites are not necessarily situated within the foraging range of nursery colonies, and most colonies do not have large areas of protected hunting grounds nearby (Fig. 22-4).

### 1.2.3    Maternity roosts: renovations and human attitudes

Nursery colonies of Greater Mouse-eared Bats often occupy old buildings such as castles and churches, which often need renovations. Renovations of a building with a maternity roost at the wrong time or in a wrong way can result in the bats abandoning the roost. Therefore, the well-being of nursery colonies depends on the attitudes of house owners, residents, architects and constructors towards the bats.

## 1.3    The management priority

There are two conservation issues that may conflict with the management of Greater Mouse-eared Bats.

In some places, Barn owls *Tyto alba* have started to use the same attics as nursery colonies of the Greater Mouse-eared Bat. Barn owls prey upon Greater Mouse-eared Bats and are presumed to be the reason for the bats abandoning their roost in some cases. In our study area, this happened in the church of Wolfsmünster in 2002. Barn owls are listed as endangered in the latest red list of threatened birds in Bavaria (Fünfstück *et al.*, 2003), which makes removing the owls from bat roosts problematic. In some cases, constructions within the roost have been successful in partitioning the attic, so that the owls no longer have access to the bats: this makes it possible for the two species to co-exist in the same place.

Forestry in Bavaria is tending increasingly towards more natural practices, aiming at more natural forests with different layers of vegetation, rather than single-aged stands with bare ground. This development is positive for many species. However, it results in a decrease in the extent of foraging habitat for the Greater Mouse-eared Bat, and has been suggested as one reason for the stagnation or decline of bat numbers in nursery colonies in Bavaria (Meschede & Rudolph, 2004). It is unclear to what degree this may jeopardise the status of Greater Mouse-eared Bats in the area.

## 2.    RESEARCH AND SURVEY

The Greater Mouse-eared Bat has been the subject of ecological research that has rendered valuable information for the conservation of the species (see Kulzer, 2003 and Rudolph *et al.*, 2001, 2004 for recent reviews). Furthermore, the Bavarian Environmental Protection Agency has recently presented guidelines for the surveillance of Greater Mouse-eared Bats (LfU, 2003).

In our monitoring unit, the Main-Spessart district, we are well informed about population development in all nursery colonies since 1993 (with data from 1983 in some localities). However, away from the nursery colonies, we have limited knowledge of the whereabouts of many animals for a large part of the year. Also, we do not have any information about swarming sites and the condition of the feeding habitat. New research is needed to shed some light on these areas of bat ecology. Monitoring programs may have to be adapted accordingly in the future.

For example, we do not know the exact whereabouts of most males during summer, but we do know that males use attics, cavities in trees, nest boxes and similar places. We also know that females regularly travel several kilometres to visit males in their mating roosts. We presume that the number of potential summer roosts for males will increase in the future, as forest practises are changing towards more natural stands with a higher fraction of old and dead tress with holes and cavities. Therefore, male summer roosts do not have to be part of the monitoring program.

In Main-Spessart district, 21 hibernacula of Greater Mouse-eared Bats are known. Most of them are surveyed annually. In total, >150 animals are found each winter, amounting to ca. 1.5% of the summer population. It is unclear in how far the conservation status of the species in Main-Spessart depends on hibernation sites within the district. Judging from the number of individuals found each winter, local hibernacula are of minor importance to the species. This is why they are not included in this monitoring programme.

Greater Mouse-eared Bats use swarming sites in autumn, but the importance and ecology of swarming is not well understood. Usually, hibernation sites are visited during swarming, but only few hibernacula have been examined with respect to their importance for autumn swarming. According to LfU (2003), we should only survey swarming sites within Natura 2000 areas. As the importance of swarming sites in Main-Spessart is unclear but seems limited, those sites have also been excluded from the monitoring program.

To gain an insight into the landscape-level conservation needs of Greater Mouse-eared Bats, we must determine the distribution, area and condition of foraging habitat within at least a 10 km radius of all maternity roosts. According to recommendations from Meschede & Heller (2002), Dietz & Simon (2003), the Countryside Council for Wales (unpublished guidance and outlines for bat contracts) and our own experience this should be done by:

1. An initial survey that maps the distribution of suitable and potentially suitable foraging habitat (deciduous forests and mixed forests >100 years or tree diameter at breast height >30 cm, with no or little understorey and a dense layer of litter, see Fig. 22-8) using analyses of existing forest maps and aerial photographs;

2. A subsequent ground-truth phase, where all identified patches of foraging habitat are mapped in the field;

3. A condition survey where all habitat patches identified in 2. are allocated to one of four clearly defined condition classes (see Table 22-1). This allocation is done using the following two variables that are looked at in permanent 1 ha monitoring plots in each of the habitat patches:
   - Proportion of forest floor not covered by vegetation;
   - Proportion of area with free airspace 0-2 m above ground.

   Furthermore, we should examine the following two variables:
   - Presence of hunting Greater Mouse-eared Bats;
   - Densities of ground-living invertebrates.

4. The monitoring plots should be randomly allocated to habitat patches and should cover 2-5% of total foraging habitat area. All monitoring plots should be permanently marked, and clear directions provided to ensure precise relocation.
5. Plots should be visited and monitored every sixth year during the vegetation period. Alternatively, aerial photos could be used to determine the condition of patches of hunting habitat during subsequent monitoring cycles.
6. Steps 1-3 should be repeated ca. every 50 years.

In this way, we will have permanent reference points for the future. We will also get information on the structure and composition of the habitats currently being used by Greater Mouse-eared Bats in the district.

During the first year, all habitat patches have to be found, surveyed and allocated to one of four condition classes with respect to ground cover and free airspace (measured as cover of shrubs and trees ≤ 2 m tall). Also, all patches are surveyed for the occurrence of foraging Greater Mouse-eared Bats and the density of large ground-living invertebrates. The classification system for condition classes may have to be adapted, if ground cover and free airspace alone do not explain the occurrence of foraging bats in the habitat patches.

All survey and monitoring should be done in close cooperation with the district administration, the responsible coordination office for bat conservation, and local bat worker groups. Furthermore, we must supply immediate and regular reports of the results to the owners, administrators or residents of buildings with maternity roosts.

*Table 22-1.* Condition classes for habitat patches. The quality of patches as hunting habitat for Greater Mouse-eared Bats decreases from 1 to 4. We presume patches in class 4 not to be suitable for Greater Mouse-eared Bats.

| | | Bare ground (%) | | | |
|---|---|---|---|---|---|
| | | > 75 | 75 - 50 | 50 - 25 | < 25 |
| Free airspace (%) | > 75 | 1 | 2 | 3 | 4 |
| | 75 – 50 | 2 | 2 | 3 | 4 |
| | 50 – 25 | 3 | 3 | 3 | 4 |
| | < 25 | 4 | 4 | 4 | 4 |

## 3.   BACKGROUND SURVEILLANCE

Current background surveillance of the Greater Mouse-eared Bat in Main-Spessart district consists of an annual survey of all known nursery colonies and hibernacula. This survey, which is conducted by the Coordination Office for Bat Conservation in Northern Bavaria in cooperation with the district administration and local bat workers, looks at both population size and habitat structure. Male summer roosts, swarming sites and foraging areas are not currently surveyed.

### 3.1   Nursery colonies

Nursery colonies are visited once every year, during daytime in July. All surveys should be done during suitable weather conditions: counts should not be conducted if the preceding night was rainy and cool, because many females may be absent from the colony in this case. The following variables are looked at (LfU, 2003):

- Number of animals present in the colony (if possible with separate figures for females and young);
- Presence of single males, mating roosts and other bat species inside the maternity roost;
- Condition of entrance and flyways, i.e. whether they are permanently open or whether there has been a change of condition;
- Renovations pending or necessary because of woodworm infestation. House owners, administrators, residents or other responsible persons are contacted to clarify this;
- Occurrence of sanitary problems (dung heaps, odours, invertebrates associated with dung or dead animals, parasites); and
- Degree of disturbance, including signs of predator presence (Barn Owl, Beech Marten *Martes foina*, House Cat *Felis silvestris catus*)

## 3.2    Hibernacula

Hibernation sites are visited once every winter, during daytime in the period December-February. All visits are made during suitable weather conditions, i.e. at sub-freezing temperatures, because most animals will be in their hibernaculum in these conditions. The following variables are looked at (LfU, 2003):

- Numbers of hibernating Greater Mouse-eared Bats;
- Numbers of individuals of other bat species;
- Accessibility of the site for bats;
- Pending or necessary renovations;
- Accessibility of the site for humans (including survey personnel);
- Degree of disturbance: signs of human activity (vandalism, dumping, tourism, caving) within roost, such as fire places, candles, garbage etc.; and
- Signs of predator presence (Beech Marten).

## 4.    THE CONSERVATION AIMS

The Greater Mouse-eared Bat in Bavaria currently has favourable conservation status. As the species is listed as near threatened in the red list of mammals in Bavaria, a population decrease would mean the status of the species could become unfavourable. We consider that a total Greater Mouse-eared Bat population in Bavaria and Main-Spessart district as great as or greater than it was in 1995 would be in favourable condition. Single nursery colonies may decrease in size or even disappear without changing the conservation status of the species, providing that the total metapopulation size in the monitoring unit remains stable.

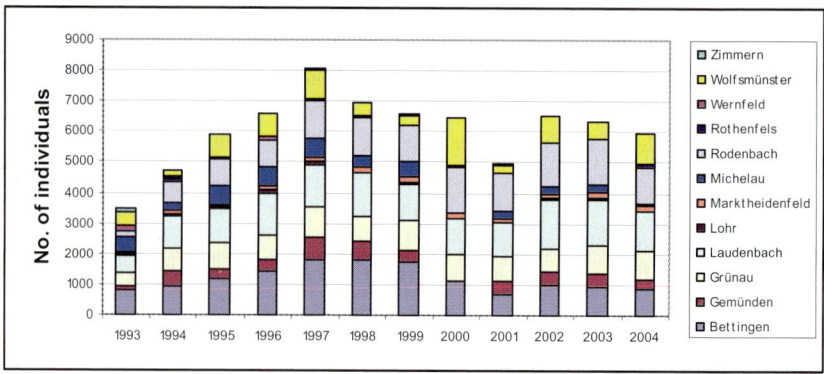

*Figure 22-5.* The number of Greater Mouse-eared Bats (females and young) in nursery colonies in Main-Spessart district during 1993–2004. Unpublished data from the Coordination Office for Bat Conservation in Northern Bavaria.

# 5.      THE CONDITION INDICATORS

## 5.1      Explanation of the indicators

*Distribution*: 12 maternity roosts are known in Main-Spessart, of which four are small. Small colonies have a relatively high probability of being abandoned, but their disappearance would not have a negative effect on the total population.

*Population size*: population increase in Bavaria until 1995; population increase in Main-Spessart until 1997, but lower numbers since then; numbers in 1995: 5 911 animals in nursery colonies in total.

*Foraging habitat quality*: Optimal foraging habitat is deciduous forest with <25% vegetation cover at ground level and free airspace between 0 and 2 m from ground. Look at patches in different condition classes, but start with class 3. If patches in worst condition class (3) pass, then all other condition classes should also pass.

*Foraging habitat extent*:

- *Lower limit*: Each female uses 30-35 ha of suitable forest in an area up to at least 10 km from the maternity roost. On average, 60% of animals in a colony are adult females. 5 900 animals x 0.6 x 35 ha/individual = 123 900 ha. Therefore, 124 000 ha of foraging habitat are needed <u>in the district in total</u> within 10 km from existing colonies at minimum population level. The area of foraging habitat <u>around each colony</u> is determined by the maximum number of females observed in the roost, because females can use different roosts in different years and each roost should keep its capacity to function as a refuge for females from other roosts. However, the numbers of females per roost should be adapted to the area of foraging habitat surrounding the roost. Increasing the number of females temporarily should mean the bats have to use a larger area for hunting during this time. Therefore, when refugees from other roosts merge with a given colony, the range of hunting females should expand (presumably from 10 to 15 km as observed in different studies) in this case, and we should monitor the larger area.

- *Upper limit*: Optimal foraging habitat for Greater Mouse-eared Bats is not optimal as habitat for many other species. Therefore, we could set an upper limit of habitat extent for 8 000 animals (maximum recorded in Main-Spessart in 1997). 8 000 animals x 0.6 x 35 ha/individual = 170 000 ha. However, as the development in forestry is towards less and less foraging habitat in class 1, this should not be a problem.

*Table 22-2.* Condition indicator table.

| Condition indicator table for population maintenance | | The population of the Greater Mouse-eared Bat in Main-Spessart district will be in favourable condition when |
|---|---|---|
| Distribution | Lower limit | Nursery colonies are present in > 7 localities in the district in any year |
| | Upper limit | None set |
| Population size | Lower limit | Total population size in the district is >5 900 animals (post-breeding) in nursery colonies in any year |
| | Upper limit | None set |
| Foraging habitat quality | Lower limit | Within 10 km of existing nursery colonies, all patches of foraging habitat belong to at least condition class 3 |
| | Upper limit | None set |
| Foraging habitat extent | Lower limit | Within 10 km from existing nursery colonies, >124 000 ha of foraging habitat (condition classes 1–3) exist in total in the district and within 10 km from each nursery colony, >30 ha of foraging habitat per female exist, where the relevant number of females in each colony is determined by the maximum number of animals found between 1993 and 2004 |
| | Upper limit | None set |
| Maternity roost quality | Lower limit | Roost unchanged or only slightly changed and Entrance open and unchanged and Signs of predators as few as or fewer and Attitudes of owner/user/resident of roost building as positive as or more positive By comparison to the situation in 1995 |

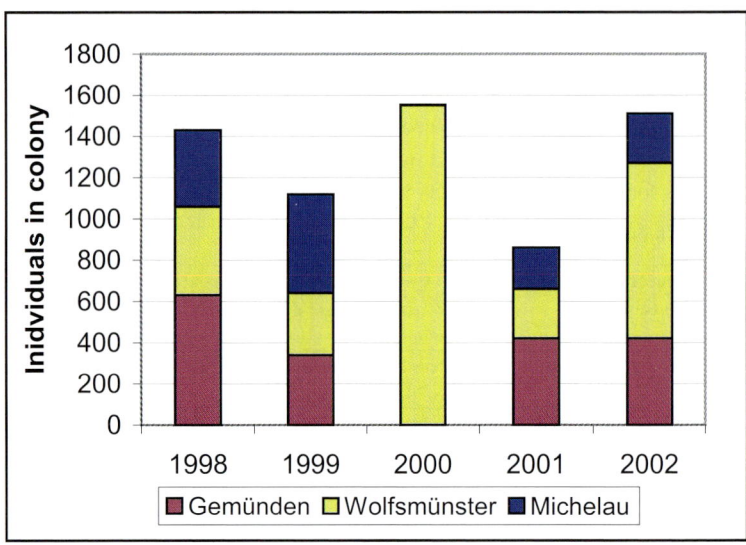

*Figure 22-6.* In 2000, the females abandoned roosts in Gemünden and Michelau and presumably moved to Wolfsmünster.

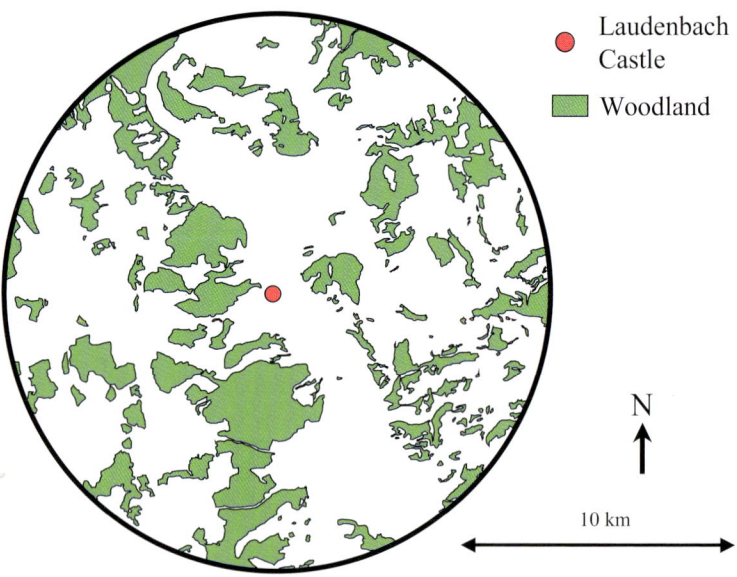

*Figure 22-7.* Presumed minimum foraging area of the females in the nursery colony in Laudenbach Castle. The woodland patches had not been allocated to different condition classes at the moment of writing.

# 6.    MONITORING

The monitoring proposed for the Greater Mouse-eared Bat in Main-Spessart district uses the data collected during the annual surveys of nursery colonies in summer. With these data, we can estimate the summer population of the species in the district and compare these to the targets for distribution and population size (Tab. 22-2).

Foraging habitats should be surveyed every sixth year, corresponding to the cycle of reporting within the Natura 2000 framework. The easiest way to do this is to use aerial photography and GIS-analyses to compute the area of suitable hunting grounds. Aerial photographs are taken every third year in Bavaria.

The alternative method, looking at changes in the condition class of patches of foraging habitat in the field, presumably is too expensive. If this is done anyway, then we should focus the monitoring on those patches that were allocated to condition class 3 during previous surveys. If these patches still pass as suitable (i.e. belonging to classes 1-3), all patches previously in classes 1 and 2 should also pass and do not have to be surveyed, assuming a linear development of deterioration of patch quality from class 1 through 2 and 3 to class 4. In areas where such a linear development is not the case (depending on forestry practices involving clear-cutting), all patches of foraging habitat should be surveyed.

*Photo by Michael Schneider*

*Figure 22-8.* Typical foraging habitat of the Greater Mouse-eared Bat at Schönartsberg, 5 km north-east of Laudenbach (here in spring aspect): deciduous wood-land with little ground cover and lots of free air space.

# 7.    RESULTS AND ANALYSIS

The number of Greater Mouse-eared Bats (females and young) in nursery colonies in Main-Spessart district increased until 1997, when more than 8000 animals were observed in maternity roosts. Since then, total population size has been lower but relatively stable (Fig. 22-5). The number of bats within nursery colonies can differ widely between years. Females also move between roosts, as was the case in our study area in 2000, when the females abandoned roosts in Gemünden and Michelau and presumably (no marked animals) moved to Wolfsmünster (Fig. 22-6). Judging from the total number of animals, other colonies may even have been involved here. Greater Mouse-eared Bats have been found to use hunting grounds several kilometres from their maternity roost. Different studies found different distances, but 10 km seems to be a reasonable average.

The presumed foraging area of the females in the nursery colony in Laudenbach Castle is depicted in Fig. 22-7. The woodland patches around Laudenbach have not been allocated to different condition classes yet. Fig. 22-8 depicts a high-quality patch ca. 5 km north-east of the colony. Theoretically, the hunting grounds of the colonies in Main-Spessart overlap considerably (Fig. 22-4). As radio telemetry studies have not been conducted in the district, we do not know where the actual hunting grounds of each colony are situated, but it is obvious that most of the district is used by bats. Natura 2000 sites alone are not big enough to satisfy the foraging needs of all colonies in the district.

# 8.    DISCUSSION

There are three issues that are not currently covered by the surveillance programme for Greater Mouse-eared Bats in Bavaria but that are of central importance in a monitoring system:

1. How will we know when Greater Mouse-eared Bats would have favourable conservation status?
2. How large should monitoring units be to take account of the metapopulation structure of the Greater Mouse-eared Bat colonies?
3. How will we know whether there is enough hunting habitat available or whether the quality of this habitat is deteriorating over time?

These questions must have a bearing on future research, woodland management, and monitoring and reporting.

## 8.1    Implications for future research

Future research has to answer the following questions:

1. What exactly is good foraging habitat, and how can habitat patches be classified?
2. How much habitat do we need? In total, per female, and in how large an area (10 or 15 km) around nursery colonies?
3. How large should monitoring units be? Or, in other words, which colonies should be treated as one metapopulation?

4. Where are the bats in winter? Are local hibernacula with only a few individuals important from a conservation perspective?

5. Where, and of what importance, are male roosts, interim roosts and swarming sites associated with nursery colonies?

## 8.2    Implications for woodland management

Although nursery colonies are protected as Natura 2000 sites, they are dependent on the surrounding woodland that is not protected. The management of these forests has to take into account the needs of neighbouring colonies as well as single males in their mating roosts, which is not done today. In Main-Spessart district, all woodland should be managed sympathetically to the needs of Greater Mouse-eared Bats and other bat species. An open question is, which direction the forestry policy in Bavaria will take in the future and whether this will favour Greater Mouse-eared Bats or not.

## 8.3    Implications for monitoring

As is obvious from our analyses, the district scale seems too small for meaningful monitoring of Greater Mouse-eared Bat metapopulations. Biogeographical boundaries are more appropriate than administrative ones. We suggest that regional coordination offices should be responsible for tracking population trends of Greater Mouse-eared Bats on the relevant biogeographical scale, while district administrations take responsibility for surveying local nursery colonies, contacting house owners, mobilising of local bat conservation groups, and surveying foraging habitat.

An increased involvement of local bat groups and other non-governmental conservation organisations seems to be necessary for the long-term success of bat conservation. Our suggested monitoring of foraging habitats could become rather expensive. Political decisions could jeopardise the monitoring and surveillance of Greater Mouse-eared Bats, if funding is redirected to species at higher risk. Local bat groups, who work for relatively little money, can guarantee long-term contacts with the owners of maternity roosts and in this way counteract changes in attitudes that could be damaging for the bats. Within Bavaria's intensive ongoing surveillance programme, almost 90% of all known nursery colonies are surveyed each year. As one result of this, we now get early warnings of any planned renovation or similar measures, and most renovation-related threats to colonies have been eliminated. However, without regular surveys, renovations of roofs and attics could again become a serious threat to nursery colonies (Boye, 2003).

## 9.    ACKNOWLEDGEMENTS

We are grateful to Peter Boye for help with the literature, Doris Grellmann for discussions about monitoring and conservation practice, and Bernd-Ulrich Rudolph for comments on the project and the manuscript.

# 10. REFERENCES

Biedermann, M., Meyer, I. and Boye, P. (2003). Bundesweites Bestandsmonitoring von Fledermäusen soll mit dem Mausohr beginnen. - *Natur und Landschaft* 78: 89-92. (In German with English summary).

Boye, P. (2003). National report on bat conservation in the Federal Republic of Germany 2000 - 2003 – Report within the Eurobats-agreement.

Boye, P., Hutterer, R., Benke, H., Braun, M., Heidecke, D., Heidemann, G., Meinig, H. and Schlapp, G. (1998). Rote Liste der Säugetiere (Mammalia). - In: Binot, M., Bless, R., Boye, P., Gruttke, H. and Pretscher, P. (eds.), Rote Liste gefährdeter Tiere Deutschlands. *Bundesamt für Naturschutz, Bonn-Bad Godesberg*. pp. 33-39. (In German).

Chiroptera Specialist Group (1996). *Myotis myotis*. In: IUCN 2003. 2003 IUCN Red List of Threatened Species.

Dietz, M. and Simon, M. (2003). Konzept zur Durchführung der Bestandserfassung und des Monitorings für Fledermäuse in *FFH-Gebieten im Regierungsbezirk Gießen*. - *BfN*-Skripten **73**: 87-140. (In German).

Fünfstück, H. -J., von Lossow, G. and Schöpf, H. (2003). Rote Liste gefährdeter Brutvögel (Aves) Bayerns. - Schriftenreihe Bayer. *Landesamtes f. Umweltschutz* **166**: 39-44. (In German).

Geiger, H. (2003). Übersicht über die Ergebnisse der Länderabfrage "Mausohrmonitoring". - *BfN-Skripten* **73**: 28-35. (In German).

Krebs, J. R. and Kacelnik, A. (1991). Decision making. - In: Krebs, J. R. and Davies, N. B. (eds.), *Behavioural ecology*. Blackwell Scientific Publications, Oxford. pp. 105-136.

Kulzer, E. (2003). Großes Mausohr *Myotis myotis* (Borkhausen, 1797). - In: Braun, M. and Dieterlen, F. (eds.), *Die Säugetiere Baden-Württembergs. Band 1*. Verlag Eugen Ulmer, Stuttgart. pp. 357-377. (In German).

LfU (Bayerisches Landesamt für Umweltschutz) (2003). Kartieranleitung für die Arten der FFH-RL (Ersterfassung und Monitoring). Entwurf März 03. Großes Mausohr *Myotis myotis* - Bayerisches Landesamt für Umweltschutz, Augsburg. (In German).

Liegl, A., Rudolph, B.-U. and Kraft, R. (2003). Rote Liste gefährdeter Säugetiere (Mammalia) in Bayern. - Schriftenreihe Bayer. *Landesamtes f. Umweltschutz* **166**: 33-38. (In German).

Meschede, A. and Heller, K.-G. (2002). Ökologie und Schutz von Fledermäusen in Wäldern. Schriftenreihe für Landschaftspflege und Naturschutz Heft 66 - Bundesamt für Naturschutz, Bonn-Bad Godesberg. (In German with English and French summaries).

Meschede, A. and Rudolph, B.-U. (2004). Landesweite Auswertungen. - In: Meschede, A. and Rudolph, B.-U. (eds.), *Fledermäuse in Bayern*. Verlag Eugen Ulmer, Stuttgart. pp. 58-96. (In German).

Rudolph, B.-U. (2004). Gefährdung und Schutz. - In: Meschede, A. and Rudolph, B.-U. (eds.), *Fledermäuse in Bayern*. Verlag Eugen Ulmer, Stuttgart. pp. 356-383. (In German).

Rudolph, B.-U., Hammer, M. and Zahn, A. (2001). Das Forschungsvorhaben "Bestandsentwicklung und Schutz der Fledermäuse in Bayern". - Schriftenreihe Bayer. Landesamtes f. Umweltschutz 156: 241-268. (In German with English summary).

Rudolph, B.-U. and Liegl, A. (1990). Sommerverbreitung und Siedlungsdichte des Mausohrs *Myotis myotis* in *Nordbayern*. - *Myotis* **28**: 19-38. (In German with English summary).

Rudolph, B.-U., Zahn, A. and Liegl, A. (2004). Mausohr *Myotis myotis* (Borkhausen, 1797). - In: Meschede, A. and Rudolph, B.-U. (eds.), *Fledermäuse in Bayern*. Verlag Eugen Ulmer, Stuttgart. pp. 203-231. (In German).

Schober, W. and Grimmberger, E. (1998). *Die Fledermäuse Europas* - Franckh-Kosmos, Stuttgart. (In German).

Schönmann, H., Kuchenmeister, B. and Kunkel, M. (2001). Fledermäuse. Flora und Fauna im Landkreis Main-Spessart, Band 3 - Bund Naturschutz in Bayern, Marktheidenfeld. (In German).

# CHAPTER 23

# MONITORING CHOUGHS *PYRRHOCORAX PYRRHOCORAX* ON THE CASTLEMARTIN PENINSULA

BOB HAYCOCK

*Countryside Council for Wales, The Old Home Farmyard, Stackpole, Pembroke, SA71 5QD*
*B.Haycock@ccw.gov.uk*

CLIVE HURFORD

*Countryside Council for Wales, Plas Penrhos, Ffordd Penrhos, Bangor, Gwynedd, LL57 2BQ*
*clive.hurford@serapias.net*

## 1.     BACKGROUND INFORMATION

The study area is situated on the Castlemartin Peninsula, on the southern edge of the Pembrokeshire Coast National Park. It is a fairly exposed Carboniferous Limestone coastal plateau, approximately 16 km in extent from Brownslade and Linney Burrows to St Govan's Head (Fig 16.1 and Fig. 23-1). The area forms part of a Natura 2000 site, and was designated because of the range of coastal habitats present, and for the maritime vegetation and associated communities and species they support. The peninsula also forms part of a Special Protection Area for birds, designated in 1993 for (Red-billed) Chough *Pyrrhocorax pyrrhocorax*.

The principal habitats in the case study area include exposed sea-cliffs, rocky slopes and headlands, maritime grassland, heath, and sand dunes.

Up to 20 pairs of Chough breed on the Castlemartin peninsula (representing about 4% of the UK population): and up to 18 of these pairs breed in the case study area, which is one of its main breeding locations in West Wales. The sea-cliffs also support important concentrations of breeding seabirds, including large colonies of Guillemots *Uria aalge*

*C. Hurford & M. Schneider (eds.), Monitoring Nature Conservation in Cultural Habitats*, 247–258.

*Figure 23-1.* Linney Head, Castlemartin coast, where some five pairs of Chough nest fairly close together.

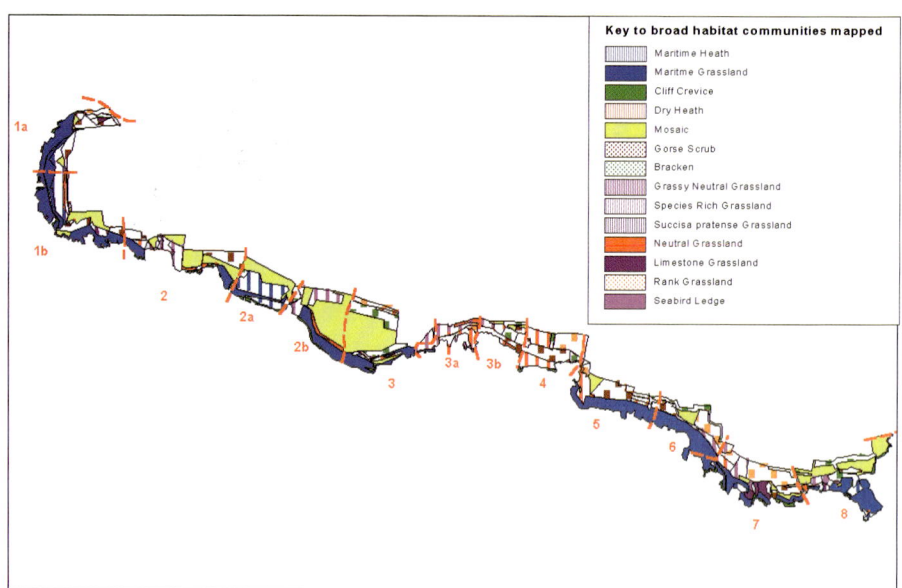

*Figure 23-2.* Part of the habitat quality map showing the sections that were selected for habitat monitoring (Wilson 2002; Davies 2002). Reproduced from Ordnance Survey mapping on behalf of Her Majesty's Stationery Office © Crown Copyright 100043571 2004-12-10.

and Razorbills *Alca torda*, and other cliff-nesting species such as Raven *Corvus corax*, Kestrel *Falco tinnunculus*, Peregrine *Falco peregrinus*, Swift *Apus apus* and House Martin *Delichon urbica*. The coastline is popular with walkers and the cliffs attract large numbers of climbers. Seasonal climbing restrictions, established by voluntary agreement, are needed to protect breeding populations of cliff-nesting species during the summer months.

Prior to 1938, the coastal vegetation along the southern coast of the Castlemartin peninsula was farmland, with coastal grazing that extended to the cliff edge. Since then, the coastline has been part of the Castlemartin military range, managed within the framework of an integrated land management plan (MOD, 2000). The limited agricultural use of the area since the Second World War has ensured that reclamation and intensification has not truncated the wide and continuous zones of sea-cliff vegetation that reflect different levels of salt in the soil.

There is a long tradition of rough grazing on the peninsula and the practices of mainly winter grazing by sheep and cattle continue, augmented by a semi-natural rabbit population. Cattle are present for most of the year, with c.350 present from June to December, rising to c.600 from January to mid-May. In November each year, cattle grazing is supplemented by c.12 000 sheep, which are transported to the Castlemartin coast from the Preseli Mountains. These sheep remain on the coastal strip until mid-May.

Rabbit numbers fluctuate quite widely, dependent on the seasonal prevalence of myxomatosis. Recently their populations appear to have declined, especially in the dunes, where the presence of rabbit viral haemorrhagic disease (RVHD) compounds the effects of myxomatosis. The local rabbit populations have yet to develop resistance to RVHD.

## 2.     CHOUGH STUDIES

Choughs have been well studied on the Castlemartin coastline. Their feeding habitat preferences have been examined (Gamble and Haycock, 1988), and over the last 20 years, their breeding and wintering populations have been regularly recorded. Since 1993 there has been annual surveillance of their breeding population size, distribution and productivity (Haycock 1993 to Haycock 2003). This surveillance programme coincided with the start of a colour-ringing programme to study post-fledging dispersal, survival and recruitment to the population (Haycock, 2002).

## 3.     HABITAT SURVEYS

Although there were earlier vegetation surveys of the Castlemartin coastline (e.g. Cooper 1987), neither of these provided information on the condition of the habitats. Consequently, three MSc students from the University of Wales Swansea carried out a full habitat quality survey as part of their MSc theses (Davies 2000; Wilson 2000; Ross 2000). This survey followed the methods described in Chapter 9, and involved mapping the distribution of the broad habitats and condition classes as appropriate (Fig. 23-2).

## 4.    HABITAT REQUIREMENTS

Choughs breed in caves and crevices along the limestone cliffs, which range from 30 to 40 m high.  During the breeding season, and particularly during the critical period when pairs are feeding their young, the adults feed close to the breeding site. In the UK this is typically within one or two kilometres of the nest (Cramp *et al. 1977 et seq.*). At Castlemartin foraging ranges from about 1.5 km to less than 200 m from the nest, due to a fairly high nesting density in optimum habitat. The current average distance between nests is about 600 to 700 m, though some sites are less than 400 m apart.

At Castlemartin, the Choughs mainly feed on soil invertebrates found among the roots of grasses and herbs, including larvae of *Tipulids*, *Coleoptera* and *Hymenoptera* (especially ants). These insects occur widely in the maritime grassland along the edge of the cliffs. The birds generally favour very short swards (<4 cm) (Cramp *et al. 1977 et seq.*). At Castlemartin Gamble and Haycock (1988) recorded Choughs feeding mainly in swards of less than 5 cm and noted that they often used interfaces between vegetated and rocky, bare ground – typically dominated by one or more of Thrift *Armeria maritima*, Sea Plantain *Plantago maritima* or Red Fescue *Festuca rubra*.

*Photograph by Clive Hurford*

*Figure 23-3*.  An adult Chough feeding in a short grassy sward.

After fledging, the juveniles remain with their parents for up to five weeks. During this period, the family parties feed mainly, though not exclusively, in the maritime grassland, again usually within a kilometre or two of the nest site.

Thereafter, the young birds leave their parents and disperse away from the breeding territory. They usually form flocks, of variable size and age composition, typically including some older (1-3 years) birds that have not yet nested. These social gatherings vary in size but flocks can comprise as many as 40 or 50 birds – feeding and roosting communally. They tend to feed in and around the areas of dune grassland at the western end of the Range in late summer. They often feed in this area during autumn and winter too, though are equally likely to be found feeding within maritime grassland, or heath on the cliff-tops or nearby coastal slopes, each day flying to and from overnight roosts in the cliffs.

# 5.   THE CONDITION INDICATORS

*Table 23-1.* The condition indicators for the Chough population at Castlemartin.

| Condition indicator table | To maintain the Chough *Pyrrhocorax pyrrhocorax* population at Castlemartin in optimal condition where: | |
|---|---|---|
| **Population size and distribution** | **Lower limit** | 10 territory-holding pairs attempt to breed. Occupied nest-sites should occur in each of the eight monitoring sections distributed from Range East to Range West (from St Govan's Head to Berryslade). |
| **Habitat extent** | **Lower limit** | Extent of maritime grassland mapped in 2000 (Davies & Wilson) |
| **Habitat quality** | **Lower limit** | > 40% of the maritime grassland in monitoring sections 1a, 1b, 2b and 5 (Fig. 23-2) should be less than 3 cm high |
| **Site-specific habitat definitions** | | |
| **Maritime grassland** | Vegetation where the combined cover of *Plantago maritima*, *Armeria maritima* and *Festuca rubra* exceeds 50% within any 50 cm radius | |

## 5.1   Reasons behind the selection of the condition indicators

Chough survival can be quite variable: productivity can be very low in years with poor invertebrate populations (including periods affected by drought or stormy weather). Numbers surviving to adulthood can be affected by food shortages and the impacts of cold winter weather.

Even allowing for weather and food shortage-related problems, however, we would not expect the population level to drop below ten pairs in the absence of changes in habitat quality. This figure is based on long-term surveillance data (including several UK decadal Chough population surveys since the 1960s). At the time of writing, ten pairs would still meet the current UK SPA qualifying level.

If the Castlemartin population fell below the SPA qualifying level (1% of the UK population), then we would be prompted to check whether the Chough population was going against the national trend, or whether it was simply a matter of other areas increasing the opportunities for them. In 2004, the Castlemartin population was almost double the lower limit, due to increased survival and recruitment. The combination of a decade or more of relatively mild winters, maintenance of regular grazing patterns, and protection of nest sites has contributed to this situation. This lower limit is likely to be reviewed every six years, and compared with population data from Pembrokeshire, the rest of Wales and future UK decadal surveys.

The principal habitats used by the Choughs at Castlemartin are caves and cliff crevices for breeding, maritime grassland for feeding throughout the breeding season, and dune grassland between late summer and winter months. They are also known to feed in other habitats, such as winter stubbles on arable land, but these tend to be used less frequently than the cliffs and dunes.

Of the three main habitats used by the Choughs, the condition of the maritime grassland is considered to be most critical, as this supports both the adults and young throughout the breeding season (from nest building through to fledgling dispersal).

This short maritime grassland is maintained by a combination of wind exposure, salt deposition and sheep grazing. The salt is deposited in sea spray, mostly during storms driven by the prevailing south-westerly winds: this is a limiting factor for many of the more aggressive plant species that would otherwise colonise the habitat. Sheep grazing plays an important role in keeping the sward low, and making it possible for the feeding Choughs to access invertebrates in the soil in short (2-4 cm) turf.

## 6.    THE CHOUGH MONITORING METHOD

The established Chough monitoring programme comprises two discrete phases: the first assesses the number of breeding pairs, while the second looks at productivity and breeding success.

## 6.1    Monitoring the breeding population

The breeding population counts at Castlemartin are carried out using a method that is broadly similar to that recommended by Gilbert *et al*. (1998). This involves recording the number of confirmed, probable and possible breeding pairs. This information is collected over a minimum of two field visits between the first week in April and first week in May.

We assess productivity and overall breeding success over two or more field visits between mid May and late June (sometimes extending into early July). Between mid and late May, accessible nests are inspected and, where possible, nestlings of a suitable size (about 14-21 days old) are colour-ringed. We also collect biometric data to help determine nestling condition and possible sex. A small team of licensed workers, including local climbers, collect these data.

Subsequent visits, later in the season, help to confirm breeding success: these are made before the young wander too far from their natal areas. Colour-ringing of nestlings increases the likelihood of correctly identifying breeding success in individual territories.

Every year, we use a Global Positioning System (GPS) to record and map the nest site locations, and subsequently transfer these data into a Geographic Information System (GIS). We also carry out boat surveys to confirm the occupancy of sites that are difficult to observe from the land (currently, however, almost all sites are visible from the cliffs). To facilitate future re-location, we take digital photographs of all new nest sites. Records of the annual status of all nest sites are stored in 'Recorder', and as GIS and Excel files.

*Table 23-2.* The critical components of the species monitoring programme at Castlemartin.

---

**Attributes assessed during the species monitoring programme:**
- Number of occupied territories;
- Number of occupied nests;
- Distance between occupied nest-sites (metres/kilometres);
- Number of adults;
- Number of young fledged.

**GPS Minimum basic equipment used:**
Binoculars (10 x 50); telescope (80 mm objective lens and 30x wide angle eyepiece lens; notebook; standard recording proforma (Excel file); 1:10 000 scale map of area and relevant coastal compartments; digital camera.

**Location of data collection:**
The entire Castlemartin coastline.

**Fixed point markers:**
Mapped, permanent red markers on cliff-tops, indicating restricted cliff-climbing zones, aid nest location. This is backed up by a mapped/photo database of nest sites.

**Special considerations:**
Health and safety: observations of nest sites must be made from safe vantage points; stormy (windy or wet) days should be avoided; agreed lone-working arrangements are essential.

Disturbance must be minimised along this very well visited and highly public stretch of coastline, and timing of visits to watch the nest or ring young must take such issues into account.

It is illegal to disturb chough nest-sites without possession of a Schedule 1 License issued by the Statutory Agency (Countryside Council for Wales).

---

## 6.2     The habitat monitoring method

We monitored the condition of the maritime grassland habitat in four of the eight monitoring sections in the study area: Sections 1a, 1b, 2b and 5 (Fig. 23-2). These covered the distribution of 60% of Chough breeding territories in the study area. Within each section, we recorded monitoring points at 20 m intervals on a systematic grid across the entire area of maritime grassland.  As the maritime grassland was generally homogeneous, both in species composition and structure, we were satisfied that recording at 20 m intervals would provide a reliable monitoring result.

In habitat patches which were not broad enough to monitor on a regular grid, we continued to record at 20 m intervals, but alternated the points between the seaward edge and the landward edge of the maritime grassland until it was possible to revert back to recording on a systematic grid.  At each monitoring point, the vegetation within a 50 cm radius was assessed for:

- \>50% cover of *Plantago maritima*, *Armeria maritima* and *Festuca rubra*; and
- Being <3 cm high.

We used a drop disc (see Chapter 10) to measure the vegetation height, and used a differential GPS (accurate to <50 cm) to record the location of each monitoring point. The MSc students who carried out the habitat quality survey also monitored the condition of the maritime grassland.  The monitoring, which was carried out in early August 2000, took three days to complete.

*Photograph by Clive Hurford*

*Figure 23-4.* A family of Choughs at roost at Castlemartin, shortly after the juveniles had fledged.

# 7. THE MONITORING RESULTS

The Chough breeding population has increased along the Castlemartin coastline since the mid 1980s. Annual surveillance since 1993 has seen a gradual rise from 11-12 territorial pairs, between 1993 and 1997, to 17-18 territorial pairs (confirmed or probably breeding), between 2001 and 2004 (Fig. 23-5).

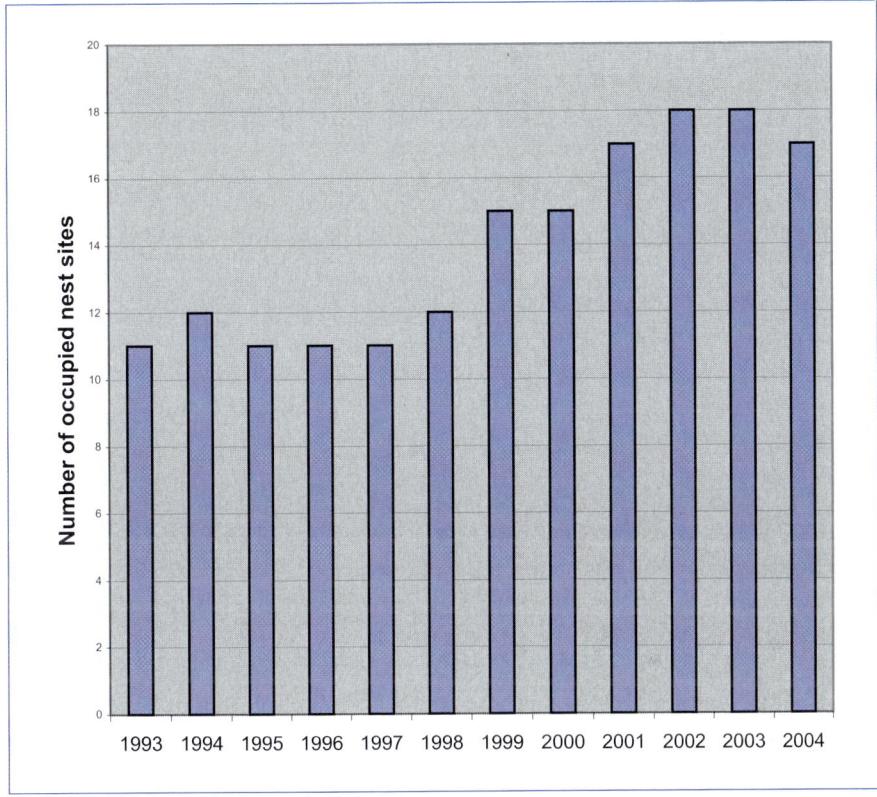

*Figure 23-5.* The Chough breeding population at Castlemartin over the period from 1993 to 2004.

In 2000, the year of the habitat monitoring, 15 pairs of Chough were resident at Castlemartin: eight in Range East and seven in Range West.

The habitat monitoring results (Table 23-3) revealed that the maritime grassland in all four monitoring sections was in optimal condition for feeding Choughs: only in Section 5 was the condition in any way marginal. Overall, the condition of the maritime grassland vegetation was assessed at 271 sample points, of which 137 passed the criteria set out in the condition indicator table. This result means that, in the summer of 2000, the Chough habitat at Castlemartin was considered to be in optimal condition.

*Table 23-3.* The monitoring results from the selected habitat sections at Castlemartin.

| Management section | Number of samples | Samples passing | Result |
|---|---|---|---|
| 1a | 60 | 29 (48%) | Pass |
| 1b | 59 | 34 (58%) | Pass |
| 2b | 52 | 30 (58%) | Pass |
| 5 | 100 | 44 (44%) | Pass |
| Overall total | 271 | 137 (51%) | Pass |

# 8.    DISCUSSION

Before monitoring the habitat, we set a provisional lower limit for >50% of the maritime grassland to be <3 cm high. However, we subsequently revised this and set it at >40%. The rationale underpinning this decision centred on the fact that we could not reasonably consider the habitat to be in unfavourable condition when the local breeding population had been steadily increasing for c. 20 years. We also took account of the fact that, if >40% of the grassland was <3 cm high, and Choughs can feed in vegetation that is at least 4 cm tall, then at least some of the vegetation that failed to meet the criteria in the condition indicator table would be in the region of 3-4 cm high and still be accessible to feeding birds.

Had we persisted with the original lower limit of >50% of the maritime grassland being <3 cm high, then management sections 1a and 5 would have been regarded as in sub-optimal condition and would have required restoration management.

## 8.1    Habitat monitoring problems

The only problem that we experienced during the habitat monitoring exercise did not become apparent until we had finished collecting the data. On returning the GPS to the hire company, it transpired that the data logger had not saved any of the data. This does not compromise the validity of the monitoring result, but it does mean that we cannot return to any of the monitoring points, nor detect where any loss of habitat has occurred. However, assuming that the same method is applied in the future, a decline in the number of maritime grassland monitoring points would indicate a retraction in the extent of the habitat.

In reality, changes in the overall extent of the maritime grassland should not be a great concern, as we expect the habitat to expand or contract on an annual basis. Changes in the extent of this habitat are likely to be determined by the frequency and severity of storms the previous winter, and not by the management regime. We can expect the management to control the sward height of the maritime grassland, but not the species composition. On this basis, the condition of the habitat is linked more to the quality of the vegetation than the extent.

*Photograph by Clive Hurford*

*Figure 23-6.* This short maritime grassland, dominated by Thrift *Armeria maritima* and Sea Plantain *Plantago maritima*, would be optimal post-breeding feeding habitat for Choughs at Castlemartin.

## 8.2     Considerations for repeat monitoring events

In 2000, the availability of successionally-young dune grassland for winter feeding was not considered to be critical and so was not monitored. However, this situation could change rapidly if the local rabbit populations along the Castlemartin coast continue to decline. In future, we should extend the Chough habitat monitoring to include the areas of dune grassland favoured by feeding birds, particularly as successionally-young dune grassland vegetation is currently declining on many dune systems in Wales.

Also, on this occasion, the academic year dictated the timing of the habitat monitoring. In future, the maritime grassland should be monitored earlier, in late May or early June, in the period immediately before the adults are feeding young. Due to military training activities, Castlemartin Range is closed to the public on most weekdays. As a consequence, the repeat habitat monitoring will have to be carried out either at weekends or during weekday evenings.

## 9.     REFERENCES

Cooper E. 1987. Vegetation maps of British sea cliffs and cliff-tops. Unpublished Report to the Nature Conservancy Council.

Cramp, S. (ed.) 1977-93. *Handbook of the Birds of Europe, the Middle East and North Africa: The birds of the Western Palearctic.* Oxford University Press. Oxford.

Davies, E.J. (2000). Is the Common Standards Model a Robust Method of Evaluating the Condition of the Vegetated Sea Cliff Communities of the Castlemartin Site of Special Scientific Interest? MSc thesis. University of Wales Swansea.

Gamble, R. and Haycock, R.J. 1988. A chough feeding survey: Castlemartin R.A.C. tank range and Stackpole Warren, Stackpole National Nature Reserve. Unpublished Report to the Nature Conservancy Council. (Summarised in Choughs and Land-use in Europe, proceedings of an International Workshop on the Conservation of the chough in the EC. 11th – 14th November, 1988. Edited by Eric Bignal and David Curtis, Scottish Chough Study Group).

Gilbert, G., Gibbons, D.W. and Evans, J. 1998. Bird Monitoring Methods – a manual of techniques for key UK species. Published by the RSPB in association with British Trust for Ornithology, The Wildfowl and Wetlands Trust, Joint Nature Conservation Committee, Centre for Ecology and Hydrology and The Seabird Group.

Haycock, R.J. 2002. Pembrokeshire Chough Study Group chough colour-ringing project: preliminary observations and a plea for more re-sightings. Pembrokeshire Bird Report, 2001. Wildlife Trust South and West Wales.

Haycock, R.J. 1993 to 2003. Choughs in south Pembrokeshire; monitoring of nest sites 1993 to 2003. Unpublished Annual Reports to Pembrokeshire Ornithological Research Committee (PORC) Chough Study Group.

Ministry of Defence, 2000. Castlemartin Range Integrated Land Management Plan, nature conservation component. Unpublished.

Ross, M. (2000). Are the Heathland Communities at the Castlemartin AFTC in Favourable Condition? A Site Evaluation of the Common Standards Model. MSc thesis. University of Wales Swansea.

Wilson, S. (2000). An appraisal of the utilisation of Common Standards Monitoring methods to assess the condition of Castlemartin's Vegetated Sea Cliff communities. MSc thesis. University of Wales Swansea.

# CHAPTER 24

# NARROW-MOUTHED WHORL SNAIL *VERTIGO ANGUSTIOR* AT WHITEFORD BURROWS

ADRIAN FOWLES[1] & DAN GUEST[2]

*Countryside Council for Wales, Plas Penrhos, Ffordd Penrhos, Bangor, Gwynedd, LL57 2BQ*
*A.Fowles@ccw.gov.uk[1]*
*D.Guest@ccw.gov.uk[2]*

## 1. BACKGROUND INFORMATION ON WHITEFORD BURROWS

The dune system of Whiteford Burrows on the north coast of Gower, West Glamorgan, Wales (Fig. 16.1), is part of the much larger Carmarthen Bay Dunes/Twyni Bae Caerfyrddin cSAC. Most of the site is owned by the National Trust and, since 1965, has been leased to the Countryside Council for Wales, who manage it as a National Nature Reserve.

The dunes have been grazed by a variety of domestic stock since the Middle Ages, particularly sheep in winter when up to 2000 ewes are known to have been present. Grazing pressure has been substantially reduced in modern times and for the past 15-20 years has consisted of around 40-50 commoners' ponies that have access to an extensive area of dune and saltmarsh. Rabbit grazing is also an important factor in the maintenance of short sward dune habitats.

The Narrow-mouthed Whorl Snail *Vertigo angustior* was discovered here in 1983 (Preece & Willing, 1984) and subsequent surveys found that it was confined to a narrow band within the dune-saltmarsh transition zone (Fig. 24-1), where it is locally abundant over a length of approximately 1.5 km.

## 1.1 Key factors influencing the management decisions at Whiteford Burrows

The ecotone occupied by *Vertigo angustior* at Whiteford Burrows is influenced by the water table underlying the adjacent dune system, grazing pressure from ponies on the saltmarsh (Fig. 24-2) and periodic flooding by high tides. Sea level change may cause this ecotone to migrate inland in the future, but at present the main management issue is concerned with vegetational succession in response to alterations in grazing pressure.

259

*C. Hurford & M. Schneider (eds.), Monitoring Nature Conservation in Cultural Habitats, 259–270.*
© 2007 *Springer.*

The ponies range widely over the dunes and adjacent saltmarsh but studies in 1983 found that they preferentially graze the saltmarsh transition zone during the hours of darkness. As the ponies are owned by commoners, CCW has little control over the numbers of stock present and it is conceivable that grazing pressure could reduce even further in the future. This is likely to be detrimental to the habitat occupied by *Vertigo angustior* and in the past part of this area became dominated by Common Reed *Phragmites australis*, while other sections are susceptible to invasion by Alder *Alnus glutinosa* scrub and Sea Rush *Juncus maritimus*. Faced with this uncertainty CCW is considering introducing its own herd of ponies to parts of the dunes, using temporary fencing – though this will be difficult to implement in areas under tidal influence.

## 1.2    The management priority at Whiteford Burrows

The site is important for several western Atlantic dune habitats and the following habitats have been identified on Annex I of the EC Habitats Directive as being of European importance: Embryonic shifting dunes (Habitat 2110), Shifting dunes with *Ammophila arenaria* (2120), Fixed coastal dunes (2130), Dunes with *Salix repens* ssp. *argentea* (2170). In addition the site has been identified as containing the following Annex II species: Fen Orchid *Liparis loeselii* var. *ovata*, Petalwort *Petalophyllum ralfsii*, and Narrow-mouthed Whorl Snail *Vertigo angustior*. The management priority for the dune system as a whole is to maintain the natural dynamics of dune succession whilst preventing over-stabilisation of the important grassland and dune slack habitats. Management of the saltmarsh transition zone to maintain favourable condition for *Vertigo angustior* is not in conflict with any of these aims and will contribute to the conservation of the plant communities. Uncommon plant species such as Sharp Rush *Juncus acutus* and Marsh Mallow *Althaea officinalis* occur here but both species are able to persist under moderate levels of grazing. However, the transition zone is the type locality for the spider *Baryphyma gowerense*, otherwise known in the UK only from a small number of fens in Wales and East Anglia: this species probably inhabits pockets of decaying vegetation so might be affected by a reduction in the amount of plant litter as a result of increased grazing pressure.

## 2.      RESEARCH AND SURVEY

*Vertigo angustior* is a small (c.2 mm) snail, which at this site occupies the transitional zone between dunes and saltmarsh where *Iris*-dominated marsh occurs on freshwater seepages inundated by sea water on the highest spring tides. The habitat here is intermediate between the dry grassland of the adjacent dunes and the wetter sections of lower-lying saltmarsh (Fig. 24-4). Freshwater marsh with higher water tables is avoided by the snail. Dense vegetation which shades out the ground layer (e.g. dominant *Juncus maritimus* or Meadowsweet *Filipendula ulmaria*) is unsuitable and the highest densities of the snail are associated with herb-rich grassland with varying amounts of fine-leaved grasses, especially Red Fescue *Festuca rubra*, particularly where there is a persistent litter

*Photograph by Adrian Fowles*

*Figure 24-1.* Optimal *Vertigo angustior* habitat a Whiteford Burrows in April 1993.

*Photograph by Adrian Fowles*

*Figure 24-2.* Ponies grazing the saltmarsh transition zone at Whiteford Burrows.

layer. The snail grazes on micro-fungi and algae on dead vegetation in damp, warm situations close to the ground.

The *Vertigo angustior* population at Whiteford Burrows had been the subject of considerable survey effort prior to the monitoring phase. In 1993 extensive sampling was carried out in order to identify the area of habitat occupied and the vegetation types supporting the greatest density of snails (Killeen, 1993). *V. angustior* was found to be present along one kilometre of the saltmarsh transition zone where there is "open, marshy grassland dominated by Yellow Flag *Iris pseudacorus* and low-growing herbs such as Common Fleabane *Pulicaria dysenterica* and Silverweed *Potentilla anserine* where the vegetation is predominantly short and there is a variable proportion of bare, friable soils covered by sparse litter" (Killeen, 1993). Typical abundances were in the order of 1200 snails per square metre in favoured vegetation types. *V. angustior* avoided areas of tall, rank vegetation or where there was evidence of waterlogging, being replaced in these situations by *V. antivertigo*.

In July 2000 we undertook a habitat survey of the transition zone mapping out all areas of vegetation dominated by *Iris pseudacorus*. The vegetation was further subdivided into six condition classes according to the known preferences of the snail.

*Table 24-1.* The habitat definitions used for the habitat survey phase of the project (see Fig. 24-3).

| Habitat type | Definitions used in mapping |
|---|---|
| Iris-dominated vegetation | Any vegetation where *Iris pseudacorus* is present at a density of >5 plants per 50 cm radius |
| Optimal habitat | Vegetation where within a given 50 cm radius search area the following criteria are met: <br> • >10 plants of *Iris* are present; <br> • Either *Lotus pedunculatus* makes up 10-60% of the vegetation cover, or *Pulicaria dysenterica* and/or *Filipendula ulmaria* are present, the latter at less than 50% cover; <br> • *Juncus subnodulosus* (and other tall rushes) account for less than 50% of the ground cover; and <br> • *Juncus maritimus*, *Samolus valerandii*, *Ranunculus sceleratus*, *Oenanthe lachenalii* and *Schoenoplectus tabernaemontani* are absent. |
| Brackish | Stands of '*Iris* dominated vegetation' where one or more of the following species are present within any given 50 cm radius search area: *J. maritimus*, *S. valerandii*, *R. sceleratus*, *O. lachenalii* and *S. tabernaemontani*. |
| *Filipendula* dominated | Stands of '*Iris* dominated vegetation' where *F. ulmaria* accounts >50% of the vegetation cover. |
| *Juncus subnodulosus* dominated | Stands of '*Iris* dominated vegetation' where *J. subnodulosus* accounts for more than 50 % of the vegetation. |
| Species-poor *Iris* | A 'catch-all' category covering all other forms of '*Iris* dominated vegetation' recorded. |

## Vertigo angustior habitat condition map showing sections used in sampling

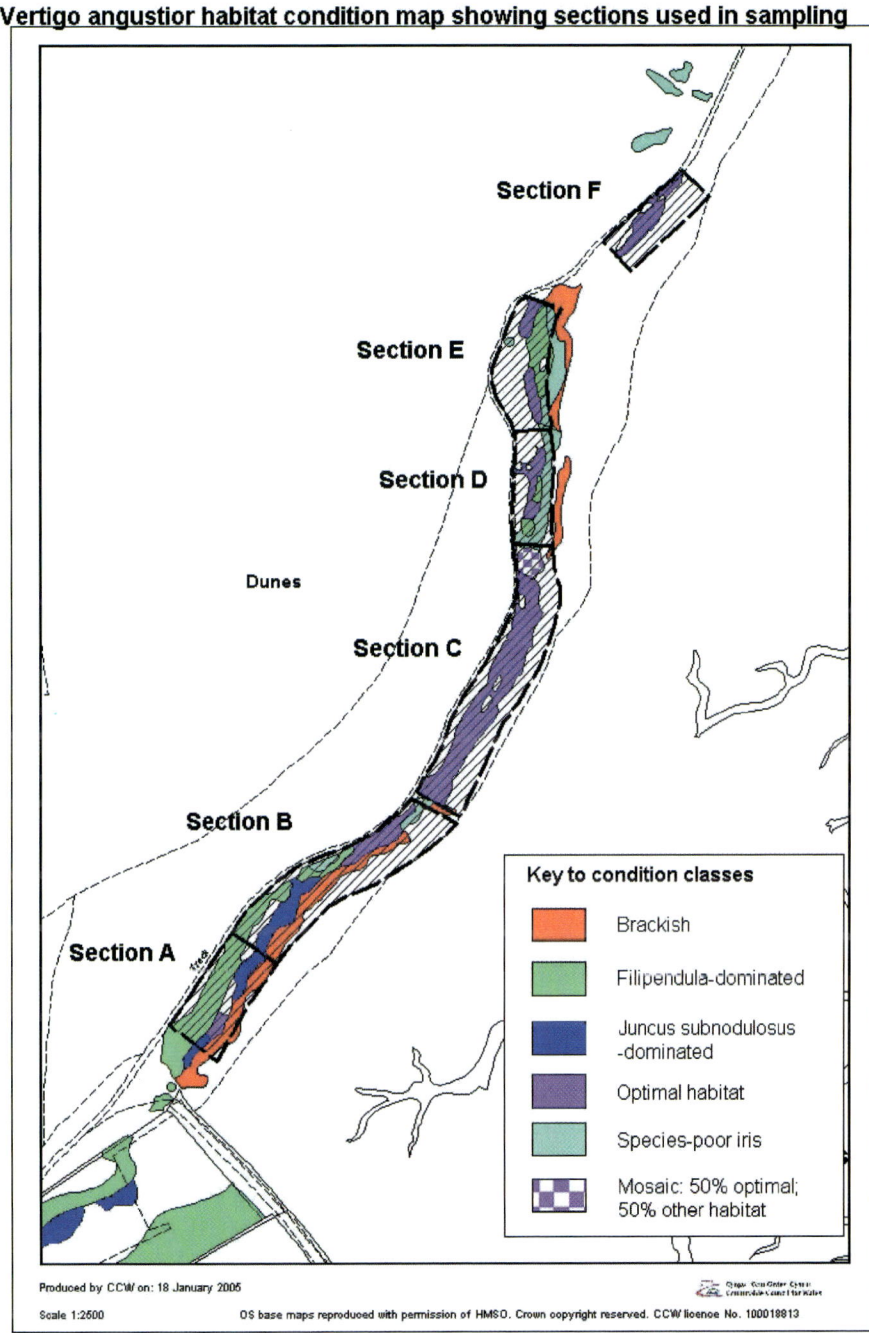

*Figure 24-3.* The habitat quality of map *Vertigo angustior* habitat at Whiteford Burrows. Reproduced from Ordnance Survey mapping on behalf of Her Majesty's Stationery Office © Crown Copyright 100043571 2004-12-10.

# 3.       BACKGROUND SURVEILLANCE

Three permanently marked, 5 x 5 m, surveillance plots were established on the transition zone in February 1994. Plots were divided into 25 x 1 sq. metre quadrats, which were in turn sub-divided into four 50 x 50 cm cells. Two cells in each of the five 1 m rows were randomly selected for sampling (Fowles & Hurford, 1995). The results up to 1996 are summarised by Fowles (1998). Sampling was undertaken annually for six years until 1999 when the surveillance programme was abandoned. By this time we had a clearer idea of the micro-habitats that produced dense populations of the snail, and a better understanding of the problems encountered in trying to undertake surveillance on an animal whose behaviour responds significantly to changes in humidity. In drier conditions the snails retreat to refuges in the upper soil layers whilst in moist conditions they emerge to browse on micro-fungi growing on plant litter. Co-ordinating sampling with standard meteorological conditions was impossible and as a result surveillance was more often than not reflecting the climate rather than changes in the snail population.

In 1997 the Countryside Council for Wales and the National Museums & Galleries of Wales collaborated to set up a Ph.D. to investigate the ecology of *Vertigo angustior* (and *V. geyeri*). The research on *angustior* was carried out at Whiteford Burrows in 1998 and 1999 (Sharland, 2000) and involved detailed studies of population dynamics, life-history patterns and habitat preferences (Cameron, 2003). However, the study also examined sampling efficiency and discovered that cutting and removing vegetation at ground level (for subsequent laboratory analysis), effectively yielded double the number of snails compared to the removal of dead plant litter alone. Excavation of a sample turf down to 2cms of soil depth increased the sample size further. Based on this research Sharland recommended the use of turf removal as a reliable method of obtaining population estimates, although this method is considerably more labour intensive in terms of the laboratory time required to extract snails from sand particles.

# 4.       THE CONSERVATION AIMS

The *Vertigo angustior* population on Whiteford Burrows is considered to be one of the strongest in Europe. We do not know if it was once more widespread at this site but the population can be regarded as healthy and hence the conservation aim is to maintain the range of the snail at its current extent. We focused on areas that had been identified as containing stands of optimal habitat as revealed by previous autecological research, seeking to manage the saltmarsh transition zone such that optimal habitat occupies an appropriate proportion of the 1.5 km length. The snail occurs at lesser densities in sub-optimal habitat between the favourable stands of vegetation and it should continue to do so if grazing pressure is such that the optimal stands are maintained. Reduced grazing pressure would lead to changes in vegetation structure and composition such that the extent of optimal habitat would decline and intervening sub-optimal habitat would no longer support the snail.

# 4.1    The condition indicators

In company with the site manager, CCW's monitoring ecologists and invertebrate ecologist met on site in May 2001 to discuss the findings of the recent research and to apply them to the development of a monitoring strategy. We scrutinized areas known to support high densities of the snail and compared their environmental characteristics and vegetation composition with adjacent sub-optimal and unsuitable areas to help us focus on the most appropriate performance indicators. Some candidates were eventually discarded because of sampling difficulties or ambiguous results, but in general it was relatively easy to identify the required indicators to the agreement of all present.

*Table 24- 2.* The condition indicator table for the Narrow-mouthed Whorl Snail *Vertigo angustior* on Whiteford Burrows NNR.

| **Condition indicators** | **To maintain the *Vertigo angustior* population on Whiteford Burrows NNR in favourable condition where** | |
|---|---|---|
| **Population range** | **Lower limit** | *Vertigo angustior* is recorded as present in Section C, plus any three of the remaining five Sections (see Fig. 24-3), during a 15-minute sampling period in each section. |
| **Habitat extent** | **Lower limit** | Current extent of *Iris*-dominated marsh in 2000 (see Fig. 24-3) |
| **Habitat quality** | **Lower limit** | In sections A - F (see Fig. 24-3), the proportion of the vegetation recorded as optimal *V. angustior* habitat is as follows:<br><br>Section A = 20%; Section B = 25%; Section C = 12%; Section D = 25%; Section E = 20% and Section F = 15%. |
| | **Upper limit** | All of the vegetation is optimal *V. angustior* habitat |
| **Definition of Iris dominated marsh** | In any 50 cm radius,<br><br>*Iris pseudacorus* is present at a density of >5 plants | |
| **Definition of optimal *V. angustior* habitat** | Within any 50 cm radius<br><br>> 10 *Iris pseudacorus* plants are present; and the cover of *Lotus pedunculatus* is between 10 and 60%<br><br>And where<br><br>*Juncus maritimus* is <10% cover<br><br>And<br><br>*Samolus valerandii, Ranunculus sceleratus, Schoenoplectus tabernaemontani* and *Oenanthe lachenalii* are absent | |

## 4.2    Explanation of the condition indicators

After six years of surveillance, a comprehensive survey and detailed research into population dynamics for the Ph.D., the most controversial decision taken was to drop population abundance from consideration as a condition indicator in favour of presence/absence data. This was deemed necessary because of the variability in sample size caused by changes in weather conditions at the time of sampling. Turf extraction might have allowed us to overcome this obstacle but a pragmatic assessment of the resources likely to be available for monitoring led us to conclude that this was too labour intensive. Research has also shown that *Vertigo* species undergo large annual fluctuations in population size (Cameron, 2003: Pokryszko, 1990) and that the peak breeding period varies considerably from year to year. We felt that there was insufficient information on behaviour under different weather conditions and fluctuations in snail density to allow us to set a meaningful target for abundance.

Having decided on presence/absence as the indicator value for the snail population, we needed to ensure that range was being maintained and hence we mapped all patches of optimal habitat and used these as our sampling stations. Loss of the snail from the current area of highest density was considered unacceptable so presence here is mandatory. We also decided that loss (or a decline below levels of detection within the sampling period) would be permissible in up to half of the remaining patches, given that patch condition may alter from year to year in response to environmental parameters, before condition was deemed unfavourable.

*Figure 24-4.* The saltmarsh transition zone at Whiteford Burrows in April 1993.

*Vertigo angustior* on Whiteford Burrows is closely associated with stands of vegetation in which *Iris pseudacorus* is frequent. The snails frequently occur within the dead leaf sheaths at the base of the plants and are believed to graze on micro-fungi and algae growing on the *Iris* leaves. It is therefore important to manage the transition zone such that *Iris* remains as a significant component over a substantial area. Within the *Iris* marsh, however, there are clearly areas that are more favourable to *V. angustior* and these coincide with herb-rich vegetation in which Greater Bird's-foot Trefoil *Lotus pedunculatus* is conspicuous or *Pulicaria dysenterica* and/or *Filipendula ulmaria* are present. These herbs may not be of direct relevance to *V. angustior* but their presence may indicate certain moisture regimes that suit the snail. Snail density declines when the vegetation becomes rank or the water table is too high. These conditions are indicated within the *Iris* marsh by the presence of Brookweed *Samolus valerandii*, Celery-leaved Buttercup *Ranunculus sceleratus* and Parsley Water-dropwort *Oenanthe lachenalii* or where *Juncus maritimus* becomes established.

## 5. THE MONITORING PHASE

It was anticipated that the snail population would be in favourable condition as grazing pressure has remained relatively constant in recent years and therefore dramatic changes in habitat quality were not expected. Monitoring sought to confirm this by establishing that the thresholds for the condition indicators were achieved.

## 5.1 Description of monitoring methods

Adult *V. angustior* are most abundant during late spring and summer and monitoring should take place during this period. Snail presence was established by sieving/shaking plant litter over a white tray and examining the debris for the presence of adult *angustior*. With experience, *angustior* is a distinctive snail but care should be taken not to confuse it with other *Vertigo* species present (*antivertigo, pygmaea,* and *substriata*) and juveniles of other species (e.g. *Clausilia bidentata, Cochlicopa lubrica, Lauria cylindracea*). The sampling period at each section should be restricted to fifteen minutes in total and if *angustior* is found within the fifteen minute field sample then the surveyor can move on to the next section.

The upper limit of the transition zone at Whiteford is clearly delimited by a permanent track, above which little or no suitable *Vertigo angustior* habitat is to be found. This track provided an ideal marker for the upper edge of the habitat monitoring plot. Vegetation sampling was carried out at 5 m intervals along a series of transects originating at the track and running westwards out into the saltmarsh. Each transect was 35 m long and hence six sampling points per transect were recorded. The transects themselves were spaced at 5 m intervals along the track creating a semi-regular grid of sample points. At each stopping point we asked the following questions:

- Is the vegetation within a 50 cm radius *Iris*-dominated marsh? And if so
- Does the vegetation satisfy our definition of 'optimal *Vertigo* habitat'?

Within each section, a variable amount of habitat is suitable for *Vertigo angustior* depending on the width of the transition zone, which is primarily influenced by local topography. As a result, many of the sample points fall within areas that that are unsuitable, usually because they have a high water table or are brackish. Taking these factors into consideration, we determined that the following pass rates were required for the habitat to be in favourable condition: Section A = 20%, Section B = 25%, Section C = 12%, Section D = 25%, Section E = 20% and Section F = 15%.

## 6.    RESULTS AND ANALYSIS

*V. angustior* was recorded as present in each of the sampling sections within the prescribed survey period and hence this attribute passed our test for favourable condition. Similarly, the pass rates for the number of points recorded as optimal achieved each of the targets, although in Section A it was equalled and not exceeded. We conclude, therefore, that *Vertigo angustior* is currently in favourable condition on Whiteford Burrows but Section A will need to be kept under surveillance to ensure that sufficient optimal habitat is maintained.

*Table 24-3.* The monitoring results from Whiteford Burrows.

| Section | *V. angustior* presence | No of habitat samples | Sample points passing as optimal habitat | % passes |
|---------|-------------------------|-----------------------|------------------------------------------|----------|
| A | + | 60 | 12 | 20 |
| B | + | 132 | 41 | 31 |
| C | + | 174 | 27 | 16 |
| D | + | 54 | 17 | 31 |
| E | + | 96 | 24 | 25 |
| F | + | 66 | 12 | 18 |

Litter sampling for the presence of *V. angustior* took 2.5 hours to complete. The vegetation sampling proved somewhat more time-consuming, taking just over six hours to complete the 602 sample points. This relatively rapid progress was made possible a) by recording pass / fail data only at each point in '*Iris*-dominated marsh' and b) by sticking rigidly to a rule where, at points that were considered too marginal to call with confidence, the surveyor would err on the side of caution and fail the point. No attempt was made to identify which criteria if any, were met at each sample point. This approach meant that little time was spent at most sample points: most clearly failed on at least one criterion or clearly passed on all the criteria, only a relatively small minority of marginal quality points requiring more diligent recording to establish their condition

# 7. DISCUSSION

Detection of minute snails amongst plant litter requires concentration and more snails are undoubtedly found with experience. During monitoring, three surveyors with different levels of experience, ranging from novice to experienced malacologist, independently sampled sections B-E with the following results: (B) 2/15/31, (C) 6/42/79, (D) 1/37/36, (E) 3/36/26. It is noticeable that the surveyor with some prior experience of mollusc sampling rapidly improved detection rates such that after half an hour he was as effective as the experienced malacologist. This relates both to the ability to identify suitable microhabitats to collect litter from and to find these tiny snails within the litter sample on inspection. A 'search image' is required for both activities and the evidence suggests that this can be acquired fairly easily with minimal training.

Both observers with prior experience found *V. angustior* in abundance during the fifteen-minute sampling period in each section, but the novice observer managed to find few specimens in the time period. This suggests that novice observers may fail to record the species when population densities are low and hence it is advised that snail sampling is only undertaken by trained or experienced observers.

These results demonstrate that *V. angustior* is present in each of the sections but they give no indication of the strength of the population. As high densities of the snail can be recorded on Whiteford Burrows, we would have greater confidence in the results if a numeric threshold for abundance was added to the attributes. However, as discussed above, snail behaviour under different climatic conditions, in combination with the large annual fluctuations in populations in size recorded for V*ertigo* species in general, create problems for setting numeric thresholds. The population may be perfectly healthy but sampling results fail their targets because of these factors. However, on the basis of the results mentioned above, it would seem that an experienced or trained surveyor should be able to record at least 20 snails during the sampling period in each section at an appropriate time of year.

The habitat monitoring was able to focus on a small number of distinctive species, which even inexperienced field workers could be expected to identify with confidence with a minimum of training. The inclusion of cover targets in the condition indicators was more controversial as, for most plants, these can change throughout the growth season and their assessment has been shown to be associated with relatively high levels of observer error. The attributes were, however, considered essential to define optimal condition habitat for the snail and, to minimise errors in recording, future habitat monitoring must be undertaken in May and recorders must be trained in making cover assessments to further ensure consistency.

Future habitat monitoring would further benefit from the use of high accuracy GPS to pinpoint the individual samples. The loss of time that precise relocation of individual sample points would inevitably incur, could perhaps be offset by omitting alternate rows of samples. The elongated sampling pattern this would produce (with 5 m between samples along individual transects but 10 m between each transect) would be more in keeping with the natural patterning of the vegetation, which is compressed into narrow bands along the transition zone.

The monitoring exercise has identified those areas of the dune-saltmarsh transition zone at Whiteford Burrows that are believed to constitute optimal habitat for the species and has determined a suite of vegetation characteristics that define this state. The site manager now has a greater awareness of the most important areas for *Vertigo angustior* on the site and has visual cues to enable him to periodically assess whether or not habitat condition is being maintained. This is particularly important at Section A, where the area of optimal habitat is restricted, and management intervention may be required to maintain the extent of optimal habitat in the near future. However, the grazing patterns of domestic stock can vary from year to year and hence all sections need to be kept under surveillance and monitored at regular intervals.

## 8.    REFERENCES

Cameron, R.A.D. 2003. Life-cycles, molluscan and botanical associations of *Vertigo angustior* and *Vertigo geyeri* (Gastropoda, Pulmonata: Vertiginidae). *Heldia,* **5**: 95-110.

Fowles, A.P. 1998. Implementing the Habitats Directive: *Vertigo angustior* Jeffreys in Wales. In: *Molluscan conservation: a strategy for the 21st Century.* Journal of Conchology. Special Publication No. 2. Eds. I.J. Killeen, M.B. Seddon, & A.M. Holmes, pp. 179-190. Conchological Society of Great Britain and Ireland.

Fowles, A.P. & Hurford, C. 1995. *Monitoring populations of the whorl snail Vertigo angustior on Oxwich & Whiteford Burrows NNRs, Gower, West Glamorgan.* Species & Monitoring Branch Report. **94/5/1**. Countryside Council for Wales.

Killeen, I.J. 1993. *The distribution and ecology of the snail Vertigo angustior at Oxwich and Whiteford Burrows NNRs, Gower, South Wales.* CCW Contract Science. **20**. Countryside Council for Wales.

Pokryszko, B.M. 1990. The Vertiginidae of Poland (Gastropoda: Pulmonata: Pupiloidea) - a systematic monograph. *Annales Zoologici,* **43**: 133-257.

Preece, R.C. & Willing, M.J. 1984. *Vertigo angustior* living near its type locality in south Wales. *Journal of Conchology,* **31**: 340.

Sharland, E. 2000. *Autecology of Vertigo angustior and Vertigo geyeri in Wales.* CCW Contract Science. **392**. Countryside Council for Wales.

# CHAPTER 25

# MONITORING THE HEATH FRITILLARY *MELLICTA ATHALIA* IN THORNDEN AND WEST BLEAN WOODS

TOM BRERETON

*Butterfly Conservation, Manor Yard, East Lulworth, Wareham, Dorset, BH20 5QP, UK*
*tbrereton@butterfly-conservation.org*

## 1.    INTRODUCTION

The Heath fritillary *Mellicta athalia* is one of Britain's rarest butterflies, and is listed as vulnerable in the UK Red Data Book. It is protected under schedule 5 of the 1981 Wildlife and Countryside Act and has been identified as a priority species for conservation action in the UK Government's Biodiversity Action Plan (BAP) (Anon, 1995). During the twentieth century, the Heath Fritillary declined severely in range (10 km resolution), with the rate estimated at 90% (Asher *et al.*, 2001).

The first national survey was completed in 1980, which located 31 colonies in the UK chiefly in two localised areas of south-west and south-east England: Exmoor, and Blean Woods in Kent (Warren *et al.*, 1981). Further surveys in subsequent years have discovered new sites, though the total number of UK colonies has never exceeded 50 (Wigglesworth *et al.*, 2004).

*C. Hurford & M. Schneider (eds.), Monitoring Nature Conservation in Cultural Habitats*, 271–284.
© 2007 *Springer.*

In 1980, four-fifths (17 of 25) of the colonies in the Blean Woods complex were in Thornden/West Blean Woods, confirming this part of the forest as the single most important station for the butterfly in the UK. During the 1980s and early 1990s the Heath Fritillary progressively declined in abundance across Thornden/West Blean, and was close to extinction in 1992 with just two colonies remaining. Concerted efforts were made from the mid-1990s to reverse the decline, through an active programme of targeted annual management. This chapter describes a monitoring study established by Butterfly Conservation, to assess the effectiveness of the management in restoring the Heath Fritillary population.

## 1.1    Background information on Thornden and West Blean Woods

Thornden and West Blean Woods form the central part of the Blean Woods complex, which extends over 2800 hectares on gently sloping land north of Canterbury in east Kent, south-east England. This woodland complex is of high importance for nature conservation and represents the second largest block of ancient broadleaved woodland in southern Britain. More than half of the woodland has been designated as a Site of Special Scientific Interest (SSSI), whilst nearly 20% of the western section has been designated a National Nature Reserve (NNR). Approximately 520 hectares was designated a candidate Special Area for Conservation (SAC) in 1995, because it contained more than 50% of the UK resource of Oak *Quercus robur* woodland with Hornbeam *Carpinus betulus* coppice. A Royal Society for the Protection of Birds (RSPB) nature reserve makes up a substantial proportion of the western part of the Blean Woods complex, whilst in the east, there is the 122 hectare East Blean Woods NNR, managed by the Kent Wildlife Trust. Other conservation and land management bodies owning or managing the wood include the Woodland Trust and the Forestry Commission. Thornden and West Blean Woods was under mixed private ownership until purchased by the Kent Wildlife Trust in 2003.

Though best known for Heath Fritillaries, Thornden Woods/West Blean is important for breeding woodland birds and supports populations of Nightingale *Luscinia megarhynchos* and a number of conservation 'Red-listed' (most threatened) species including Turtle Dove *Streptopelia turtur* and Nightjar *Caprimulgus europaeus*. The coppice and high forest habitats are also important for Dormice *Muscardinus avellanarius*, Wood Ants *Formica rufa* and moths.

The main management at Blean Woods over much of the twentieth century has been continued coppicing and steady removal of Oak standards. In 1950, for example, the forest was a mixture of Oak high forest with coppice underwood (60%) and Sweet Chestnut *Castanea sativa* coppice with Oak standards (37%) (Holmes and Wheaten, 2002). In the 1970s and 1980s large areas were cleared and replanted with non-native conifers (Warren, 1985a; D. Hoare, Tilhill Forestry, pers. comm.). With an increasing emphasis on the need to incorporate biodiversity objectives in important lowland woodlands as part of the UK BAP, and the majority of the wood is now in conservation ownership, extensive removal of non-native conifers is likely over the coming decades.

## 1.2     Key factors influencing the management decisions on the site

In Thornden/West Blean Woods, the Heath Fritillary is restricted to rides and sheltered woodland clearings, especially newly cut coppice, which contain the larval foodplant Common Cow-wheat *Melampyrum pratense* growing in unshaded positions. Woodland clearings generally only remain suitable for Heath Fritillaries for a few years (2-9 summers growth), due to re-growth and shading. Therefore, fresh habitat needs to be created year after year, within reach of the butterfly, to balance natural extinctions with colonisations at a whole wood scale. In this woodland complex, not every new woodland clearing will contain suitable habitat conditions, and hence more clearings need to be created than are likely to be colonised in order to maintain a stable population.

The main factor threatening the Heath Fritillary at Thornden/West Blean Woods has been changing woodland management practices, resulting in reductions in ride management and coppicing. More than half of the wood was coniferised in the 1970s and 1980s, with plantings of Corsican Pine *Pinus nigra* ssp. *laricio* and Norway Spruce *Picea abies* replacing less economically productive coppice (coups with lower stool density or suffering Birch *Betula* invasion). This temporarily increased the availability of clearing habitat for the Heath Fritillary, but substantially lowered the long-term potential of the woodland for the butterfly (Warren, 1985a & b).

With the decline and fragmentation of coppicing, newly cut clearings have frequently been too isolated from existing colonies to be colonised, or have been created too infrequently to prevent local extinctions within the wood.

Coppice has gradually been less widely practised in the wood over the latter half of the twentieth century due to the decline in its economic value. The future economic viability of coppicing remains a concern. Further changes to the timber market in recent years, have meant that instead of making a profit from coppicing, there is now an economic cost to conservation bodies and other landowners to remove coppice.

## 1.3     The management priority

The national importance of the Blean Woods complex for Heath Fritillaries was discovered by survey work in 1980 (Warren *et al.*, 1981). This survey and subsequent research was highly successful in highlighting the rarity of the butterfly, the national importance of Blean Woods, and the high priority for local landowners and managers to cater for the needs of the butterfly in management.

At Thornden Woods/West Blean there was initially slow progress in managing the woodland to help conserve the Heath Fritillary, due to initial resistance by the site managers to SSSI notification and a delay in notification because of lack of resources within the Nature Conservancy Council. Following SSSI notification in 1989, a management agreement was drawn up between the NCC and the private landowners, with positive woodland management to maintain and where possible expand the Heath Fritillary being the "key management objective" (English Nature, unpublished). An

*Photograph by Tom Brereton*

*Figure 25-1.* Typical Heath Fritillary habitat in good condition: a second summer's growth coppice coup containing abundant Common Cow-wheat.

active programme of targeted coppicing (in special management areas) and ride management, including high forest thinning, was devised.

Many of the specialist species (e.g. Turtle Dove, Nightingale) at Blean Woods also depend on coppicing for survival. Fine-scale management prescriptions have recently been recommended for Dormice conservation that highlight a potential management conflict with the Heath Fritillary (Bright & Morris, 1989). However, this is only likely to be problem on small, isolated sites, and in a large woodland complex such as at Blean where both species have co-existed over long time periods, local site managers do not perceive a management conflict or a need to manage at a finer scale than traditional coppicing for Dormice and the Heath Fritillary (M. Walter, RSPB, pers. comm.).

## 2.      SURVEY AND RESEARCH

Detailed research was carried out during the 1980s to determine the status and ecology of the Heath Fritillary in the Blean Woods complex (Warren 1985b, 1987 a-c). The research identified that the species requires Common Cow-wheat-rich coppice clearings, aged 2-9 years after cutting. Mobility studies confirmed that in woodland the butterfly remained in, or close to, discrete colonies in suitable clearings, with adjacent mature coppice and trees forming a barrier to dispersal.

The research solved the crucial problem of how to manage woodland habitats to benefit the butterfly, stressing the vital need to create an annual supply of coppice clearings near to existing colonies and connected by rides to ensure a high probability of colonisation. It also drew attention to the need for close monitoring, as a) the species status can rapidly decline, and b) it was necessary to determine the effects of the recommended management prescriptions, which had only been trialed for a few years.

## 3. SURVEILLANCE

Periodic surveillance of the Heath Fritillary at Thornden/West Blean Woods was carried in four years during the 1980s and in 1992 (Warren, 1989: Brereton *et al.*, 1998). In each period three key condition indicators of the butterfly population – the number of colonies, the population size of each colony, and the extent of breeding habitat in each colony. The surveys highlighted an alarming decline from 17 colonies in 1980, to 12 colonies in 1984, eight in 1989 and just two by 1992. It was a source of particular concern that the butterfly was close to extinction in 1992, despite SSSI notification, and the production of a detailed management plan aimed specifically at conserving the butterfly and the continuation of small-scale coppicing over parts of the wood. The main reason for this decline was the reduction and fragmentation of traditional coppicing activity.

## 4. THE CONSERVATION AIMS

Conserving the nationally important Heath Fritillary population was identified as a key objective in the Thornden/West Blean Woods SSSI Management agreement. In 1994, the UK BAP was produced which gave a clear target to restore the UK Heath Fritillary to its 1980 status by 2005, giving renewed impetus of the importance of extensive positive management for the butterfly at Thornden/West Blean Woods. During the mid-late 1990s an extensive programme of conservation management work was carried out in the wood specifically to benefit the Heath Fritillary, enhanced through grant funding from two sources (English Nature Management Agreements and the Forestry Commission's Coppice for Butterflies Challenge Scheme). Although the level of coppicing remained similar to levels in the late 1980s and early 1990s, the key difference during the mid and late-1990s was that the coppicing was targeted adjacent to existing Heath Fritillary colonies and there was an associated extensive programme of wide ride management, including thinning of High Forest areas rich in Common Cow-wheat (Davis & Warren, 1999). The ride management was carried out to improve linkage between coppice clearings. The momentum for this work was due in no small part to the work of Steve Davis, of English Nature's Kent Team.

# 4.1    The condition indicators

*Table 25-1.* The condition indicator table for the Heath Fritillary at Thornden/West Blean Woods.

| Condition indicators | Targets | |
|---|---|---|
| | Lower | Upper |
| **Species** | **Limit** | **Limit** |
| Number of colonies | 13 | 20 |
| Number of 1 km squares occupied | 8 | 12 |
| Total extent of breeding habitat occupied (ha) | 5 | 20 |
| Relative abundance index (number per hour x flight area at peak period) | 400 | 10,000 |
| Total population (number on peak day x 3) | 1200 | 30,000 |
| Number of new colonies | 2 | 6 |
| Number of large colonies* | 1 | 10 |
| **Habitat**** | | |
| Number of coppice clearings cut previous winter | 3 | 6 |
| Number of rides or clearings aged 2-4 years since cutting | 10 | 18 |
| Number of coppice clearings aged 2-4 years since cutting containing Common *Cow-wheat Melampyrum pratense* (rank abundance 3-5***) located adjacent to existing Heath Fritillary colonies | 10 | 18 |

\* Colony sizes are: Large = >200 adults during the peak flight period;
\*\* Condition indicator, when butterfly monitoring has not been carried out.
\*\*\* *M. pratense* Rank Abundance Scale: 0 Absent, 1 Rare - a few spikes only, 2 Scarce - a few patches present, 3 Frequent - patches always in view, 4 Common - ground cover more than 10%, 5 Abundant - ground cover more than 40%

# 4.2    Explanation of the condition indicators

The condition indicators for the Heath Fritillary at Thornden/West Blean Woods (Table 25-1) have been developed from the baseline survey and research work conducted during the 1980s and early 1990s. The indicators provide four key measures of condition:

- Population status (number of colonies);
- Population size (relative abundance index);
- Geographic spread/range (measured at 1 km square resolution); and
- Extent of breeding habitat occupied (measured in hectares).

The main condition indicator is the number of colonies present in any one year, as this directly measures progress in the UK BAP-related target of restoring populations to the 1980 level by 2005. Applying the UK BAP target to Thornden/West Blean Woods would give a target of 17 colonies by 2005. This target is considered unrealistic given that coniferisation produced a temporary surge in suitable habitat in the late 1970s/early 1980s. Consequently, Butterfly Conservation (the UK Government's 'lead partner' for BAP implementation) has set a more realistic site-specific target of at least 13 colonies: this excludes those 1980 colonies that occurred exclusively in conifer clearings. An upper limit of 20 colonies per annum has also been set, as creating too many suitable habitats in a single year may lead to lack of suitable areas that are available to be cut in subsequent years.

Though the number of colonies present in each year is the chief indicator, the other attributes listed in Table 25-1 are also important measures of condition. They provide a safeguard to ensure that achieving favourable condition requires more than conserving the butterfly in a series of small colonies (in terms of both the number of individuals and area occupied) in a geographically restricted part of the wood. This is particularly important, as previous studies have demonstrated the importance of maintaining extensive networks of butterfly populations ('metapopulations') to long-term persistence. A further condition indicator included is the number of new habitat patches colonised each year, to provide ongoing evidence that natural extinctions are being balanced by colonisations. A lack of clearings in any one-year would provide an early warning that management was failing to create suitable habitat in the right location. Surveys of the adult butterfly are recommended each year, but if this is not possible, the number of coppice clearings aged 2-4 years since cutting, containing Common Cow-wheat and located adjacent to existing Heath Fritillary colonies can be used as a surrogate condition indicator.

## 5. THE MONITORING PHASE

In 1996, Butterfly Conservation, in conjunction with English Nature and Tilhill Forestry (the site managers), formulated an annual monitoring programme for the Heath Fritillary in Thornden/West Blean Woods. The programme had three main aims: (1) to determine the effects of positive management for the Heath Fritillary at Thornden/West Blean Woods; (2) to aid future targeting of annual management (coppicing/rides) to benefit the butterfly; and (3) to assess conservation progress towards the site-specific (BAP-related) target of 13 colonies by 2005.

### 5.1 Description of monitoring method

Monitoring of adult butterflies was carried out annually from 1996-2003 by professional staff from Butterfly Conservation, with help from local volunteers in some years. The surveys were completed in late June or early July, with the aim being to visit during the week of peak numbers. This peak varied by two to three weeks between years due to the weather, and local specialists were consulted to ensure the timing of visits was optimal. With good weather, two experienced surveyors could complete the monitoring in less than two days.

In each year, all known colonies, rides and woodland clearings that had been cut over the previous seven years were assessed for their suitability for Heath Fritillaries. This involved recording the size of clearing, and the extent of Common Cow-wheat growing amongst bare ground/short vegetation in an unshaded position. Tilhill forestry provided information on the location of cut compartments.

If Heath Fritillaries were present in the habitat patch, a timed count (Warren *et al.*, 1984) was completed to determine colony population size and spatial area occupied. Timed counts are the preferred method for rapid monitoring of rare species, especially those that have temporally and spatially dynamic distributions in extensive habitats. In a timed count, the extent of the flight area (where the butterflies are flying) is determined by a casual walk over the whole habitat patch. The number of butterflies per minute of search effort is then recorded along a systematic W-shaped (zigzag) walk of the whole of the flight area. The walk passes through both low and high butterfly density areas so that adult abundance is not over or under-estimated.

The results are used to obtain an estimate of adult density (expressed as the number of adults seen per hour x flight area in ha). The adult density is then standardised to generate a predicted relative abundance index at the peak flight period (estimated peak number per hour) by calibration with local transect data. This index can be used to directly compare sites and years. The Relative Abundance Index can be converted to an absolute measure of population size (Peak Population) using a formula derived from timed count and mark recapture calibration studies (Warren *et al.*, 1984). The Peak Population is calculated as 2 x Relative Abundance Index + 4.8. Given potential inaccuracies, the Peak Population estimate is better used to classify the population into one of three categories: large, medium or small (Warren *et al.*, 1984). The Total Population (total number emerging) can be roughly calculated as three times this peak number. After the field monitoring, the following were produced annually:

- A map showing the location and extent of colonies;
- A table of the number, area (ha) and abundance index of each colony; and
- An assessment of the overall condition of the Heath Fritillary population in the wood.

The condition of the Heath Fritillary population was assessed by Butterfly Conservation following the generic condition categories developed by English Nature and JNCC to assess the condition of SSSIs (JNCC, 1998): *Favourable*; *Unfavourable – recovering*; *Unfavourable - no change*; *Unfavourable – declining*; *Partially destroyed*; *Destroyed*.

The map and the table were sent to English Nature and Tilhill Forestry as feedback on conservation progress and as an aid to targeting Heath Fritillary management over the following winter.

## 6.      RESULTS AND ANALYSIS

### 6.1      Changing status and condition assessment

The fortunes of the Heath Fritillary population in Thornden/West Blean Woods improved considerably during the mid/late 1990s, coinciding with the most extensive period of targeted positive conservation management work. From just two colonies in two 1 km squares in 1993, there were 14 colonies present, located in ten 1 km squares by

2000. The total population was nearly 8000, with the total colony area extending over nearly 8 hectares. The Heath Fritillary population was classed as *Unfavourable – recovering* from 1996-1998, but achieved *Favourable* condition status (Table 25-2, Figure 25-2) by 1999, and this status was maintained in 2000 and 2001. In 2002, the condition was assessed as *Unfavourable - declining*, due to the lack of at least one large colony, whilst further evidence of a deterioration included a reduction in (1) the number of new colonies, (2) the total population (by nearly four-fifths) and (3) the extent of habitat occupied. In 2003 and 2004, there was a progressive deterioration in the condition of the Heath Fritillary population (failing on two and three attributes respectively), and condition was similarly classed as *Unfavourable - declining*. By 2004, in distribution terms, the butterfly was at its lowest ebb in the wood since the start of annual monitoring, with the 2005 BAP-related target looking unlikely to be achieved.

*Table 25-2.* Annual monitoring results 1996-2004 and overall condition assessments.

| Condition indicators | 1996 | 1997 | 1998 | 1999 | 2000 | 2001 | 2002 | 2003 | 2004 |
|---|---|---|---|---|---|---|---|---|---|
| Number of colonies | 6 | 7 | 10 | 14 | 14 | 13 | 14 | 11 | 8 |
| Number of 1 km squares occupied | 6 | 7 | 9 | 10 | 9 | 9 | 9 | 8 | 6 |
| Extent of breeding habitat (ha)* | 2.07 | 3.62 | 5.6 | 7.2 | 7.4 | 7.74 | 6.22 | 3.93 | 4.4 |
| Relative Abundance Index | 33 | 420 | 436 | 1223 | 1250 | 996 | 223 | 334 | 274 |
| Total Population (emergence) | 99 | 1260 | 1308 | 3669 | 3750 | 2988 | 669 | 1002 | 822 |
| Number of new colonies | ? | 1 | 3 | 5 | 4 | 4 | 3 | 1 | 1 |
| Number of large colonies | 0 | 2 | 2 | 2 | 4 | 3 | 0 | 0 | 1 |
| **Condition assessment**** | **U-r** | **U-r** | **U-r** | **F** | **F** | **F** | **U-d** | **U-d** | **U-d** |

Values in red font highlight a failure in a condition indicator
* Occupied by Heath Fritillary
**Condition assessment categories: U-r = *Unfavourable – recovering*; F = *Favourable*; U-d = *Unfavourable – declining*

## 6.2    Habitat and colony persistence

Between 1996 and 2004 a total of 27 discrete colonies were established, though the actual number of habitat patches colonised was greater than this (n = 36), as a number of the colonies spread into adjacent new clearings between years. Recently cleared coppice (aged 2-4 years since clearance) was the main habitat (52% of colonies) utilised (Table 25-3), followed by thinned Oak high forest (21%) and rides (7%).

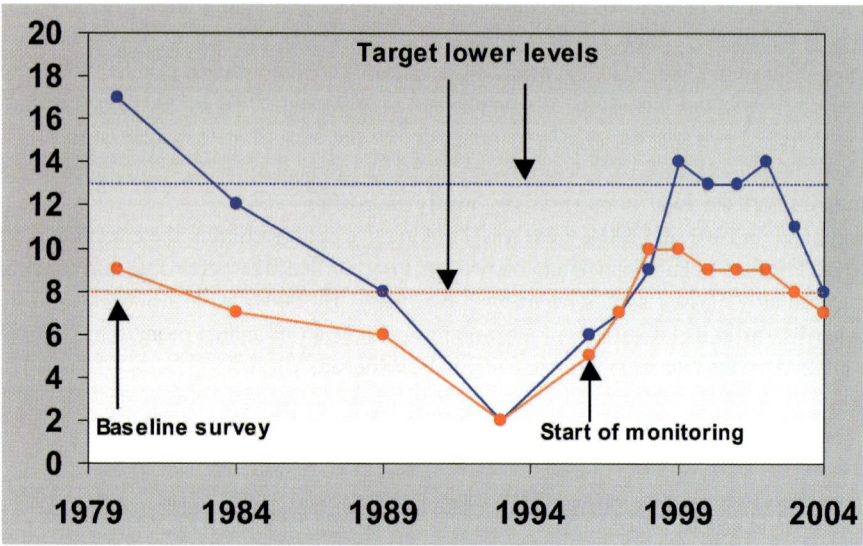

*Figure 25-2.* Changing status of the Heath Fritillary at Thornden/West Blean Woods in relation to condition indicator targets. Blue points refer to the number of colonies in a given year; red points are the number of 1 km squares occupied (a measure of range). Dashed lines are the lower target levels for each condition indicator.

Coppicing was the main management activity (45% of colonies, n = 18) that created suitable habitat conditions for new colonies to establish. Wide ride/thinning also played an important role in creating suitable habitat, with 40% of colonies (n = 15) forming on clearings created in this way, though this management mainly created suitable non-ride habitat (mini coppice plots/thinned high forest) rather than grassy woodland rides. Colonies also formed on land created by clearfelling of conifers (n = 3) and pylon lines (n = 1).

Between 1994 and 2003, 35 coppice coups were cut of which 17 (nearly 50%) were colonised by the Heath Fritillary. In addition, four smaller coppice sections were created as part of the wide ride/thinning programme. Coppicing was carried out in nine 1 km squares, with Heath Fritillaries occurring in eight of these. Of 8.5 km of wide ride/thinning work carried out between 1994 and 2003 in eleven 1 km squares, 3.13 km (37%) was colonised by Heath Fritillaries, across eight 10 km squares.

The majority of butterfly colonies persisted for 3-5 years, with the available data indicating that on average, colonies were shorter-lived in coppice than other habitats. Colony persistence was more variable in ride habitats, ranging from one to at least nine years.

*Table 25- 3.* Number of colonies by habitat and longevity.

| Habitat | Number of colonies | Colony longevity (years) | |
|---|---|---|---|
| | | Mean | Range |
| Pylon lines | 1 | n/a | 9+ |
| Coppice | 22 | 4 | 2-6 |
| Thinned High Forest | 9 | 4.2* | 1-9+ |
| Rides | 7 | 6* | 1-9+ |
| Clearfell | 3 | 4.7* | 3-6+ |

* Likely to be an underestimate, due to the short time period of monitoring

## 6.3     Impact of targeting of coppicing and ride management

The improvement in fortunes in the Heath Fritillary population in the late 1990s could not be explained solely by the increase in coppicing because:
- There was no increase in the level of coppicing (number of plots) over the 1990s and early 2000s (correlation coefficient of no. of coppice plots with year; 1990-1999 $r=-0.1$, $n=10$, $p=0.78$ *n.s.*).
- Nor was there a change in the level of coppicing before or during the monitoring period (mean no. plots per annum 1990-1995=4.2, 1996-2003=2.89, $t=0.88$, $p=0.42$, *n.s.*).
- Also, there was no direct correlation between the level of coppicing and the number of Heath Fritillary colonies (correlation coefficient of the annual number of clearings aged 2-4 since cutting and the number of Heath Fritillary 1993-2003 $r=-0.48$, $n=9$ years, $p=0.18$, n.s.).

However, the late 1990s increase in Heath Fritillaries coincided with improved targeting of coppicing close to existing colonies, which increased the level of occupancy of coppice coups. Between 1990 and 1992, 18% (2 of 11) of cut coppice coups were colonised by Heath Fritillaries, compared to 75% (9 of 12) between 1998 and 2000, with the change being statistically significant (Yates corrected $\chi^2=7.45$, d.f. 1, p<0.05). In 1996 and 1997 (prior to targeting), there were no Heath Fritillary colonies located in commercially managed coppice coups, even though there were 18 scattered through the wood which had been cut over the previous four years.

# 7.    DISCUSSION

## 7.1    Monitoring strategy

The monitoring programme devised proved highly successful in assessing the changing condition of the Heath Fritillary population at Thornden/West Blean Woods in relation to management. Three distinct phases in the condition at the wood were readily identifiable, each related directly to the level of targeted annual habitat management work carried out. From 1996 to 1998, the condition of the population improved, as extensive wide ride management work was carried out. From 1999 to 2001, the Heath Fritillary achieved favourable condition status, as the programme of ride management and thinning continued and coppicing was targeted successfully adjacent to current colonies. However, from 2002 the momentum to conserve the Heath Fritillary faltered, as less ride management, thinning and coppicing was carried out and correspondingly the condition of the population progressively deteriorated. Alarmingly, by 2004, the condition of the population was little better than what it had been at the start of monitoring eight years previously, despite a huge amount of good work in the interim. However it should be borne in mind that without the regular management over this period the butterfly would almost certainly have become extinct in the area.

The study has demonstrated the key role that monitoring data can play to aid the successful targeting of habitat management work. The change in occupancy of coppice from 18% in the early 1990s (without targeting) to 75% by the late 1990s (by targeting using monitoring data) is testimony to this. The study has also  highlighted the importance

*Photograph by Tom Brereton*

*Figure 25-3*. Heath Fritillary butterflies at Thornden and West Blean Woods.

of monitoring the species as well as the habitat, because Heath Fritillaries (due to their low dispersal powers) do not occupy every clearing in suitable habitat condition. Up to date information on the location of colonies proved essential to the success of the management targeting. Finally, the study has highlighted the importance of annual monitoring to help keep managers focused on the conservation priority, and to avoid complacency as the condition status can soon change. A management success can (as happened at Thornden/West Blean) quickly be reversed.

## 7.2    Management implications

The monitoring study has reinforced the message that successful conservation of the Heath Fritillary in large woodland complex requires a combination of:
- Coppicing, to create the preferred habitat; and
- Ride management and thinning, to create additional suitable habitat and facilitate dispersal and colonisation.

The success of the targeted management has highlighted that it is not just the amount of habitat available but the location, suitability and degree of linkage of habitat that determines overall condition.

Over the monitoring period, colony persistence was short (mostly 3-5 years) emphasising the importance of creating suitable new habitat every year to balance extinctions with colonisations. There was some evidence to indicate that colonies were becoming shorter-lived in coppice, with colony longevity over the study 2-6 years, compared to 3-9 years described by Warren in the 1980s (Warren, 1985b, 1987b, 1991). This shortening of colony lifespan has not been documented at the west end of the Blean complex (M. Walter, RSPB pers. comm.), but if real could possibly be a consequence of climate change, with a warming climate leading to a more rapid vegetation re-growth following cutting. The management implication is that more clearings may need to be created than previously prescribed, though it also follows that coppice rotations may be shortened if grow-back is quicker. More research is needed in this area to clarify the situation.

The recent monitoring results show that management needs to be stepped up once more to achieve favourable condition for this crucially important woodland for Heath Fritillaries. There is considerable hope for the future as a conservation body, the Kent Wildlife Trust, has recently purchased the woodland.

# 8.    REFERENCES

Anon. (1995). Biodiversity: The UK Steering Group Report. Volume 2: Action Plans. HMSO, London.

Brereton, T.M., Warren, M.S. and Roberts, R.E. (1998).  Action for the Heath Fritillary: status, monitoring and conservation progress 1996 & 1997. Butterfly Conservation, Wareham.

Bright, P. & Morris, P. (1989). A Practical Guide to Dormouse Conservation. The Mammal Society, London.

Davis, S. & Warren, M. S. (1998). The heath fritillary butterfly (Mellicta athalia): a review of recent population changes in relation to management in the Blean Woods, Kent. Unpublished paper from the conference, Ecology and History of the Blean. Canterbury, Kent.

Holmes, W. and Wheaten, A., (2002). The Blean: the woodlands of a cathedral city. White Horse Press, Canterbury.

Joint Nature Conservation Committee. (1998). A Statement on Common Standards Monitoring. JNCC, Peterborough.

Warren, M.S., Thomas, C.D. and Thomas, J.A. (1981). The Heath Fritillary.  Survey and Conservation Report.  Report for the Joint Committee for the Conservation of British Insects.

Warren, M.S., Thomas, C.D. and Thomas, J.A. (1984). The Status of the Heath Fritillary Butterfly Mellicta athalia Rott. in Britain.  Biological Conservation 29 pp. 287-305.

Warren, M.S. (1985b). The Status of the Heath Fritillary Butterfly Mellicta athalia Rott. In relation to changing woodland management in the Blean Woods, Kent. Quarterly Journal of Forestry, **79**, 174-182.

Warren, M.S. (1985b).  The Ecology and Conservation of the Heath Fritillary (Mellicta athalia). Unpublished Report - Nature Conservancy Council.

Warren, M.S. (1987a). The ecology and conservation of the heath fritillary butterfly Mellicta athalia. I. Host selection and phenology. Journal of Applied Ecology, **24**, 467-82.

Warren, M.S. (1987b).  The ecology and conservation of the heath fritillary butterfly Mellicta athalia. II. Adult population structure and mobility. Journal of Applied Ecology, **24**,    483-98.

Warren, M.S. (1987c).  The ecology and conservation of the heath fritillary butterfly Mellicta athalia. III. Population dynamics and the effect of habitat management.   Journal of Applied Ecology, **24**, 499-513.

Warren, M.S. (1989). National Heath Fritillary Survey and Progress Report, 1989.  Unpublished Report to WWF (World Wide Fund for Nature).

Wigglesworth, T. Bourn, N. Brereton T. and Bulman C. (2004). Status of the Heath Fritillary in the Blean Woodlands 2004. Butterfly Conservation, Wareham.

# PART V

## WOODLAND MONITORING

# CHAPTER 26

# WOODLAND MANAGEMENT

## A historical overview

KEITH KIRBY

*English Nature, Northminster House, Peterborough, PE1 1UA, UK*
*Keith.Kirby@english-nature.org.uk*

## 1.   INTRODUCTION

Consider a fairly typical wood in southern England. Most of the trees in it will probably be less than 50 years old, and almost all less than 100 years; a consequence of the major fellings that took place across the country in the 1914-18 and 1939-45 wars. Rather than single-stemmed trees, there are multiple stems growing up from a single stump (stool), a legacy of hundreds of years of coppice cutting. We walk across an earth bank marking the boundary between the old coppice wood and land that had formerly been common grazing. The latter has grown up as dense woodland in the last 50 years since it became uneconomic to keep cattle on it, but amongst the young growth are some huge old trees that used to stand out in the open. They were formerly pollarded and are full of dead wood. The Bluebells *Hyacinthoides non-scripta* are less common in this section of the wood, reflecting in part the differences between ancient woods – sites continuously wooded since medieval times – and land that has grown up from grassland only in the last few hundred years.

This example illustrates how the current composition and structure of a wood is the product of the interaction between the site's environment (soil and climate), its long-term land-use history and its management (or lack of it) over the last century. The choices we make about woodland management must take account of the past.

*C. Hurford & M. Schneider (eds.), Monitoring Nature Conservation in Cultural Habitats*, 287–292.
© 2007 *Springer.*

## 2.    THE ORIGINS OF WOODLAND MANAGEMENT

Not all woods across Europe show the effects of management quite so sharply, but in only a few remote areas are there forests that are totally free from human impact. There is active debate about what 'natural woodland' in the temperate and boreal zone would look have looked like – was it a predominantly dark, closed canopy cover across the landscape, or much more like a parkland or savannah kept open by large herbivores? Whatever the outcome of this debate, over most of Europe the wildwood has long gone; what we now value for nature conservation are the woods that have evolved under several thousand years of land management.

Early peoples modified the natural forests in a variety of direct and indirect ways: through clearance of trees to create open farmland for cultivation or for their grazing animals; through reducing the populations of large herbivores such as Aurochs *Bos Taurus primigenius*, Wild Horse *Equus caballus* and Bison *Bison bonasus*; through fire, particularly in coniferous and Mediterranean forests; and through active management of trees and woods.

The size and shape of poles found in Neolithic trackways and in Bronze Age archaeological excavations suggest that, at least locally, woods were already being managed in prehistoric times. By the time of the Roman Empire woodland management must have been well organised to support the major demands for fuel for industries such as iron-working, glass-making, potteries and of course bath-houses!

By the medieval period our picture of woodland management becomes clearer because of written records. These must be interpreted with some caution, because they were often made by lawyers or others with a particular point to make. Nevertheless they illustrate:

- Concerns about woodland ownership, implying that woods were a valued commodity;
- The trading of a wide range of woodland products, sometimes over large distances, suggesting differences in the amounts and quality of wood being produced;
- Distinction between different land-use types, for example between meadows and forest, and between coppices and wood-pastures;
- An appreciation of good and bad management practices, as indicated by disputes about the breaking of fences and illegal grazing of areas.

Many woods were managed as coppices: the trees were cut, several stems re-grew from the stump, and when these were big enough to be useful they were cut again. The process on some sites has been repeated for centuries. Grazing animals had to be largely excluded from coppices if there was to be good re-growth from the stools. In other sites, however, trees and domestic herbivores might co-exist if the trees were cut above the browse height. The wildlife from the natural forests had to adapt to the changing woodland structures and landscape patterns that these practices created. Some may have increased in abundance, for example birds that thrived in the dense coppice regrowth; some decreased, for example species with a dependence on dead wood, which was much reduced because it was a valuable resource; and some went extinct either locally (Beaver *Castor fiber* and Wild Boar *Sus scrofa* in the UK) or across the whole of Europe (Aurochs).

*Figure 26-1.* Failed coppice due to browsing by Muntjac *Muntiacus reevesi*, leading to dominance by Pendulous Sedge *Carex pendula*.

*Figure 26-2.* Successful mixed coppice structure at Ham Street woods.

## 3.    THE DEVELOPMENT OF MODERN MANAGEMENT SYSTEMS

As socio-economic conditions changed, so woodland management evolved to meet the new demands. In the nineteenth century there were major shifts from wood to coal or coke as fuel. Sawmill technologies developed that made it easier to cut large trees into planks for building work, rather than harvesting small trees and using them whole. Oaks being grown to provide timber for the navy were overtaken by the shift to steel-hulled ships. Paper consumption rose dramatically, fuelling increased demand for pulp from conifers.

These trends continued through most of the twentieth century. Land management became more polarised, with farmland managed more intensively for food production and woods for timber production. Marginal lands and mixtures of land-uses such as wood-pastures tended to be squeezed out of the system. Forestry has tended to become more mechanised, to rely more on fast-growing single species stands (often of trees not native to the region), and to involve larger scale harvesting.

The consequences of changing land and forest use for nature conservation have varied across Europe, depending on the nature of the forests, and their extent and importance to a country's economy, but often include the following:

*   Fragmentation and isolation of key woodland habitats and species by a 'hostile' agricultural environment has increased. Even where forests remain dominant in the landscape, intensification of their management may contribute to isolation of, for example, old growth species.
*   Changes to the woodland structure. In former coppice woods the fine-scale mosaic of open space and dense young stands was often lost as management was abandoned. Wood-pastures and their associated veteran trees (important for deadwood beetles and lichens) have been converted either to young plantations or to farmland. Old-growth high forest stands in the boreal zone have been felled and replaced by more uniform, younger, even-aged stands.
*   Tree species composition has become more uniform, with conifers tending to be favoured over broadleaves in most circumstances because of their faster, more uniform growth.

## 4.    CURRENT TRENDS

During the latter part of the twentieth century there have been moves in many countries back towards forestry practices that are more sympathetic to nature conservation. The importance of multiple objectives for sustainable forestry is recognised in national and regional policies, and in the development of independent certification schemes for woodland management such as those under the umbrella of the Forest Stewardship Council. Groups such as the Continuous Cover Group in Britain or Pro-Sylva on the continent encourage 'close to nature' approaches to management that do not involve clear-cutting or other major management interventions. There are some signs that

*Photograph by Keith Kirby*

Figure 26-3. Browse-line and grass dominance following browsing by Fallow Deer *Dama dama.*

recent changes to the Common Agricultural Policy in the European Union may improve the integration between forestry, nature conservation and agriculture.

## 5.    USING THE PAST TO GUIDE NATURE CONSERVATION MANAGEMENT

The whole question of whether we need to or should manage for nature conservation is being debated widely. If we leave woods alone will we produce the sort of rich variety of wildlife that we want? We should certainly explore, in so-called 'strict reserves' or 'minimum intervention areas', what happens when we try.

However most woods have been actively managed in the past and are only a fraction of their former extent. Therefore, we cannot assume that the species and habitats that we value will survive if we let natural processes take their course everywhere. The end product will certainly be different to what may have been in the former wildwood because of the loss of key species, the presence of introduced species and a changing climate.

More often we will seek to manipulate the management of the forest in various ways to favour particular elements of the system, or at the least ensure their survival. In doing so we must identify how they were related to past and current management, bearing in mind that the key management practices may be those operating 10, 50 or even 100 years ago. What we encourage in one patch of woodland or forest should take account of how other similar areas are being treated within a region. If the trend is towards woods becoming more shaded as coppice cutting is abandoned then ensuring that open space is maintained in some woods becomes a priority; if most woods are being actively cut then a higher priority may be to ensure that some areas are left alone to accumulate deadwood.

We cannot avoid making management decisions – even doing nothing is a positive decision to allow the wood to change in response to the effects of past interventions. Just as past management has shaped how the woods are now, so our management will determine what our successors inherit. We must therefore take what we value from the past and try to ensure it survives into the future.

# 6.     REFERENCES

Kirby, K. J. & Watkins, C. (eds). 1998. The ecological history of European forests. CABI, Wallingford.

Peterken, G. F. 1993. Woodland conservation and management (2nd edition). Chapman and Hall, London.

Peterken, G. F. 1996. Natural woodland. Cambridge University Press, Cambridge.

Rackham, O. 2003. Ancient woodland (revised edition). Castlepoint Press, Dalbeattie.

# CHAPTER 27

# ISSUES SPECIFIC TO MONITORING BROAD-LEAVED WOODLAND

CLIVE HURFORD

*Countryside Council for Wales, Plas Penrhos, Ffordd Penrhos, Bangor, Gwynedd, LL57 2BQ*
*clive.hurford@serapias.net*

## 1.    INTRODUCTION

In the first instance, and before engaging in a potentially difficult decision-making process, we should consider whether monitoring is necessary.  If we do not intend to manage the wood, or have not developed a management strategy, then setting up a surveillance project is a more appropriate option.  Many of the difficulties associated with woodland monitoring stem from not having clear management aims.  The following sections make the assumption that we are committing, or intending to commit, resources towards actively managing a wood.

## 2.    DECIDING WHAT TO MONITOR

In many ways, how we approach monitoring in broad-leaved woodland differs little from the approach recommended for other habitats with a history of cultural management, i.e. first we decide what we want the management to achieve, and then we monitor in the appropriate areas to see if we have achieved it.  The major difference between woodlands and other terrestrial habitats is the size and spatial distribution of the dominant plants, which exacerbates the problems associated with recording vegetation cover.   We have to be aware of this as we develop the condition indicators for the wood.   In most woods, the monitoring will focus on the following attributes (adapted from Kirby *et al.*, 2002):

- Area;
- Structure;
- Species composition;
- Regeneration potential;
- Quality indicators; and
- Associated species (fauna and flora).

*C. Hurford & M. Schneider (eds.), Monitoring Nature Conservation in Cultural Habitats*, 293–300.

*Photograph by Clive Hurford*

*Figure 27-1.* Upland Sessile Oak *Quercus petraea* woods in Wales, like this one at Blackmill, typically support high breeding densities of insectivorous passerines such as Pied Flycatchers *Ficedula hypoleuca*, Redstarts *Phoenicurus phoenicurus* and Wood Warblers *Phylloscopus sibilatrix*. This is not sustainable in the long-term, however, as prolonged periods of intensive sheep grazing have resulted in relatively uniform stands of old trees. As these trees die there will be a net loss of breeding sites for the birds. Note the lack of an understorey and regeneration.

Monitoring the overall extent of a wood is not usually a problem, as this will be visible on remote images. Therefore, the difficult issues for monitoring broad-leaved woodland are associated more with assessing the quality of the habitat.

## 3.      INCREASING THE RELIABILITY OF THE MONITORING PROJECT

The most reliable way to monitor many woodland attributes is to use measures of abundance, either within a monitoring plot or at sampling points throughout the wood. The structural complexity of woods and the size of the trees can make it difficult for recorders to get the sort of overview that is possible in grassland, which affects their ability to make estimates of cover. This problem can often be overcome, however, by using alternative measures.

For example, in assessing the structure of a wood, we may well be interested in monitoring gaps in the canopy – this, as noted above, is difficult. However the gaps are usually created by fallen trees, and simply counting the number of fallen trees in the monitoring area is likely to achieve a far more consistent result than trying to estimate the area of gaps in the canopy. Furthermore, information on the number of fallen trees may go some way towards answering the question of whether there is enough deadwood present. Estimating the canopy cover of non-native trees is also prone to observer error. However, if we set a upper limit for non-native canopy-forming trees, expressed as, for example, no more one in ten canopy-forming trees should be non-native', we will get a reliable monitoring result from checking the ratio of native : non-native trees at our monitoring points. Similarly, rather than estimating the cover of understorey trees we could set a lower limit for the number of understorey trees in our monitoring area. This too will deliver a consistent result. After all, we do not need to know how many there are, only that there are enough.

We can assess regeneration by setting a lower limit for the number of canopy gaps (the locations of fallen trees) with viable native saplings within a 5 m radius. In this case, we could also discriminate between different species to safeguard against future changes in composition through selective browsing. Associated species of conservation value and invasive species could be assessed using presence and absence data or against a lower limit for abundance, as appropriate.

These are all 'measurable' options for monitoring attributes related to woodland structure, species composition, local distinctiveness, regeneration, and invasion by non-native species.

*Photograph by Clive Hurford*

*Figure 27-2.* Clear-felling in the early 1900s has resulted in many young, densely shaded, even-aged stands in Wales. Also, the presence of Beech *Fagus sylvatica* in these woods, which is not a native species in this part of Wales, suggests that some planting has taken place in the past.

# 4.    WHERE TO MONITOR

Remote images, such as aerial photographs, can provide reliable information on the overall extent, and to some degree the composition, of the woodland canopy. However, they cannot tell us anything about species in the understorey, associated species of interest, or problems related to regeneration. Therefore, we can only use remote methods in combination with ground-based data.

In smaller stands of woodland, say 5-10 ha, we could choose to monitor across the whole of the wood, perhaps on a systematic grid. This is not necessarily the most efficient option, but it is one way of monitoring a small wood. The method is similar to that recommended for other terrestrial habitats (Chapter 13), but differs primarily in the size of the monitoring points and how we record at them. In broad-leaved woodland, we would record points at 50 m intervals and, within a 25 m radius of each point, use measures of abundance to monitor the attributes. The case study for the Beech *Fagus sylvatica* forests at Biskopstorp (Chapter 29) was developed along these lines.

## 4.1    Monitoring in selected areas

In larger stands of woodland, we could use a random sampling approach, but this is an inefficient way of detecting what may be relatively rare events (e.g. a tree-fall, or small fell) particularly if we know which parts of the wood are being managed and hence which are most likely to change in a given period. A more efficient way of dealing with large areas of woodland is to adopt a selective approach that uses our knowledge of the site and the management strategy to focus our monitoring effort. Hence our monitoring effort will be influenced by a number of factors, such as the long-term management strategy, the source of perceived threats to the conservation value, and the distribution of rare and threatened associate species.

Deciding where to monitor is aided, in the first instance, by a map of the site (preferably supplemented by a recent aerial photograph). By outlining the areas targeted for management (or which have been subject to some major unplanned change such as windthrow), and incorporating information on the projected timing of the management, we can plan where we need to monitor and when. After carrying out the management, we can either:

- Monitor across the whole of the management unit, using either a systematic or random sampling method; or
- Set up a small number of monitoring plots in the management unit.

If we take the latter option, two or three 50 x 50 m monitoring plots in each managed area should suffice. Often we should already know the answer to some of the questions that we might ask, for example relating to structure and invasive species, as our management is likely to have targeted these attributes. Therefore a fairly simple quick record that the management has gone as expected may be all that is required for these. The more detailed monitoring can then focus more on issues relating, for example, to a) the success and nature of the regeneration or b) the distribution of associated species. We would use measures of abundance to record the attributes throughout the plot, and would

*Photographs by Clive Hurford*

*Figure 27-3.* Clockwise from top left, *Usnea florida, Usnea articulata, Lobaria pulmonaria*, and *Teloschistes flavicans.* Each of these species of lichen is easy to identify and is a 'pure air' indicator that occurs on trees.

assume that our monitoring result reflected the condition of the habitat elsewhere in that management unit.

We need not worry too much about being able to find the plots again, as in future monitoring cycles we would probably want to focus our monitoring effort in other parts of the wood. It can take considerably longer to find plots in woodland than to record them, so unless we want to use the plots to gather surveillance data, why bother? We can use the time that we have saved to record more plots and ensure that we have the precision of result that we need.

### 4.1.1     Taking account of the management strategy

Traditionally, many larger areas of woodland were managed on a rotation basis, ensuring that there could be a timber harvest every year. Inadvertently, this benefited the diversity of the woods, by creating a mosaic of habitat patches that could accommodate species with a variety of different requirements, e.g. disturbed ground, direct sunlight and shaded areas. However, the declining level of management activity in many broad-leaved woodlands since the early 1900s has discriminated against species that require light or disturbance, perhaps for the first time in thousands of years. In these woods, our management effort should focus on creating a more varied age structure through selective thinning, opening rides, and the removal of non-native species.

Similarly, our monitoring effort should focus on the stands of woodland that are being managed, as these are the areas where we are looking to make conservation gains, through the creation of open space. We can make some assumptions about the stands not being managed: these are most likely to be relatively young, even-aged stands, densely shaded and with low levels of disturbance.

The length of time it takes different stands to respond to management must also be taken into account. In woods managed as high forest, we should not monitor for regeneration within 3-4 years of carrying out the management, because viable saplings will not have had an opportunity to develop. The monitoring could therefore give a misleading impression of the condition of the wood. By contrast, coppiced stands should be monitored within a year or two of coppicing, as signs of intensive browsing activity will be evident by the end of the first growth period.

*Photograph by Clive Hurford*

*Figure 27-4.* Ancient oak woods, like this one at Ty Canol, are rare in Wales: few lowland woods escaped clearfelling in the early 1900s. This wood is known for its exceptional lower plant flora.

At the other extreme, changes in the levels of deadwood may take decades to develop through natural processes; therefore we may well have to create it. However, think carefully before targeting mature stands of broad-leaved woodland, or even mature trees, for management. These are relatively scarce, and the restoration of species associated with older trees and deadwood could well depend on them persisting in these areas until the younger stands in close proximity begin to mature.

## 4.1.2    Taking account of perceived threats

If we are concerned about invasion by non-native species, we should consider which areas of the wood are under most threat. Typically, it will be those in the vicinity of areas either a) already infested or b) adjacent to non-native or infested stands outside the site. We can target these areas for monitoring, on the basis that any future invasion is most likely to happen here. If there is no evidence of invasion in these parts of the wood, we can make some assumptions about those further away from the source.

Other threats may be similarly concentrated: enrichment from fertiliser overspread may be greater at the edges of a wood. A few threats may be more widespread, for example the effects of air pollution on lichens, but even with these it may be possible to identify the most sensitive locations in the wood (Luigi Nimis *et al.*, 2002; Richardson, 1992).

## 4.1.3    Monitoring associated species

Ultimately, the conservation value of a wood will be determined by the species that are, or should be, associated with it, whether they are mammals, birds, butterflies, higher plants, bryophytes or lichens. If these species are not present, then we should not consider the wood to be in optimal condition.

From a monitoring perspective, if the species of interest are distributed throughout the wood, then we should detect them wherever we decide to monitor. If the species have special requirements, such as lichens that are restricted to older trees, or plants that require disturbance, or butterflies that require sunlight, we can monitor in the areas where the species are most likely to occur. In most cases, we will know where these areas are without needing additional survey information. However the species may not always be present at the time of the visit – butterflies may not fly if the temperature is too low, birds may not be singing in the middle of the day – or identifying the species may require specialist knowledge. In such situations a two-stage monitoring approach may be helpful: a check that the relevant habitat conditions exist and, on a longer cycle, a survey of the actual species abundances.

# 5.    IN SUMMARY

If we know, from monitoring the key attributes in the critical period after management that a) our management strategy is achieving its aims, b) the wood is not being invaded by non-native species, and c) the rare and threatened species are still present, we can probably consider the wood to be in optimal condition.


# 6.    REFERENCES

Kirby, K., Latham, J., Holl, K., Bryce, J., Corbett, P. & Watson, R. (2002). Objective setting and condition monitoring within woodland Sites of Special Scientific Interest. English Nature Research Reports, No. 472. English Nature. Peterborough.

Luigi Nimis, P., Scheidegger, C. & Wolseley, P. (eds.). (2002). Monitoring with Lichens – Monitoring Lichens. Kluwer Academic Press. Dordrecht.

Richardson, D.H.S. (1992). Pollution monitoring with lichens. Naturalists' Handbook 19. Richmond Publishing Co. Ltd. Slough.

# CHAPTER 28

# MONITORING ISSUES SPECIFIC TO WESTERN TAÏGA FORESTS

ANDERS HAGLUND

*Ekologigruppen AB, Dalslandsgatan, 7 EP, SE-118 58, Stockholm, Sweden*
*anders.haglund@ekologigruppen.se*

## 1.     INTRODUCTION

This chapter focuses on the Western Taïga (Habitat 9010), with particular emphasis on the situation in Sweden. The Western Taïga, a priority Natura 2000 habitat, accounts for the vast majority of coniferous forest in the Boreal region and is widespread throughout it. In Sweden alone, almost one million hectares of Western Taïga are protected within the Natura 2000 network. It is a complex habitat dominated by old-growth coniferous forests, and takes in several sub-habitats, including secondary forest types such as the young broad-leaved forests that develop after large-scale

disturbances. To monitor the Western Taïga, we must divide the habitat into different ecological subgroups. The following groups are recognised in Sweden:

1. Forests requiring active management, e.g.:
- Natural old pine forests (except those on rock outcrops and lichen-dominated ground);
- Natural old boreal deciduous forests;
- Recently burnt areas and younger successional stages that have developed after fires, such as young deciduous stands.

2. Forests with minimum intervention management, e.g.:
- Natural old spruce forests;
- Natural pine forests on rock outcrops and lichen-dominated ground;
- Natural old wet coniferous forests (swamp forests);
- Natural old mixed forests (which may, in certain cases, require management).

*C. Hurford & M. Schneider (eds.), Monitoring Nature Conservation in Cultural Habitats*, 301–308.
© 2007 *Springer.*

## 2.     WHAT ASPECTS OF THE WESTERN TAÏGA CONCERN US?

The most important ecological attributes of the Western Taïga focus on the occurrence of old trees of different species, deadwood of different sizes, and species composition. The majority of the fauna and flora associated with the forest type are dependent on these structures. We don't need to worry too much about deadwood because, as long as there is no extraction of timber from the forest, we would expect deadwood to be plentiful. The situation with respect to deadwood will be clarified during a baseline survey that precedes the Natura 2000 monitoring programme. Similarly, as long as we know from this baseline survey that we are dealing with an old forest, we can assume that the trees in the forest will grow older in time, so we do not worry about this either.

Therefore our main concerns focus on tree species composition, regeneration, and the populations of demanding and highly specialised associated species.

### 2.1     Tree species composition

In its natural state, the Western Taïga consists of a mosaic of fire refuge areas, such as swamp forests dominated by Spruce *Picea abies*, and burnt areas at different stages of succession, dominated by deciduous forests, Pine *Pinus sylvestris* forests or mixed coniferous forests. Many of the tree species need large-scale disturbances such as fire (Fig.28-5) or large windfalls to aid their regeneration. Neither Pine, one of the key species in the Western Taïga, nor the deciduous trees occurring in the Western Taïga, can in the long-term compete with Spruce in undisturbed forests on richer soils. Our main worry is that, without regular disturbance by fire or selective logging of Spruce, the majority of the forest will in time be dominated by Spruce and consequently lose much of its natural biodiversity.

### 2.2     Regeneration

While the absence of large predators such as Wolf *Canis lupus* and Lynx *Lynx lynx* has, in many parts of Sweden, contributed to a large population of browsing animals, the main reasons for this are related more to hunting regulations and forestry practices. The commoner browser species, i.e. Elk *Alces alces* and Roe Deer *Capreolus capreolus*, have a strong preference for Pine and some broad-leaved tree species, e.g. Goat Willow *Salix caprea*, Rowan *Sorbus aucuparia* and Common Aspen *Populus tremula*. Young specimens of these broad-leaved trees are totally absent over large areas of Sweden. As a result, species that are not favoured by the browsing animals, e.g. Spruce and Birch *Betula* spp., are indirectly favoured on the regeneration sites. This is also a concern for us.

*Photograph by Clive Hurford*

*Figure 28-1.* Parts of the ancient Pine *Pinus sylvestris* forest at Tyresta show no signs of human management activities.

*Photograph by Clive Hurford*

*Figure 28-2.* Old growth Spruce *Picea abies* forest at Tyresta.

## 2.3    Specialised species

Many invertebrates are dependent on fire and live almost exclusively on a constant supply of recently burnt trees. These fire-adapted species will die out if they are not provided with burnt trees on regular basis. As forestry workers extinguish natural fires soon after they start, many of these species are critically endangered.

Many of the mammals and birds of the Western Taïga need large areas of old forest. Capercaillies *Tetrao urogallus*, White-backed Woodpeckers *Dendrocopos leucotos*, and Three-toed Woodpeckers *Picoides tridactylus* need several hundred hectares of Western Taïga to maintain stable populations and can only survive in large reserves. Many of the reserves containing Western Taïga in Sweden are so small, however, that we are not confident that they can support viable populations of these species. Therefore, to ensure effective long-term conservation of these species we will have to take into account the intensively managed Western Taïga in the landscape outside the reserves. Our main concern in this case is that we are not certain that the existing reserves are large enough, or that the quality of the habitat outside the reserves will be maintained to a suitable standard.

In south-western parts of Sweden, many species of lichen that are sensitive to air pollution have decreased dramatically. Since many species of lichens are dependent on the habitat this is a major conservation concern.

Finally, many of the reserves in southern parts of Sweden have been managed intensively in the past and it is not uncommon to find drainage ditches present in wet coniferous forests. A concern is that these draining ditches will degrade the swamp forest habitats.

## 3.    MANAGEMENT OF THE WESTERN TAÏGA

When we refer to active management of the Western Taïga we normally mean the use of fire to produce regeneration or disturbance. This was the natural form of habitat regeneration before humans began logging operations. If we want to preserve the biodiversity of the habitat, logging cannot replace fire as a management tool. However, in many areas, burning represents a hazard for forests or buildings outside the reserves and so we cannot risk using fire. In these areas, we can consider selective logging as a means of creating the desired species composition. Other options may include culling Elk and/or Deer, and filling draining ditches in the swamp forests.

## 3.1    Making difficult decisions

We will need to make well-informed management decisions to preserve the biodiversity of the pine forests and deciduous trees in the Western Taïga. Whether the aim of the management is to preserve species dependent on fires, or to create large coherent patches of Western Taïga, these decisions should always be made on a landscape level. However, before making a management decision we will need to a) know the distribution of the habitat subgroups, b) analyse the history of fires in the area, and c) know the species composition of the Pine-dominated forests and the mixed coniferous forests. This information will be provided by the Natura 2000 baseline survey.

Conservationists have little experience of actively managing the habitat (and fire is a dangerous management tool). Furthermore, there is also a strong resistance to active management in the counties. After all, why should we burn the old forests in the reserves, when we have so few old forests left, especially when the species that depend on burnt forests do not really need old trees? Most fire-dependent species are not so fussy, and do not need very large trees: mature trees will suffice. Therefore, in most reserves, we can avoid setting targets for the existing areas of old growth forest by concentrating on regenerating the parts of the forest that do not score highly on age, amount of deadwood, or the other key attributes.

In the long-term, however, we also have to manage the old-growth forest to maintain the species composition. Therefore, in our larger reserves, 5-10% of the areas with Pine forest or deciduous forest should be burnt during each 10-year period. This strategy of "prescribed burning for nature conservation purposes" (where a certain amount of wood is left on a regeneration area before burning) should be carried out at least every third year on a landscape level. This should be conducted by co-operating with surrounding landowners.

One decision that is very unfamiliar to most people involved with nature conservation is the regulation of the browser population. This will be done in co-operation with local hunters. The problem of regulating browsing is complicated by a conflict of interest, however, as the local hunters do not want smaller prey populations to hunt.

## 4. DECIDING WHAT TO MONITOR

As in all monitoring at the site level, first we decide what we want the management to achieve, and then we monitor in the appropriate areas to see if we have achieved it. The major difference between woodlands and other terrestrial habitats is the size and spatial distribution of the trees. There could be hundreds of metres between old individuals of certain broad-leaved trees. Furthermore, the size of the habitat is almost overwhelming. If we spent one hour a year per hectare monitoring the Western Taïga forest in Sweden, we would use up the whole national nature conservation budget. Therefore the monitoring must be highly focused on the following attributes:

- Extent of the subgroups of Western Taïga, with special focus on those subgroups which require active management, including burning;
- Species composition, with special focus on Pine and broad-leaved species sensitive to browsing;
- Regeneration, with special focus on Pine and broad-leaved species sensitive to browsing;
- Associated species, with special focus on "fire" species and species that need large areas of high quality forest.

In some cases we would also monitor the amount of deadwood

*Photograph by Clive Hurford*

*Figure 28-3.* Managed Western Taïga forest with even aged trees and little deadwood.

## 4.1    Monitoring or surveillance

When monitoring sites, we do not intend to measure the quality and composition of the Spruce-dominated and mixed coniferous forest. On a national level, however, we need information on the status of attributes like deadwood and species composition. Therefore the site monitoring is complemented with a surveillance programme covering a selection of sites, including both intensively managed and unmanaged forests (i.e. nature reserves). We will also use a surveillance programme to track changes in the occurrence of lichens sensitive to air pollution: this programme will cover a selection of sites. The national surveillance programme will also incorporate data from the site monitoring, particularly with respect to the extent of burnt areas in the landscape and the extent of the habitat subgroups in the reserves.

## 4.2    Monitoring methods

Remote images, such as aerial photographs, can provide reliable information on the overall extent of the Western Taïga. In Sweden, we collect remote images every 24th year, using infrared colour film, which makes it easy to separate the deciduous trees from the conifers. These photographs are taken from no higher than 30 000 m to ensure that the images have an acceptable resolution. However, while it is relatively easy to monitor the overall extent of the subgroups dominated by broad-leaved trees on remote images, it is far more difficult to separate the Pine forests from Spruce forests. Similarly, it is very difficult to separate the broad-leaved tree species that are sensitive to browsing from other broad-leaved trees. Furthermore, remote sensing cannot tell us anything about associated species of interest and problems related to regeneration. Therefore, the remote images must, to some extent, be complemented by fieldwork. However, the map produced by the remote sensing exercise will help us to decide where to monitor and provide at least some information on the composition of the woodland canopy. If we outline the subgroups that are targeted for management (or which have been subject to some major unplanned change such as windthrow), and incorporate information on the projected timing of the management, we can plan where we need to monitor and when.

With respect to management interventions such as the filling of draining ditches, we will often already know the answer to some of the monitoring questions. Therefore, simply recording whether the management has been done and whether the dam is secure will often be enough.

Within an appropriate period of carrying out other types of management, we could set up a small number of monitoring points or plots in the management unit. These recording points or plots should be geographically spread over the area, but the exact location can be chosen selectively. We would use measures of abundance to record the attributes throughout the plot, and would assume that our monitoring result reflected the condition of the habitat elsewhere in that management unit.

*Photograph by Clive Hurford*

*Figure 28-4.* An area of some 300 ha of old growth Western Taïga at Reivo in northern Sweden was burnt accidentally when fire spread from a nearby stand in 1966. This area was left to recover naturally, and the effects of the fire were still evident when this photograph was taken in 2002.

Taking into account the large area covered by the habitat, the monitoring of the Western Taïga has to be efficient. The method for measuring species composition and deadwood of forest stands is very simple and quick. The main tool is a relascope, which consists of a piece of plastic with a small gap. The relascope is held at a certain distance from the eye, regulated by a piece of string. On looking through the gap of the relascope, all trees that exceed the size of the gap are counted, giving a figure for the standing volume per hectare of each species. We measure deadwood by recording the length of the fallen or standing dead trees that fall within the gap of the relascope. Depending on the variation between the measuring sites, 4-6 measuring points should suffice for each stand. If deadwood is targeted, the number of points should be at least 6-8.

The regeneration of burnt areas is measured by counting the number of saplings more than 2 m high in circular plots of 100 m$^2$ and then comparing the total with the target value. If the target value is not reached, browsing damage to all trees lower than 4 m will be recorded in the plots. We should not monitor for regeneration within 12-18 years of carrying out the management, because viable saplings of broad-leaved trees will not have had an opportunity to develop and exceed the browsing height of Elk.

We will not worry too much about being able to re-locate the measuring points or plots, as in future monitoring cycles we will probably want to monitor in other parts of the forest. As it can take considerably longer to re-find plots in forest habitats than it does to record them, unless we want to use the plots to gather surveillance data, we should avoid the problems associated with precise replication. As with broad-leaved woodland, we can use the time saved to record more plots and increase the precision of the monitoring result.

## 4.3   Monitoring associated species

Ultimately, the conservation value of a coniferous forest will, as for the deciduous nemoral woods, be determined by the species that are associated with it, whether they are mammals, birds, invertebrates, higher plants, wood-living fungi, bryophytes or lichens. If these species are not present, then we should not consider the forest to be in optimal condition. As for wood-living fungi and bryophytes, we are quite confident that these species will survive as long as they are provided with suitable habitat such as old trees or large woody debris. In the Western Taïga we have chosen to focus the monitoring on birds that depend on large coherent forests, and on fire-dependent invertebrates.

Capercaillies are monitored by counting the number of male individuals at the lek on a five-year rotation. We will count woodpecker species, such as the White-backed Woodpecker and Three-toed Woodpecker, on standardised 10 km routes, as used in the national surveillance programme. If the population of the species is low, standard point and line transects will be used.

We can monitor most fire-dependent beetles by counting the number of trunks that have feeding-traces or hatching-holes of the species. Provided that the monitoring is done three years after burning, it will be relatively easy to find and identify the unique traces of most species. This can be a very quick and qualitatively good method of assessing the populations of key species.

# CHAPTER 29

# BEECH *FAGUS SYLVATICA* FORESTS AT BISKOPSTORP

ÖRJAN FRITZ

*Department of Southern Swedish Forest Research Centre, Swedish University of Agricultural Sciences, Box 49, S-230 53 ALNARP, Sweden*
*orjan.fritz@ess.slu.se*

## 1.  INTRODUCTION

Biskopstorp, situated in the nemoral region of SW Sweden, is an area of some 865 ha that, with the exception of some small lakes and bogs, has almost complete forest cover. Together with the neighbouring nature reserve Vapnö Mosse, a large bog including surrounding forests, it forms an area of almost 1100 ha.  The area constitutes the single most important forested area for biodiversity in the county of Halland, and is one of the most important in southern Sweden. About 30% of Biskopstorp is deciduous broad-leaved forest, with Beech *Fagus sylvatica* dominating the central and eastern parts of the area, and Oaks *Quercus* spp. dominating the western and southern parts.

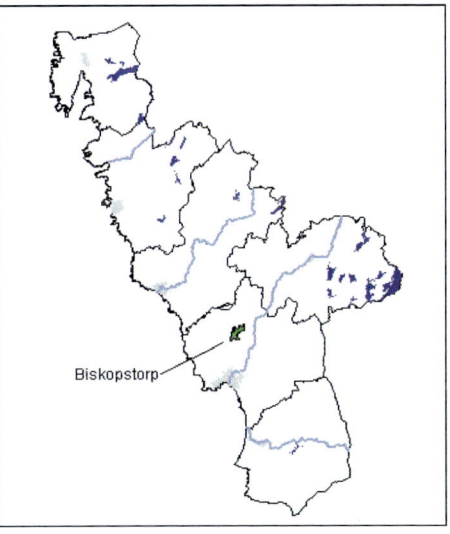

Biskopstorp

Despite a relatively species-poor vascular plant flora and mollusc fauna (primarily because of a soil and bedrock poor in lime/calcareous minerals), the overall biological diversity connected to the old Beech and Oak forests is of international importance in a European context.  At the time of writing, at least 136 species on the official Swedish Red-list (Fritz, 2004: Gärdenfors *et al.*, 2000), and a further 87 species of regional interest have been found at Biskopstorp. The most important groups are epiphytic lichens  (Fritz, 2004) and wood-inhabiting insects (Abenius, 2004: Andersson. 2001: Jansson, 2004) with epiphytic bryophytes (Fritz, 2004) and wood-inhabiting fungi (Heilmann-Clausen, 2005)

*C. Hurford & M. Schneider (eds.), Monitoring Nature Conservation in Cultural Habitats, 309–322.*
© 2007 *Springer.*

also well represented. The high conservation value of the forest can be attributed mostly to:

- A concentration of old Beech and Oak stands on many different substrates:
- Long continuity of broad-leaved deciduous forests;
- Sub-oceanic conditions, with an annual mean precipitation of ca.1200 mm; and
- The broken topography, i.e. rocky hillside areas, with different kinds of exposure in several directions, which results in diverse conditions of light, humidity, atmospheric depositions etc., has made it difficult to fell trees on a large scale.

Pollen and macro fossil analyses (Björkman, 2002: Hannon, 2002: Karlsson, 1996) have shown that the forest has a long continuity, with broad-leaved species dominating over thousands of years. The last 1000 years have seen dramatic changes, however, with increasing human impact. Beech expanded in the early Middle Ages and, during the 18th century, the remnants of forests dominated by Oak, Lime *Tilia cordata* and Hazel *Corylus avellana* appear to have been cut down, probably to increase the extent of Beech forest for pigs. During the 20th century, the more accessible areas were planted with Norway Spruce *Picea abies*, which made up approximately 35% of the forested area in 1996.

*Photograph by Örjan Fritz*

*Figure 29-1.* Beech forest habitat at Biskopstorp.

In the early 1990s, plans were developed to protect the whole area as a nature reserve. After negotiations with landowners, most of the area had been secured by 2004, though the official designation of the area has yet to be finalised. The exact border of the reserve may change somewhat before the official designation.

In the secured areas, a restoration programme has already started. The main object is to restore the entire area to forest dominated by Beech, Oak and other deciduous species. The plan is to convert the Spruce plantations to deciduous forests over a 30-year-period. Bengtsson (1999) and Karlsson (2000) provide a detailed framework for achieving this.

## 1.1    Key factors influencing the condition of the forests at Biskopstorp

There are key four factors that threaten the conservation value of the broad-leaved forests at Biskopstorp:

- Poor continuity of successional phases;
- Invasion by Norway Spruce;
- Changing management practices; and
- Atmospheric depositions of airborne pollutants.

Most species, and especially the red-listed species, at Biskopstorp are associated with old Beech and Oak trees, and can be found on large living trunks, as well as on thinner 'late-grown' trunks and deadwood (snags, logs). The broad-leaved forests are mostly in the old succession phase, in the age span of 200-300 years. However, direct succession from single-layered old forests (created by thinning management) to regenerating stands dominated by primary successors can cause problems for substrate-dependent specialist species.

Invasion by Norway Spruce constitutes a major problem. Norway Spruce is very competitive and acts aggressively in the broad-leaved forests at Biskopstorp. It can convert virtually all types of deciduous forest, including Oak, Aspen *Populus tremula*, Alder *Alnus glutinosa*, Birch and Beech, into Spruce-dominated forests. As the Spruce invades a stand it creates a habitat with a changed microclimate, making it cooler, moister and darker, and causes regeneration problems for the deciduous species. It also has a negative impact on many red-listed species, by shading out those living on broad-leaved trees. The Norway Spruce is not a native species of the area and most Spruce plantations consist of plants with foreign provenances (mainly from southeast Europe). Furthermore the nature conservation value of Spruce in this area is very low: it harbours very few specialised and demanding species in this part of south-west Sweden – in contrast to the situation further north in the country.

The cessation of grazing by cattle has created a lack of disturbance in the Oak areas, where now both Beech and Norway Spruce are expanding. The resulting dark and closed Oak forests will not produce viable Oak offspring.

Atmospheric deposition impacts on both soil and water quality and has negative effects on certain groups of organisms, mainly epiphytes, mainly as a result of the low pH in precipitation. These depositions originate mainly from sources outside Sweden.

## 1.2    The management priority at Biskopstorp

A preliminary management plan exists for the proposed nature reserve. The plan focuses on:
- Restoring the existing deciduous forests by suppressing the Norway Spruce; and
- Converting the Norway Spruce plantations into deciduous stands to create continuous blocks of broad-leaved forest.


## 2.    RESEARCH AND SURVEY

Since the mid-1990s, many research projects and surveys have been carried out to obtain information on the biodiversity of the Biskopstorp area, mainly by students from local universities (at Alnarp and in Lund) and by personnel from the local administration board of Halland. These studies have focused on the woodland key habitats (wkh) and include:
- Pollen and macro fossil analyses,
- Assessing the age of trees in the most valuable broad-leaved stands (Fahlvik, 1999: Niklasson, 2002);
- Assessing Beech mortality (Jacobsson, 2002) and
- Distribution surveys for epiphytic lichens, bryophytes, wood-inhabiting insects and fungi etc.

In addition, there is an ongoing project to establish the detailed management history of the area since the 17th century.

Of particular relevance to this case study, there is also an ongoing project to identify epiphytic mosses and lichens that could be used as indicator species for the red-listed species in the most valuable Beech stands in the Biskopstorp area (Fritz, 2001). This project is also being carried out elsewhere in Halland. In all, 45 localities comprising 132 stands in the county have been visited since 1994, mostly stands of Beech. In each stand, all of the stems, logs and snags have been searched for these indicator species and red-listed species. The results to date suggest that the indicator species work well; the more indicator species you find, the more red-listed species are present. As many of the red-listed species are small and difficult to find, the indicator species can be useful surrogates for assessing the condition of each stand.

# 3.    BACKGROUND SURVEILLANCE

Over the period 2000-2003, a baseline surveillance project was carried out at Biskopstorp. This comprised five studies:

1. An inventory of woodland key habitats. This was done to obtain spatial information on the area and geographical distribution of the most interesting forested habitats in Biskopstorp. In all 47 woodland key habitats were found, mostly Beech habitats. Furthermore 17 areas with nature values close to woodland key habitats also were found and registered.
2. An inventory of surveillance plots. In total, 115 plots, each with a 20 m radius, were randomly distributed across Biskopstorp. Within each plot, we gathered data on several attributes, and most importantly we recorded all epiphytic and epixylic species of lichen and bryophyte, as well as the forest type, tree and structural data. Indicator species and red-listed species were noted with more accuracy.
3. An assessment of deadwood. In about 75% of the plots, we investigated the amounts of deadwood of different sizes and qualities as well as some additional forest data.
4. Surveillance of the seven most valuable Beech stands, comprising a total area of about 12 ha. Red-listed and indicator species of epiphytic lichens and bryophytes were investigated on all trees, living and dead (standing as well as fallen) in these stands, providing information on population size, fertility and vitality.
5. An inventory of all surveys on red-listed species and regionally interesting species in Biskopstorp, with the number of localities and (if epiphytic) number of tree-occurrences (as an indicator of population size).

These studies are presented on maps and figures in two reports by the local administration board of Halland (Fritz, 2004: Martinsson, 2004). These exercises will be repeated on a ten-year cycle.

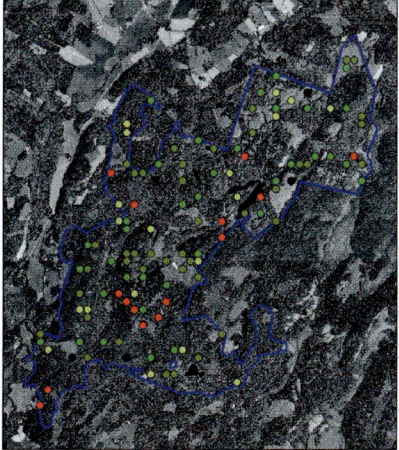

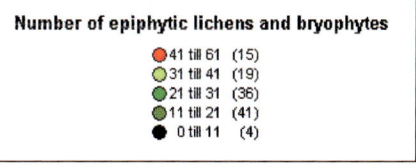

**Number of epiphytic lichens and bryophytes**

- 41 till 61  (15)
- 31 till 41  (19)
- 21 till 31  (36)
- 11 till 21  (41)
- 0 till 11   (4)

*Figure 29-2.* The number of epiphytic lichens and bryophytes associated with the different stands of woodland at Biskopstorp.

Photographs by Örjan Fritz and Clive Hurford
*Figure 29-3. Pyrenula nitida* and *Lobaria pulmonaria*, two of the species of lichen in the condition indicators for the Beech forests at Biskopstorp.

Building on this information, in 2004:

- Tree age data were collected from all plots situated in deciduous broad-leaved stands:
- There was a study of wood-inhabiting fungi on Beech logs carried out in several of the most valuable and interesting stands, using the same method as the EU-project Nat-Man (Heilmann-Clausen 2005): this will make it possible to compare the results from different investigated Beech localities around Europe; and
- There was surveillance of Beech regeneration in areas where the County Administration had either carried out management or had plantations of broad-leaved trees.

## 4.    THE CONSERVATION AIMS

A number of conservation issues exist in this extensive and diverse area. However, the most important ones concern the broad-leaved forests, i.e. the Beech (Annex I habitat 9110) and Oak forests (Annex I habitat 9190). The overall aim is to ensure optimal condition, now and in the future, in the Beech and Oak forests of the area.

## 4.1 The condition indicators for the Beech forests at Biskopstorp

*Table 29-1.* The condition indicator table refers to the Beech forests of Biskopstorp (Habitat 9110).

| Condition indicators | To restore the Beech *Fagus sylvatica* forest at Biskopstorp to optimal condition where | |
|---|---|---|
| **Overall extent** | Lower limit | To reflect the extent in the 1800s. |
| **Quality** | Lower limit | Within >50% of the 20 m radius samples there is:<br><br>At least one mature or late-grown Beech tree (>150 years old) and no canopy-forming Spruce present.<br>**And**<br>Beech regeneration with young trees<br>**And**<br>> 20 m³/ha of coarse woody debris ≥10 cm in diameter<br>**And**<br>Viable spruce seedlings are absent<br>**And**<br>3 or more indicator and / or red-listed species are present |
| **Definitions** | | |
| **Beech forest** | Deciduous broad-leaf forest dominated by Beech (>50% of canopy-forming trees) where mature spruce are scarce or absent. | |
| **Mature Beech tree** | Beech trees with a diameter at breast height (DBH) of >40 cm | |
| **Late-grown Beech tree** | Thin, almost always infected with fungus, suppressed trunk of old age | |
| **Beech regeneration with young trees** | Seedlings and saplings with >2 younger trees 1-20 cm DBH present | |
| **Viable Spruce seedlings** | Spruce seedlings > 2 m high | |
| **Indicator & red-listed species** | Lichens: *Biatora sphaeroides, Lecanora glabrata, Leptogium lichenoides, Lobaria* spp., *Normandina pulchella, Nephroma* spp., *Pyrenula nitida, Parmeliella triptophylla, Peltigera* spp. *Thelotrema lepadinum,* Bryophytes: *Antitrichia curtipendula, Homalothecium sericeum, Neckera* spp., *Porella* spp., *Zygodon* spp. | |

## 4.2 Rationale behind the selection of condition indicators

In general, the optimal condition for the Beech-dominated forest at Biskopstorp depends on the extent, age structure, amount of deadwood, and species composition of the habitat.

A restoration target for the extent of the Beech forest is outlined on a map in the preliminary management plan, and this will be used to prompt the restoration of the area. This takes into account several factors:

- The former distribution of Oak and Beech forests, the extent in 2004 was considerably less than in 1650 or 1850 (Malmström, 1939);
- Edaphic and topographic conditions;
- Increasing tree species diversity in pure Beech forests; and
- Practical management aspects, such as planting of Oak on larger clearcuts, but Beech on smaller clearcuts and under shelterwood of Spruce.

Age structure has also been identified as an important attribute, as many of the red-listed bryophytes and lichens are associated with mature Beech trees, which are more than 150 years old but not necessarily big trees. Thin, late-grown, trunks seem to be an important substrate for epiphytes at Biskopstorp.

Amount of deadwood is also important, e.g. as substrate for biological biodiversity and nutrient recycling in the forest ecosystem. The investigation of plots in Biskopstorp resulted in an average of 21.3 m³/ha of deadwood pieces >5 cm in diameter in woodland key habitats. The average in Biskopstorp area as a whole was lower, 11.8 m³/ha (Martinsson, 2004). However, very few plots in the study exceeded the target for deadwood in Biskopstorp area.

Studies from semi-natural Beech forest in nemoral Europe resulted in a span from almost nothing to 550 m3/ha of deadwood. Also, there was considerable variation in deadwood volumes over time. The total deadwood volume depended on forest type, reserve age and living wood volume. Extraordinary volumes could be found in collapsed formerly managed even-aged stands (Christensen & Hahn, 2003). Because of the nutrient-poor soils, we would generally expect to find lower volumes of deadwood in the less productive sites at Biskopstorp than we might see in many other European nemoral Beech forests. A few un-thinned and very old stands in Biskopstorp may, however, have considerable volumes of deadwood, but we have no evidence to confirm this as yet. The durability of the Beech logs in Biskopstorp also seems to be shorter than average. The common wood-inhabiting fungi *Fomes fomentarius* seems to break down the wood at a faster rate in Biskopstorp than in other places. An amount of at least 20 m3/ha of coarse woody debris (CWD) seems to satisfy a range of wood-inhabiting organisms at stand level (de Jong *et al.*, 2004), and could be the preliminary target for deadwood in Biskopstorp (for CWD ≥10 cm in diameter). Besides the amount of deadwood, there are, of course, a number of other variables concerning deadwood that may be important. Both the stage of decay, and the exposure to wind and sun, for example, will influence which species can use the deadwood. Also, on the national level there is a monitoring target for deadwood, which will be measured with a new and fast field method (using relascopes in plots) developed for the Western Taiga (Habitat 9010). This method should be tested and evaluated in the Beech habitat as well. However, if gap-phase dynamics are prevailing in the Beech forest habitat, we shouldn't expect to find ageing Beech stands with large volumes of deadwood in every plot on every occasion. There will be a dynamic between different phases (innovation, aggradation, biostatic and degradation phases). The long-term natural or desirable proportions between the phases in this area are not known at this moment.

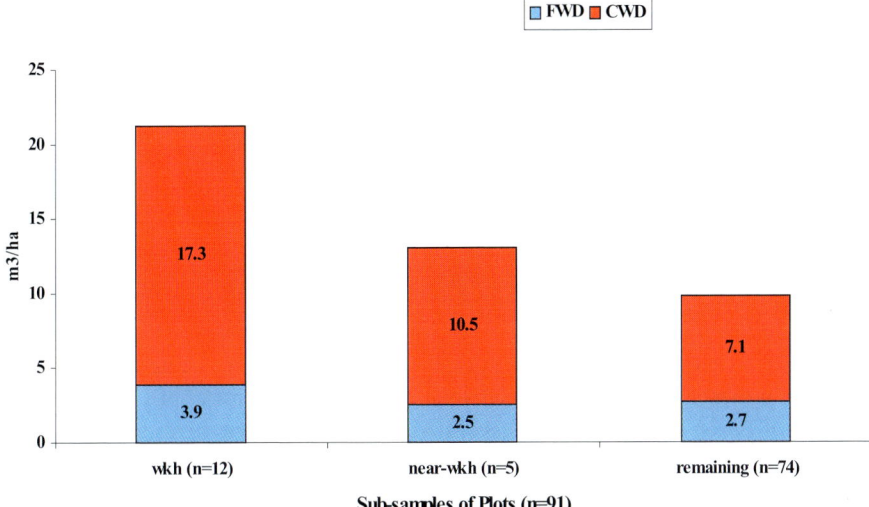

*Figure 29-4.* Amount of deadwood in sub-samples* (n=91) of 20 m radius plots (n=74) in the Biskopstorp area 2001-2003 (Martinsson 2004). The deadwood includes all types of tree species, deciduous and conifers. Definitions:
FWD = Fine woody debris (pieces 5-9 cm in diameter),
CWD = Coarse woody debris (pieces at least 10 cm in diameter).
 wkh = woodland key habitats

Near-wkh is an area of forest judged to be intermediate in nature conservation value between wkh (high value) and the remaining forest (low value). In near-wkh, either indicator species are present and/or the structure implies some (moderate) nature conservation value. If not destroyed, near wkh has the potential to develop into a woodland key habitat in the near future, i.e. in years or decades.

*Some plots were heterogeneous and contained t wo forest habitats, often with different nature conservancy values, e.g. an old beech forest (wkh) and a middle-aged spruce plantation (not a wkh nor a near-wkh).

Although mature Beech trees are widely distributed, and regeneration of viable saplings is relatively common, we are concerned about the general age structure of the stands. There are few trees in the 50 to 100 year age class, which suggests that it may be difficult for the red-listed species, which favour the more mature trees, to persist as the older trees die. This discontinuity of age classes has arisen as a result of historic management practices, such as thinning. There is no certain way of overcoming this problem in the short term, and experimental management is needed to increase the opportunities for the rarer species to persist.

Regeneration is not perceived to be a problem in the area, either for broad-leaved species or conifers. The regeneration of viable Beech seedlings and even saplings is evident throughout areas of deciduous forest, with little or no indication that browsing is a problem. The major threat to broadleaf regeneration is competition by Norway Spruce. In 2002, Spruce saplings were numerous throughout the areas of broad-leaved forest. These saplings need to be removed periodically to prevent a gradual loss of broad-leaf canopy. The proximity of the Norway Spruce plantations to the broad-leaf woods means that, until the plantations have been replaced by native broadleaf stands, this is likely to be an ongoing management activity.

We do not know for certain how many Beech saplings in each plot will be enough to guarantee restoration. Nor do we know the mortality rate of Beech saplings in the area or how many mature trees will result from the regeneration. For the time being we will have to settle with presence of regeneration and younger trees. However, in time, the ongoing surveillance of Beech regeneration in Biskopstorp should shed some light on how many saplings, or groups of saplings, is sufficient to restore the forest.

Species composition is assessed by the co-existence of a mixed suite of red-listed and indicator species. These are bryophytes and lichens that are typically associated with the trunks of old deciduous trees. Most of these species appear to work well in both the Beech and Oak forests in the area, with only a small number restricted to Beech. The species listed in the condition table should be seen as good examples of indicators. There are other species that could provide similar information, but these are either more difficult for non-experts to identify or generally less prominent in the forest in Biskopstorp.

With respect to how many of these species we should expect to find at a monitoring point: at Holkåsen (4.4 ha), which is the oldest Beech forest in the area, there are more than 20 red-listed lichen species present, but these are present on only a minority of the old trees (Larsson, 2000). The species occur aggregated throughout the forest habitats. So while we should not expect to find the species in all plots with old broad-leaved trees, the co-existence of three or more of these species within any 20 m radius does suggest a healthy situation.

Mature Spruce is generally undesirable these habitats and the targets reflect this. However, in one small area in the north of Biskopstorp a mature mixed conifer (Spruce and Pine *Pinus sylvestris*) and Beech forest of spontaneous (and natural) origin will be allowed to develop with minimum intervention.

There remains the question of whether the "healthy indicators" given above should be the same everywhere in Biskopstorp? As such, the condition indicator table for Biskopstorp must be seen as the starting point for monitoring against targets. Due to differences, for example, in stand history, edaphic and topographic conditions, biodiversity content, it may follow that, with greater knowledge of each stand, the appropriate indicators may vary. As a matter of good practice, any future evaluation should review whether:

- The species that we have chosen are appropriate indicators;
- The number of species that we require at each sample point is a reasonable expectation; and
- The target for the number of monitoring points with these species is appropriate.

On the basis of current knowledge, however, the criteria in the condition indicator table above would seem reasonable.

The methods intended to be used in the Swedish baseline survey and monitoring of Natura 2000 habitats are, at the time of writing, not yet finalised (Abenius *et al.*, 2005). The final outcome will inevitably influence the monitoring programme in Biskopstorp - even if it is a special area.

# 5.    THE MONITORING PHASE

Monitoring focuses on the extent of the broad-leaved forest, the co-existence of typical species, regeneration of Beech and Oak, the presence and status of old trees, the amount of deadwood, and the absence of Spruce (both as mature trees and as viable seedlings).

The extent of the Beech and Oak-dominated forests will be monitored using aerial photographs, as broad-leaf forest is easily distinguished from conifer plantation. The extent of the broad-leaved forest will then be compared with the target in the management plan.

We will use a selective approach to monitor the regeneration of Beech, Oak and Spruce, with plots situated in the areas least likely to pass the targets in the condition indicator table. If the forest passes the targets for structure and quality in these areas, we will assume that the other areas of forest would also pass.

Two approaches have been considered for monitoring the species composition of the forests. The first involves recording a) the frequency of mature Beech trees, b) the frequency of mature Spruce, and c) the co-occurrence of more than three indicator and/or red-listed species at contiguous, systematically distributed, 20 m radius monitoring points. This is possible because the individual broad-leaved stands are relatively small. We have also considered an alternative method. This involves dividing each stand into five or six sections (A-F) and setting a lower limit for the number of mature Beech trees and an upper limit for the number of mature conifers in each section. The search for the five red-listed and indicator species would target the mature *Beech* trees in each section. This method may be more efficient than using contiguous 20 m radius monitoring points, as the result would be known as soon as a section failed to meet the criteria in the condition indicator table, allowing the recorders to abandon the monitoring in that stand and move on to the next. Both methods will be tested before a final decision is made.

# 6.    MONITORING RESULTS

The data gathered during the numerous survey, surveillance and research projects at Biskopstorp have shown, without any doubt, that the current condition of the Beech forest habitat, as well as the Oak habitat, is sub-optimal. We have no immediate need to carry out monitoring to establish this because:

• The extent of the Beech forest is clearly less than it was in ca. 1850;

- Mature Spruce trees are still present in many of the Beech and Oak stands;
- Viable Spruce seedlings are relatively common and widespread in the Beech and Oak stands; and
- The areas of forest still dominated by Spruce will not support the red-listed and indicator species assemblage.

In the short-term at least, these results suggest that our resources should be prioritised for managing the forests, rather than for doing monitoring that would serve to confirm only what we already know. The monitoring will become increasingly important in the future, however, and in the first instance we must test and finalise the methods for assessing regeneration, species composition and the amount of deadwood in the Beech stands. When we have done this, we can begin a monitoring programme that initially assesses the condition of the stands that have already been managed. We can then build gradually on this as more and more stands come under management control.

## 7.     DISCUSSION

In the first instance, the monitoring results should drive both the current management regime and the restoration programme in Biskopstorp. On achieving the targets in the condition indicator table, it will be possible to maintain the habitat in optimal condition into the future (and contribute to Favourable Conservation Status, as defined in the EC Habitats Directive).

What then is "optimal condition" for Beech forests? The ecology of natural Beech forest, and its disturbance and regeneration regimes, in Europe is still not fully understood, primarily because most have been impacted by human activities. As a consequence, it is not clear what direction European Beech forest will take if it is allowed to develop without management (Peters, 1997). As research continues, new information could change our view of what healthy conditions are (and should be) in different kinds of Beech forest in Europe. We would expect the attributes and targets in the condition indicator table to be reviewed as new information becomes available.

### 7.1     Management issues at Biskopstorp

The focus on the restoration of the deciduous forests in Biskopstorp since 1994 seems to have been largely successful. Many broad-leaved stands have been cleared of Norway Spruce, improving conditions for the regeneration of Beech, Oak and other deciduous trees. Furthermore, a substantial area of Norway Spruce plantation has been removed, and replaced by Beech, Oak and Lime (Karlsson, 2000), increasing the area of deciduous forest.

Apart from the problems associated with the Norway Spruce, the major threat is the discontinuity of age distributions in the deciduous stands. Most old Beech stands are rather even-aged and will reach maximum age within the next 50 years. In the decades to come we will see more and more stands of Beech breaking down and beginning the succession again. The larger the areas of breakdown, the more probable that primary successors such as Birch, Aspen and Rowan will take over, probably to be replaced by secondary successors like Beech further on down the line. It will take centuries, perhaps, to develop a dynamic forest landscape with more balance between the different growth phases (initial regeneration, young, mature, old, breakdown). At the same time, wholesale turnover will not occur in every stand of Beech, as some have a more or less ongoing gap-phase dynamic and uneven age distribution (mostly those on steep hillsides and in rocky areas).

## 7.2    Management of Beech and Oak habitats at Biskopstorp

The effort to remove Norway Spruce from Beech and Oak stands in Biskopstorp has been largely effective, but there is more to do. There are, and will continue to be, plantations of Norway Spruce inside Biskopstorp for some 20 years or more, and the area will continue to be mostly surrounded by Spruce plantations. The violent storm of 8th January 2005 changed the restoration situation somewhat, by toppling several stands of middle-aged and almost mature Spruce and destroying others.

As such, it is inevitable that emerging Norway Spruce seedlings will continue to be a threat, not only to the Beech forest but also to the Oak and Alder forests of Biskopstorp. Therefore, the management must be seen as a long-term commitment. Similarly, re-instating Beech and Oak on former Spruce plantations (which have been cut down recently to increase the area of deciduous forest) will be a long process stretching decades, if not centuries ahead.

However, the regeneration of Beech seems to be better than for many years. Many factors have contributed to its increased regeneration capacity, for example:

- The interval between mast years in Beech has decreased and has been 1-3 years since 1990;
- Wild Boar *Sus scrofa* have increased, resulting in increased digging activity in the area;
- The Roe Deer *Capreolus capreolus* and Elk *Alces alces* populations have been kept deliberately low through hunting (to reduce browsing pressure on seedlings);
- Weather has mostly been favourable with a mild climate, i.e. with plenty of rain and increased mean spring temperatures since the early 1990s. Spring frosts do, however, still occur, for example in late May of 2004.

Regeneration of Beech has not, however, been restricted to Beech forests. Seedlings of Beech are appearing in several different forest types, not the least in Oak stands with closed canopies, where the Beech seedlings seem to thrive and out-compete the Oak seedlings. So not only Norway Spruce, but also Beech, is posing a threat to some Oak stands in the area. In order to avoid a new management problem, we may have to consider introducing cattle and/or using small fires in the Oak forests. Fire management has already been tested on a small scale, with seemingly good results, i.e. killing off the Norway Spruce and Beech seedlings, but leaving Oak seedlings mostly intact.

# 8. REFERENCES

Abenius, J. (2004). Vedlevande gaddsteklar i Halland. Länsstyrelsen Halland. Meddelande 2004:8.

Abenius, J., Aronsson, M., Haglund, A., Lindahl, H. & Vik, P. (2005). Uppföljning av Natura 2000 i Sverige. Naturvårdsverket. Rapport 5434.

Andersson, R. (2001): Förekomst av vedlevande insekter i Biskopstorp i Halland. Länsstyrelsen Halland. Meddelande 2001:16.

Bengtsson, S. (1999a): Tempererad lövskog i Halland i ett europeiskt perspektiv – ekologi, naturlig dynamik och mänskliga störningar. Länsstyrelsen Halland. Meddelande 1999:1.

Bengtsson, S. (1999b): Biskopstorp - skogstyper, ekologi och skötsel. Länsstyrelsen Halland. Meddelande 1999:20.

Björkman, L. (2002): Pollenanalytisk undersökning av en torvmarksföljd från Trälhultet i Biskopstorp-området, Halmstads kommun. Länsstyrelsen Halland. Meddelande 2002:21.

Christensen, M. & Hahn, K. (eds.) (2003). A study on deadwood in European beech forest reserves. *NatMan project WP2. Working Report 9. Deliverable 20.* (Homepage of NatMan: www.flec.kvl.dk/natman/)

Fahlvik, N. (1999): En dendrokronologisk studie över sambandet mellan beståndsålder/tillväxthastighet och förekomsten av rödlistade lavar på bok inom Biskopstorpområdet i Södra Halland. Examensarbete nr. 9. Inst. för Sydsvensk skogsvetenskap, SLU, Alnarp.

Fritz, Ö. (2001): Indikatorartsövervakning av biologiskt värdefulla ädellövskogar i Hallands län. Länsstyrelsen Halland. Meddelande 2001:25.

Fritz, Ö. (2004): Uppföljning av biologisk mångfald i Biskopstorp. Inventeringar av nyckelbiotoper, provytor och rödlistade arter. Länsstyrelsen Halland. Meddelande 2004:1.

Hannon, G. (2002): Bokskogens historia och dynamik i Biskopstorp och Dömestorp – resultat från makrofossilstudier. Länsstyrelsen Halland. Meddelande 2002:27.

Heilmann-Clausen, J. (2005). Diversity of saproxylic fungi on decaying beech wood in protected forests in the county of Halland. Länsstyrelsen Halland. Meddelande 2005:7.

Jacobsson, D. (2002): Mortalitet av bok i Biskopstorp och Frodeparkens naturreservat. Examensarbete nr 32. Inst. för Sydsvensk skogsvetenskap, SLU, Alnarp.

Jansson, N. (2004). Vedskalbaggar i 20 lövskogsområden i Hallands län 1999-2002. Länsstyrelsen Halland. Meddelande 2004:23.

de Jong, J., Dahlberg, A., Almstedt, M., Jonsson, B.-G., Hysing, E. & Silfverling, G. (2004). Mer död ved i skogen – en förutsättning för tusentals arters överlevnad. Fauna och Flora 99 (2).36-41.

Karlsson, M. (1996): Vegetationshistoria för en artrik bokskog i Halland – stabilitet eller störning? Examensarbete nr. 1. Inst. för Sydsvensk skogsvetenskap, SLU, Alnarp.

Karlsson, M. (2000): Granavveckling och föryngring i Biskopstorp. Länsstyrelsen Halland och Inst. för Sydsvensk Skogsvetenskap, SLU, Alnarp. Meddelande 2000:7.

Larsson, K. (2000). Indikatorartsövervakning av epifytiska lavar och mossor i skogliga nyckelbiotoper. Länsstyrelsen Halland. Meddelande 2000:15.

Malmström, C. (1939). Hallands skogar under de senaste 300 åren. Meddelanden från Statens Skogsförsöksanstalt. Häfte 31, nr. 6.

Martinsson, H. (2004): Död ved i Biskopstorp – En inventering utförd 2001-2003. Länsstyrelsen Halland. Meddelande 2004:16.

Niklasson, M. (2003). En undersökning av trädåldrar i halländska skogsreservat. Länsstyrelsen Halland. Meddelande 2002:28.

Peters, R. (1997). *Beech Forests.* Geobotany 24. Kluwer Academic Publishers.

# PART VI

## USING INFORMATION FROM REMOTE IMAGES

# CHAPTER 30

# USING EARTH OBSERVATION TO MONITOR HABITATS

GRAHAM THACKRAH

*Research Systems International, UK. 31 Wellington Business Park, Dukes Ride, Crowthorne, RG45 6LS*
*gthackrah@rsinc.com*

## 1.    INTRODUCTION

Mapping and monitoring habitats for nature conservation purposes has traditionally been undertaken using a mixture of ground based field survey techniques, often combined with aerial photography. The aerial photographs provide an overview of the whole site of interest, and allow field surveyors to direct their efforts on the ground by providing some knowledge of the location of the various habitats within a given site. Earth Observation (EO), through the use of aerial photography, is thus already a tool used by many conservation practitioners.

The main benefit of Earth Observation (a general term for remote images obtained from aircraft or satellites) is the objectivity it adds to the process of habitat identification and subsequent mapping. Subsidiary benefits include the broad temporal and spatial coverage that EO data provide over a site of interest.

## 1.1    Objectivity

Perhaps the strongest argument for using EO data in a habitat mapping exercise is the potential to increase the objectivity of the habitat identification process. EO data lend themselves to objective analysis; they are quantitative measurements of properties of the area of interest. Field survey methods, even when every attempt is made to ensure their objectivity, will include a certain level of operator bias when they are used to acquire field data. Whilst the interpretation of EO data is still subject to such errors, the ability exists to introduce more objectivity into the process. The subjectivity of EO data interpretation is at least subjective on a repeatable basis, because it comes after the data gathering stage, whereas in the case of field surveys, observer effects will compromise the basic data set.

*C. Hurford & M. Schneider (eds.), Monitoring Nature Conservation in Cultural Habitats, 325–331.*
© 2007 *Springer.*

## 1.2    Cost

EO sensors, and the platforms they are mounted on, are expensive items of equipment. Although many satellite platforms are launched by national or international space agencies, there is an increasing number of commercial satellite platforms being launched, particularly those with sensors acquiring high spatial resolution optical data. Data from national and internationally launched sensors for EO are usually less expensive for the end user than data from commercial EO sensors.

Processing of EO data for use in habitat identification studies is a specialist task and typically requires both specialist software and good computer hardware. EO data, and the expertise to suitably process them, can thus be expensive. The costs that are outlined in Tables 30-1, 30-2 and 30-3 are relative costs, 1 being free data and 5 being expensive. Absolute costs depend on the nature of the organisation buying the data; often, academic institutions or research organisations can apply for free data. Commercial companies can expect to pay the most for image data, though some satellite operating authorities provide their data free to anyone.

The costs of processing EO data vary according to the nature of the data. These differences are generally much smaller than that introduced by the cost of the data, even though the absolute value of EO data processing may be greater than the data costs.

## 1.3    Sensors

This section provides an overview of some of the EO sensors available in 2005. EO sensors can be generalised into a number of categories and discussions relating to a category may be applied to future sensors that become available within the categories outlined here. Advancements in sensor design and data processing hardware and software will lead to increasing technical capabilities of all categories. EO sensors fall into three broad initial categories:

- Optical sensors sensitive to wavelengths of the electromagnetic spectrum between approximately 300 nm and 2500 nm;
- Thermal sensors sensitive to wavelengths between approximately 3000 nm and 10000 nm and
- Microwave sensors, sensitive to wavelengths between approximately 2 cm and 70 cm.

Sensors may further be characterised as either active or passive. Active sensors transmit energy at a particular wavelength towards a target and detect a portion of the energy reflected from the target impinging on the sensor. Passive sensors rely on an independent source of energy (usually the sun) to illuminate the target.

Perhaps the most appropriate classification of sensors for conservation is on the basis of spatial resolution. High-resolution airborne sensors are suitable for small scale mapping exercises over sites of a few hundred hectares. High spatial resolution satellite data may also be suitable for this task, but their usually limited spectral sampling may be a disadvantage. Medium to low-resolution satellite data are more useful for regional and national scale mapping exercises.

The scales suitable for airborne survey range from sites of a few hectares to a few thousand hectares. Sites much smaller than this will simply be uneconomic to survey using EO sensors (unless components of very high conservation value are present), sites much larger benefit from the broader spatial coverage of satellite borne EO sensors. Common airborne sensors for survey work used within the UK are briefly described in Table 30-1. These include:

- Film cameras (increasingly being replaced with their digital equivalents), and imaging spectrometers such as;
- ATM (Airborne Thematic Mapper - Daedalus);
- CASI (Compact Airborne Spectrographic Imager - ITRES); and
- HyMap (Hyperspectral Mapper – Integrated Spectronics).

The imaging spectrometers are listed in order of increasing spectral resolution. When flown on an aircraft at a height of a few thousand metres these sensors will provide image data on the scale of a few metres.

Also dependent on flying height is the swath width. Airborne sensors flown at a typical range of operational heights will provide a swath from a few hundred metres to a kilometer or more. It must be borne in mind that the image characteristics of wide swath-width sensors will vary considerably from the centre to the edge of the image and this will have an effect on the ease of processing these data.

Medium scale survey work, mapping areas greater than a few thousand hectares, may be undertaken using either airborne sensors or satellite-mounted imaging sensors. Satellite-mounted imaging sensors have a swath width of several miles to several hundred miles and may image the entire Earth over a matter of days. These sensors often have archives of past image data that would be useful for historical studies of habitat development. Using airborne imaging sensors to undertake monitoring on this scale must be planned carefully, particularly in temperate areas where cloud cover is often an issue. This means that it can take a long time to image the entire area of interest. By contrast, satellite sensors often have the ability to image entire areas of interest, and can cover many thousand hectares in a single swath or satellite track.

A steadily increasing number of satellite-based imaging sensors, capable of providing image data on the scale of a few metres, is providing an alternative to airborne survey for mapping small and medium size sites. These data are typically limited in their spectral range, covering a handful of bands usually a red, green and blue (RGB) channel augmented with a near infrared channel for broad category land cover mapping purposes. Information from the infrared region of the electromagnetic spectrum is particularly useful in studies of vegetation. A sample of these is described in Table 32-2. The usefulness of these sensors for habitat mapping work may be limited (compared to the sensors mentioned in Table 30-1) due to their low spectral resolution; it is usually the case, however, that as much useful information is gathered from these as is routinely gathered from aerial photographs.

Large-scale survey work, such as mapping entire countries, is the exclusive preserve of satellite-based imaging sensors, or very extensive airborne campaigns. In temperate zones particularly, an airborne campaign covering an area such as this would take upwards of a year to complete[1] and be very expensive to undertake. A sample of sensors suitable for large-scale work is included in Table 30-3. They include sensors with a long history of successful image acquisition in orbit on a series of satellite platforms (such as Landsat, SPOT and AVHRR) as well as more recent additions (such as MODIS on board TERRA and AQUA).

An important omission so far is the non-optical sensors. RADAR sensors are mainly active sensors emitting radio waves in one of four main wavebands, X, C, L and P band (2.5 to 3 cm, 3 to 7 cm, 15 to 30 cm and 30 to 100 cm wavelength respectively). The radiation scattered back towards the sensor is measured and stored in an image product in a similar way to optical sensors. The reader is directed to the introductory texts on remote sensing for a more thorough introduction to RADAR remote sensing (see Bibliography). Features of interest to the conservation practitioner are the sensitivity of the technology to soil and vegetation moisture content and the ability to derive elevation models from advanced processing of RADAR image data.

*Table 30-1.* Characteristics of various airborne EO imaging systems.

| Sensor/Platform | Typical Spatial Resolution | Spectral Resolution | Cost / Ha (1 to 5) | Notes |
|---|---|---|---|---|
| Aerial camera | 30cm | 1 to 3 bands, | 2 | Poor spectral calibration, excellent spatial resolution and good geometric properties |
| CASI | 1 to 20m | 12 to 64 bands | 5 | Good spectral resolution, geometric issues |
| ATM | 2 to 20m | 12 bands | 5 | Good spectral resolution, including thermal channel, geometric problems |
| LiDAR | 10cm to 5m | 1 band | 4 | Elevation data only, large datasets, processing issues, good spatial coverage compared to traditional survey |
| HyMap | 1 to 20m | > 100 bands | 5 | Excellent spatial and spectral resolution, good geometric quality compared to other airborne sensors |

[1] A recent survey of the UK using airborne SAR was discussed in a presentation at the RSPSoc (Remote Sensing and Photogrammetry Society) annual conference during 2004. IFSAR DTMs for flood risk mapping in Great Britain by John Michael and Andrew Shepard. This survey, covering the whole of the UK, took over a year to complete and covered hundreds of individual flightlines. As SAR surveys are not limited by the weather, or daylight hours, an optical survey would be expected to take even longer.

## 1.4 Potential for success

The ability of EO data to provide useful information to the conservation practitioner involved in habitat identification and mapping depends on their ability to differentiate between habitats of interest. This ability is a product of several factors. The imaging sensor has to acquire image data with a resolution below that occupied by a typical habitat patch. The imaging sensor must also acquire spectral data that adequately allow two different habitats to be distinguished in the multidimensional space described by its spectral sensitivity. Given an *a priori* definition of the habitats present at a site, spectral information on each should be gathered to determine their separability using the spectral bands included on the instrument intended to map and monitor them. These data should be gathered using a portable radiometer, an expensive item of equipment that is usually better hired than bought. These are straightforward to use and instruction is usually given in their operation from the organisation hiring them out. Processing the data from such instrumentation is a specialised task but worth completing before data are purchased to avoid buying data incapable of answering the critical questions.

*Table 30-2.* Characteristics of various high-resolution spaceborne EO imaging systems.

| Sensor/Platform | Typical Spatial Resolution | Spectral Resolution | Cost /Ha (1 to 5) | Notes |
|---|---|---|---|---|
| **IKONOS** | 1 to 4m | 4 (RGB + NIR + Pan) | 3 | Very high resolution space borne sensor, limited spectral resolution, RGB and NIR bands only, plus a panchromatic high spatial resolution band |
| **QuickBird** | 61cm to 2.5m | 5 (RGB + NIR + Pan) | 3 | Similar characteristics to IKONOS |
| **SPOT HRG/HRS** | 2.5 to 10m | 1 to 3 | 3 | Longer history of image acquisition than QuickBird or IKONOS |
| **Landsat** | 25m | 7 | 1 | Good temporal record of image data, lots of expertise available in data user community, many papers published using these data, currently platform difficulties mean no more image acquisitions |

*Table 30-3*. Characteristics of various low-resolution spaceborne EO imaging systems.

| Sensor/Platform | Typical Spatial Resolution | Spectral Resolution | Cost /Ha (1 to 5) | Notes |
|---|---|---|---|---|
| **AATSR on ENVISAT** | | 7 (RGB, NIR and TIR) | 1-3 | |
| **MODIS on board EOS TERRA/AQUA** | 250 m to 1 km | 3 to 64 | 1 | Latest NASA Earth Observing satellite, excellent scientific basis, wide range of data products and processing levels |
| **MISR on board EOS TERRA/AQUA** | 275 m | 4 | 1 | Acquires image data at 9 look angles. |
| **NOAA - AVHRR** | 1 km | 4 to 6 (range spanning, R, NIR and TIR) | 2 | Long time series of data and substantial experience of processing in the scientific community |
| **SPOT VEGETATION** | 1 km | 4 (RGB + NIR) | 1 | Large-scale coverage of entire planet over short periods of time. |

## 1.5    Field work

A successful EO project relies on good fieldwork to gather reference data with which the EO data will be processed and compared. These reference data are crucial to any statements of accuracy and repeatability. The different imaging sensors have a variety of requirements for fieldwork: some field data are required for all imaging sensors, whilst others are usually only applied when using image data from a single class of imaging sensors. Basic requirements are data to perform geometric and radiometric correction of the image data. With poor geometric and radiometric correction, EO image data are of limited use in conservation. Field data of the form of spatial positioning information and radiometry of land cover types are a basic minimum requirement.

Spatial and radiometric reference information for habitats of interest are best gathered using portable radiometers, as mentioned in the previous section, and GPS systems (differential GPS or standard GPS depending on the level of spatial accuracy required). Both of these are often available for hire from national science agencies and other bodies. Many of these bodies have equipment available for hire from national pools, though often priority is given to nationally funded scientific research over commercial exploitation.

## 1.6    Data Processing

Once acquired, digital image data from airborne or spaceborne sensors require processing in some manner to provide useful information to the conservation practitioner. Examples of some of this processing are mentioned in the case studies in this chapter. Usually processing can be separated into two categories, pre-processing, required before the data can be more usefully used in conjunction with other geographical data, and post-processing, usually the stage where useful information is extracted from the image data. Later stages of data utilisation may well include substantial analysis within a formal geographic data processing environment, or GIS.

Pre-processing stages include geometric and radiometric correction. These are stages where field based measurements of geographical position (acquired using GPS or DGPS receivers) and reflectance (acquired using a field spectrometer) are compared to the pre-processed image data to give some feedback on the quality of the corrections made. Post-processing stages, including classification of spectral image data, or derivation of measures of topography from elevation data, are usually compared to field based measurements of land cover, typically habitat information. The information gathered by these comparisons is invaluable in making assessments of the quality of the information derived from EO measurements and is essential in validating their use for the purposes of monitoring for conservation.

Image data providers often carry out various pre-processing stages as part of their service. Customers of image data pay for different levels of pre-processing from raw image data (cheapest) to fully geometrically and radiometrically corrected image data (most expensive). Post processing would normally be carried out by the practitioners directly involved in the project using the field data collected for this purpose; usually this is information on a number of samples of the habitats or characteristics of interest at the site.

## 2.    BIBLIOGRAPHY

Campbell, J. B., (2002). Introduction to Remote Sensing. Taylor and Francis.
Lillesand, M. L., Kiefer, R. W. and Chipman, J. (2003). Remote Sensing and Image Interpretation. John Wiley and Sons Ltd.
Mather, P. (2004). Computer Processing of Remotely-Sensed Images: an Introduction. John Wiley and Sons Ltd.

# CHAPTER 31

# PLANNING A REMOTE SENSING PROJECT

## General advice for conservation practitioners

### CLIVE HURFORD

*Countryside Council for Wales, Plas Penrhos, Ffordd Penrhos, Bangor, Gwynedd, LL57 2BQ*
*clive.hurford@serapias.net*

## 1.    INTRODUCTION

To classify the habitats in a remote sensing image, we must provide the classification software with enough information to identify both the broad habitats and habitat condition classes of conservation interest.   This is the critical difference between quantitatively analysing remote sensing images and qualitatively analysing aerial photographs.  Without careful preparation, the chances of success are limited.  This chapter outlines the main problem areas and how best to avoid them.

Firstly, we must decide whether to use satellite images or images collected from an airborne platform; there are advantages and disadvantages to both (see Chapter 30).  In general, airborne remote sensing can provide detailed high-resolution images for mapping heterogeneous habitats on small sites, while satellite images are better suited to assessing large-scale changes across relatively homogeneous terrains.   Whichever system we choose, our  preparation will be critical to the success of the project.

## 1.1    Assessing the likelihood of success before letting a contract

Before letting a remote sensing contract, we must be confident that it will deliver the necessary information.  This means visiting the survey area and checking for the risk of confusion between similar habitats. This process should provide quantitative information, ideally from a spectroradiometer, regarding the separability of the key habitats in terms of their spectral properties.

Most successful remote sensing projects to date demonstrate that the sensor can differentiate between the more obvious habitats, such as woods and arable fields.  There are fewer examples, however, showing that the technique can be applied equally successfully to habitats comprising fine-scale mosaics or habitats with a similar structure and biomass, e.g. semi-improved grassland and semi-natural grassland.

*C. Hurford & M. Schneider (eds.), Monitoring Nature Conservation in Cultural Habitats, 333–340.*
© 2007 *Springer.*

If the survey area includes habitats with a similar structure and appearance to the main habitat of interest, we should take a series of readings from within each habitat (or habitat state) with a spectroradiometer: this will determine whether the spectral signatures of the habitats overlap. If they do, then the sensor may not be able to isolate the habitat of interest. In reality, however, it is difficult to obtain a spectroradiometer, because they are expensive to buy (in the region of €75 000 in 2001) and there is a high demand for the few that are available to hire.

In the absence of a spectroradiometer, our only alternative is to make a field assessment of whether a project is likely to succeed. If the habitat of interest has a structure, or dominant species, that is distinct from all of the other habitats in the survey area, we should be able to identify it on the image. If there are several habitats, or habitat states, of a similar species composition and structure, we will need a good scatter of training areas in each to have any chance of separating them on the image.

Our expectations should be realistic. For example, using an airborne platform the smallest pixel size that we can hope for will be of the order of 1 x 1 m, while using satellite images the pixel size is going to be in the region of 30 x 30 m. There are some commercial satellite imaging sensors that now acquire data on a smaller spatial scale than this, but these are typically of reduced spectral resolution compared to the larger scale sensors, of the order of three to four bands. The detail in the image cannot be any better than the pixel size, and each pixel will represent some aggregation of the radiance values of the vegetation in the area covered by that pixel. Consequently, we will not be able to isolate data for any habitat that occurs at a smaller scale than the pixel size.

We are not yet familiar enough with the reflectance values attached to our habitats (or habitat states) to make any sensible predictions of what the values might be before obtaining the data. So without knowing something about the site and the distribution of the habitats before we collect the image data, we are unlikely to be able to interpret the image.

## 2.      PREPARING FOR DATA COLLECTION

Vegetation monitoring usually focuses on the extent and quality of a habitat. The extent of a broad habitat type can often be discerned on aerial photographs. There are times, however, when the boundaries of a habitat are indistinct on aerial photographs, or when we are also interested in the extent of high quality vegetation within a broader habitat class. On these occasions, remote sensing images will sometimes deliver critical information that cannot be obtained from aerial photographs. To obtain that information, the following decisions need careful consideration:

- Which system to use;
- When to obtain the data;
- How to differentiate between the habitats on the image.

Chapter 32 provides guidance on the merits of the various sensors that are operational in 2005, and should help us to decide which system is most appropriate for our site. The following sections deal with when to obtain the images and how to ensure that we can interpret them.

# 3. WHEN TO OBTAIN THE IMAGE DATA

The timing of the data collection can have a major influence on the success of a project. For example, if we are interested in intertidal vegetation we will need to specify that the image data are collected when solar noon coincides with a low spring tide, or risk the images being collected during a high tide period, when most of the intertidal habitat is covered by the sea.

Terrestrial habitats need the same consideration. If we are collecting aerial photographs, for example, there will be times of the year when the chances of success are optimised. In early and mid summer, most vegetation will be green and it may be difficult to differentiate between habitats that are structurally similar. However, the growth periods of habitats often differ, and it may easier to detect differences between the habitats in late spring or early autumn. For example, *Pteridium* stands can be difficult to differentiate from grassland in late spring or early summer, but are obviously different as they die back in autumn. In fact, many upland habitats become visibly distinct in late summer or autumn, including *Molinia* grassland, *Nardus* grassland and *Calluna*-dominated habitats. By contrast, lowland grassland habitats on thin soils tend to burn off quite early in the summer, making late spring an ideal time for survey. This may influence whether we opt for aerial photographs or remote sensing images, as remote sensing methods are most effective when the sun is directly overhead, when the sky is clear and, in the case of airborne platforms, when turbulence is slight. This minimises the effects of shading and allows better comparisons between image data acquired on different dates by minimising differences in the images as the result of varying illumination conditions. Away from equatorial countries, these conditions are less likely in spring and autumn, and will be relatively scarce in winter (when the sun is always lower in the sky).

The common dilemma for the practitioner is whether to fly in summer, when we can get optimum performance from the remote sensing apparatus but it is not the best time of year to distinguish the key habitat, or to fly outside the summer period and risk flaws in the data set (see Chapter 33).

# 4. DIFFERENTIATING BETWEEN THE KEY HABITATS ON AN IMAGE

Collecting training data has two main functions: to accurately pinpoint the locations of the different habitats and allow radiometric correction, and to improve the geometric correction of the image. The intensity of the training data collection exercise will depend on the pixel size and what information you need to extract from the image data. In

*Figure 31-1.* In this image, generated in ENVI[1], we can see across-track radiance trends within each flight strip. These made it impossible to classify the habitats in the mosaicked CASI image. The major problem was that there were differences in the radiance values within the flight strips as well as between them. This artefact of the data collection process is most visible in the flight strip containing Kenfig Pool, which is bright green along its eastern edge but a paler green along the western edge.

general, geometric correction will be less of a problem using satellite data than airborne data, as satellite image data are much less affected by changes in the orientation of the platform.

However, because of the generally larger pixel size generated by satellite data, we will often be able to detect only large-scale patterns in a vegetation mosaic. Whichever platform we use, we will need to provide training points for all of the habitats that occur in stands greater than the pixel size. These training points should be taken from within homogeneous stands of vegetation, and each point should be more than the diameter of a pixel away from any habitat boundary. This should ensure that the spectral signature refers to a pure stand of the habitat. To be safe, however, the rule of thumb (particularly when using

[1] ENVI image processing software, Research Systems International Inc.

airborne platforms) should be that each training point is taken from more than two pixel diameters inside homogeneous stands of habitat that are more than four pixel diameters in area. Ideally, we will obtain a scatter of training points from each habitat and habitat condition class across the survey area. The chances of an accurate classification improve with increases in the number of points in each habitat class.

The location of each training point should be recorded with a differential Global Positioning System (DGPS) that is accurate to less than 5 m, and we should describe and photograph the habitat at each point. This will provide a valuable record of the vegetation classes used to 'train' the image.

If we are using an airborne platform, the training data set has to be well distributed to take account of: a) potentially poor image registration; b) different radiance values between and within the relatively narrow flight strips (typically 0.5 – 1 km wide); and c) the increased detail in the habitat data. In this instance, we will need a large number of training points in each habitat class in each flight strip to minimise potential problems at the classification stage (Fig. 32-3).

Even if we are using an onboard GPS to register the image, which will be possible only if the plane supports one, we should also obtain a set of training points chosen specifically to aid geometric correction (or image registration). These can be used to check the accuracy of the correction by the onboard GPS, or to correct the image data if the plane does not have an onboard GPS. Ideally, the DGPS points collected for the purpose of image registration will mark the location of permanent features that will be visible on the image, e.g. buildings, field corners and road junctions. This can be a serious problem if we are collecting images over a large area of relatively featureless terrain, such as a mountain range. In this situation, it may be better not to risk collecting airborne data without onboard GPS, as there are few, if any, working examples of these problems being overcome.

On smaller sites of less than 1000 ha, we could stake out a series of different coloured target boards (large enough to be visible in the acquired image data) and record their locations with a DGPS. These would have to be in place at the time of the flight. Therefore, before committing to this approach, we need to know the position of the flight strip boundaries well ahead of the flight. This information should be available from the pilots and mission planners.

Adjoining flight strips can be misaligned even when an image has been registered with onboard GPS, and shifts of c.10-15 m sometimes occur. These are clearly visible where a linear feature, such as a road or track, cuts across the boundary of two flight strips. As the systems become more accurate these problems should be reduced. However, until these issues have been resolved, training points for habitat classification should be situated well within homogeneous stands of habitat, i.e. for a typical airborne survey, more than 10 m from the habitat boundary in any given direction. Consequently, even when data are collected at the 2 x 2 m pixel level, it may not be possible to map the distribution of any habitat that does not occur in at least one stand of more than 20 x 20 m in the survey area.

# 5.    THE CONSEQUENCES OF POOR IMAGE REGISTRATION

If some of the linear features in an image are misaligned, similar but less obvious problems will probably be present elsewhere in the image (where convenient points for comparison do not exist).   In this situation, provided that we can still see the extent and approximate distribution of the habitats in the survey area, we may have to accept that their locations are not precise.

The major problems occur, however, if the misalignment within the image exceeds the size of the training areas.  For example, if we record training points centrally from within a 20 x 20 m stand of vegetation, an absolute misalignment of more than 10 m would put the training points outside the habitat.   So the training points for habitat X would now be situated in habitat Y, and we would end up using the spectral signature of Habitat Y to map the distribution of Habitat X.   This problem becomes more likely if the training points have been taken from smaller stands of vegetation.   If we are not familiar with the general distribution of the habitats in the survey area, and we are relying on the image to give us this information, then this type of problem is easily overlooked.

# 6.    THE CONSEQUENCES OF DEFINING TOO FEW HABITAT CLASSES

Because image classification software uses multivariate statistical analysis methods, any habitats or habitat classes that are not covered in our training data set will be allocated to those habitat classes that were used to train the image.  In practice, this means that the extent and distribution of the habitat classes on the final image will be inaccurate because our selected habitat classes willl incorporate areas of "unclassified" habitat.

On complex sites, where a large number of habitats are present, this scenario is almost inevitable (even with a relatively comprehensive coverage of training points).  However, if we are aware that it is happening then we may be able to reduce the level of error, by using stricter training areas, increasing the number of habitat classes and using other means of separating the habitats, e.g. by introducing rules relating to aspect or slope (assuming that a data elevation model (DEM) is available for the area).  For example, in upland areas we could find that dry heath vegetation and *Calluna*-dominated blanket bog share the same radiance values (as the canopy of both habitats may be formed of dense ericoid sub-shrub cover).  However, the blanket bog will occur mostly on the mountain plateaux while the dry heath will be found more often on the valley slopes.  A rule stating that all of the *Calluna*-dominated vegetation on a slope of more than 20° is dry heath could help to differentiate the habitats.

In some cases, it may be that the sensor will not be able to separate some of the habitat classes that we have identified and provided training data for.  An example might be 'species-rich' and 'species-poor' hay meadow vegetation, which are similar both in structure, appearance and productivity.  Provided we have saturated the survey area with training points for the habitats, we will soon realise that there is a problem, and we will be able to lump the classes under a single 'hay meadow' habitat class. Our image will not provide as much information as we would like, but it will be more accurate and we will at least be aware of its limitations.

## 7.     THE CONSEQUENCES OF NOT CLEARLY DEFINING THE HABITAT CLASSES

As with terrestrial-based methods, unambiguous habitat definitions are essential if we want our image to be a baseline for tracking change. In reality, the habitat definitions may be the only constants between the baseline and repeat image data, as:

- The plane will not be in precisely the same position,
- The plane will not be flying at precisely the same height;
- The data may not be collected at precisely the same time of day or on the same date;
- The growth stage of the vegetation may differ as a result of seasonal weather patterns;
- The radiance levels and atmospheric conditions will be different;
- The accuracy of the image registration will be different;
- The edges of the flight strips will be in different parts of the survey area;
- The pixel size will differ slightly; and
- The vegetation at many of your original training points will have changed, sometimes into a different habitat class.

None these factors is a major problem, as long as there are clear and unequivocal definitions of each habitat class. These definitions will allow us (or our successors) to identify a new set of training points, safe in the knowledge that the classification will be based on the original habitat classes. Ideally, data collection should be scheduled for the end of the growth period in the habitat, as this will increase the chance of achieving consistency in repeat surveys.

It is important to realise that we cannot use the original training points to inform a repeat remote sensing exercise. We will need a new set of training points, which should be collected as close to the time of image data collection as possible.

## 8.     THE REQUIRED SKILL LEVEL

Unlike terrestrial monitoring methods, which can be simplified for repeat sampling, remote sensing projects have a high skill requirement throughout all phases of the operation. As with any other form of monitoring project, initially we have to decide what information we want the image to provide. We then have to ensure that the classification software can isolate that information. This means that the ecologist setting up and collecting the image training data must be able to identify and define all distinct developmental phases of each vegetation type in the survey area.

The person collecting the training data for repeat exercises does not need the skills to define the classes of each habitat present, but will need to be able to recognise the habitat class at each training point. In effect, this means that a competent botanist will be required for every data collection event. These skills, combined with detailed site knowledge, are equally important during the image classification phase of the project (see Chapter 32).

## 9.    IN SUMMARY

Before embarking on a remote sensing project, it is good practice to:
- Have clearly stated, and carefully considered, aims for the project;
- Be familiar with the habitats and permanent features on your site before you let a contract;
- Use a DGPS to record the grid co-ordinates for a scatter of the permanent features that will be visible on the image;
- Provide unequivocal definitions for every distinct habitat and habitat state in the survey area;
- Take photographs of every distinct habitat and habitat state;
- Use a DGPS to record the location of several homogeneous stands of each distinct habitat and habitat state;
- Make sure that each DGPS point is situated within a stand of homogeneous vegetation that is at least larger than the width of two pixels, and that some points in each habitat class are located within homogeneous stands more than three or four pixel diameters in area; and
- Obtain DGPS co-ordinates from stands of habitat across as much of the survey area as possible.

Finally, we should ensure that we have access to remote sensing and GIS specialists from the outset.

# CHAPTER 32

# REMOTE SENSING OF DUNE HABITATS AT KENFIG NNR

CLIVE HURFORD

*Countryside Council for Wales, Plas Penrhos, Ffordd Penrhos, Bangor, Gwynedd, LL57 2BQ*
*clive.hurford@serapias.net*

## 1.    INTRODUCTION

Kenfig National Nature Reserve, a coastal dune system of ca. 600-ha, has been popular with local naturalists since the early 1900s, and is one of the best-studied nature reserves in Wales. The site, located in south-east Wales (Fig. 16-1), comprises a complex mosaic of vegetation types and is of international conservation importance for three habitats on Annex I of the EC Habitats Directive and for two species on Annex II.

In the early 1990s, however, there were concerns about the loss of successionally-young habitats on the site (Jones *et al.*, 1996). These concerns were underpinned by evidence from aerial photographs, which showed that the extent of open sandy habitats at Kenfig had declined from c.40% of the site in 1945 to less than 5% in 1991. Aerial photographs also indicated that, by 1991, the dynamic geomorphological processes at Kenfig had virtually stopped; the fore-dunes showed signs of erosion, there was no evidence of new sand accretion, no new dune slack development, no new sand blow-outs, and a decline in both the size and mobility of existing sand blow-outs (Fig. 32-2).

*C. Hurford & M. Schneider (eds.), Monitoring Nature Conservation in Cultural Habitats*, 341–352.
© 2007 *Springer.*

The gradual loss of open sandy habitats on the site implied an associated decline in the successionally-young stages of sand dune development, though these seral stages are less easy to discern on aerial photographs.

In 1997, the CCW monitoring team carried out a terrestrial-based monitoring project at Kenfig: this focused on the condition of the dune grassland and slacks in some of the remaining successionally-young parts of the site (Hurford & Perry, 2000). Shortly after completing this work, CCW was invited to be a partner in the Glamorgan Coastal Monitoring Initiative (Pan & Morgan, 1999); a collaborative three-year project jointly funded by the Welsh Assembly Government, University of Wales Swansea (UWS), the Countryside Council for Wales and Bridgend Borough Council (the site managers). This initiative was a pilot project to demonstrate how remote sensing technology could contribute to monitoring the coastal habitats in south-east Wales.

This collaboration presented a rare opportunity for remote sensing specialists, GIS specialists, monitoring ecologists and land managers to work together: and on a complex site that would provide a good test of the technology. Fortunately, our work coincided with other UWS remote sensing projects focused on Kenfig (Thackrah *et al.*, 1999: Sanjeevi & Barnsley, 1999), which us gave an opportunity to share our experiences as the projects developed. The following case study describes the process that we used to produce a classified Airborne Remote Sensing (ARS) image of Kenfig NNR.

## 1.1    Setting up the project

Initially, we had to decide what information we needed to obtain from the image data. An important aspect of remote sensing image data is that any image generated by a classification process is based solely on the information that you provide for it. At Kenfig, the primary conservation interest centres on two developmental phases of the internationally important 'humid dune slack' habitat (2190), namely embryo dune slacks and successionally-young dune slacks. Of secondary importance is the successionally-young phase of the 'fixed coastal dunes with herbaceous vegetation' habitat (2130). All three of these young seral phases, which host several nationally and internationally rare plant species, e.g. Fen Orchid *Liparis loesēlii*, Petalwort *Petalophyllum ralfsii* and Hutchinsia *Hornungia petraea*, are understood to be declining at Kenfig. A successful project would reveal the extent and distribution of each of these seral stages on the site.

## 1.2    Collecting the data to train the image

We collect site-based training data to help interpret what we can see in image data acquired by remote sensing techniques. In practice, this means visiting the area of study and pinpointing a series of reference points on the site that can be used to inform the classification of the image. We might carry out a similar exercise to help us to interpret aerial photographs.

This phase of a remote sensing project is critical: unless you have enough reference points to isolate the key vegetation types, other vegetation with a similar spectral signature

*Figure 32-1.* The flooded humid dune slack habitat at Kenfig Pool in winter.

*Figure 32-2.* The only remaining area of embryo slack at Kenfig in 1998. Note the clonal patches of Creeping Willow *Salix repens* in the foreground and how the Marram Grass *Ammophila arenaria* is stabilising the exposed face of the dune blow-out, effectively preventing further movement of the dune.

will be assigned to the key vegetation types during the image classification phase of the project. This will compromise the accuracy of the final image. The data collection exercise at Kenfig comprised two initial phases:

- Developing clear definitions of the distinct habitat classes present on the site; and
- Locating a series of reference points in each habitat class.

We had already identified and described the key vegetation types at Kenfig during a terrestrial-based mapping project in June 1997 (Aubrey, 1997: Besley, 1997: Hurford & Perry, 2000). These habitat definitions were adapted to form the basis for those used during the airborne remote sensing exercise. Table 32-1 lists the all of the visually distinct habitat classes identified at Kenfig, provides the key attributes of each, and states the number of training points recorded in each habitat class. Identifying and locating all of the main vegetation types at Kenfig, and not only those of high conservation interest, would allow us to make informed decisions in relation to combining habitat classes if problems arose during the image classification phase of the project.

We did not use the National Vegetation Classification (NVC) communities (Rodwell, 2000) as habitat classes for this exercise because they were not sensitive enough to isolate the critical seral phases of dune development. For example, SD14, the humid dune slack community, takes in all of the seral stages from open, species-rich vegetation to species-poor vegetation dominated by Creeping Willow *Salix repens*. Both of the internationally rare plant species at Kenfig, *Liparis loeselii* and *Petalophyllum ralfsii* are restricted to successionally-young SD14 slacks and could not survive in the more mature stands.

At Kenfig, we provided site-specific definitions for all of the visually distinct phases of habitat succession, from bare sand through to dune woodland. We used a differential GPS (accurate to <50 cm) to locate each training point and, as far as possible, we tried to ensure that each training point was situated within a homogeneous stand of vegetation greater than 5 x 5 m in area. This was not always possible, however, as a few habitat classes typically had a smaller patch size than 5 x 5 m. For these vegetation types, we saturated the largest stands with training points. This drew attention to these stands when we were classifying the image. At each training point, we checked the flora within the stated area of search (usually within a 50 cm or 1 m radius) to confirm the vegetation type, and then checked the vegetation within a 5 m radius to ensure that we were within a homogeneous stand.

We concentrated most of the training points in the habitat classes of high conservation value and in vegetation types that were structurally similar to them, as we suspected that it might be difficult to separate these. By contrast, we recorded only a small number of points in vegetation types that were locally distributed on the site and of low conservation interest, and ignored habitat types that we knew would be obvious on the image, e.g. open water, woodland and scrub. The small number of points in 'embryo slack' and 'young slack' vegetation reflects the scarcity of these habitats on the site. In fact, we used GPS co-ordinates from an earlier monitoring exercise as training points for 'young slack' vegetation: this error of judgment subsequently caused us difficulties during the image classification phase of the project.

The image data were collected by NERC using CASI and ATM sensors in May and August 1997, and our first period of data collection coincided with the time of the August flight. LiDAR data were also made available during 1999 from a flight undertaken by the Environment Agency as part of the Glamorgan Coastal Monitoring Initiative (GCMI). This ensured that we had data reflecting the condition of the habitats when the image data were collected. On this occasion, we recorded the location and vegetation data at five training points in each of the key vegetation types.

## 2. PREPARING THE IMAGE DATA FOR CLASSIFICATION

Before we could classify the image, we had to mosaic the flight lines, ensure that the image was geometrically and radiometrically correct, and 'mask' (or exclude) any areas outside the nature reserve where we had not collected training data. Without masking the image, habitats outside the reserve boundary, but not within it, e.g. improved pastures and road verges, will be allocated to the habitat class with the closest spectral signature. If this happens, it will compromise the accuracy of the classified image. However, if the raw image has been masked and is technically sound, i.e. geometrically correct and with even radiance values for the habitat classes across flight strips, then the success of a classification will depend on:

- The distribution of training points;
- The subsequent selection of the training areas; and
- Existing site knowledge.

### 2.1 Training the image with the training points

After collecting the training data, we downloaded the GPS co-ordinates and created a text file for each of the vegetation types. This listed the eastings, northings and habitat type for each ground truth point. We then used two software packages to process the image:

- ENVI, which processes the image and carries out the classification; and
- ArcView, which we used to identify the training areas on a mosaicked ATM image (Fig. 32-3) of Kenfig (imported from ENVI).

### 2.2 Selecting training areas

When the training points had been imported into ArcView, we opened up the same ATM image of Kenfig in ENVI. With both the ArcView and ENVI images on the computer screen, we identified appropriate training areas from the distribution of training points on the ArcView image, and then defined these as 'regions of interest' on the ENVI image. This was easier if we dealt with each habitat class separately, i.e. when we were selecting the training areas for closed rich grassland vegetation, we plotted only the training points for closed rich grassland on the ArcView image.

# 3.     GENERATING AND CORRECTING THE
   CLASSIFIED IMAGE

After identifying and selecting the training areas for each vegetation type, we ran the first classification. This generated an image showing the distribution of the various habitat classes within the survey area. This classification, however, was clearly inaccurate and was not a true representation of the distribution of habitat classes at Kenfig. As such,

*Table 32-1.* Brief definitions of the habitat classes identified at Kenfig and the number of training points collected in each

| Habitat class | Key attributes | No of points |
|---|---|---|
| | | |
| Embryo slack | >25% bare ground with clonal *Salix repens* | 14 |
| Young slack | < 25% bare ground with thalloid liverworts common | 32 |
| Species-rich slack | > 2 spp. of orchid in any 1m radius | 143 |
| Mature slack | Dense *Salix repens* with *Hydrocotyle* or *Carex nigra* | 111 |
| Saline slack | Abundant *Juncus maritimus* or *Glaux maritima* | 10 |
| Dense *Calamagrostis* | *Calamagrostis epigejos* dominant | 37 |
| Tussocky *Molinia* | Dominant and tussocky *Molinia caerulea* | 19 |
| Fen meadow | *Lysimachia vulgaris* abundant | 5 |
| Tall fen | *Typha latifolia* locally dominant | 2 |
| *Phragmites* swamp | Dense beds of *Phragmites australis* | 12 |
| *Eleocharis* fen | Dominant *Eleocharis palustris* | 8 |
| *Juncus* fen | Dominant *Juncus* spp. | 7 |
| *Schoenoplectus* fen | Dense stands of *Schoenoplectus tabernaemontani* | None |
| Strandline | Dominant *Eryngium maritimum* or *Cakile maritima* | 1 |
| *Ammophila* and bare sand | Tussocky *Ammophila arenaria* with >50% bare sand | 48 |
| Young grassland | 25-50% bare sand / bryophyte cover, annual spp. present | 157 |
| Closed rich grassland | Closed species-rich grassland, | 102 |
| Closed poor grassland | Closed species-poor grassland, *Arrhenatherum* rare | 70 |
| Rank grassland | *Arrhenatherum, Dactylis* (or both) abundant | 121 |
| Dense *Pteridium* | *Pteridium aquilinum* dominant | 44 |
| *Betula* scrub | Dune woodland dominated by *Betula pubescens* | None |
| *Alnus* scrub | Woods dominated by *Alnus glutinosa* | None |
| *Salix* scrub | Tall scrub dominated by *Salix* spp. | None |
| *Hippophae* scrub | *Hippophae rhamnoides* dominant | 4 |
| *Salix repens* sub-shrub | Dominant *Salix repens* on dry dunes | 3 |
| *Calluna* sub-shrub | *Calluna vulgaris* dominant | 5 |
| **Other habitats** | | |
| | | |
| Bare sand | >90 open sand – plant cover sparse | 14 |
| Deep open water | Open water >1 m deep | None |
| Shallow open water | Open water < 1 m deep | None |

this first classification was simply the starting point in an iterative process that eventually resulted in a representative image. These iterations involved image processing specialists, the appropriate habitat specialists, and the local site managers (whose intimate knowledge of the site was critical in helping to identify areas that had been allocated to incorrect habitat classes).

Our first attempt at classifying the image revealed two problems that could compromise the accuracy of the classified image:

- Our inability to predict the precise location of the flightlines meant that some habitat classes were not represented within all of the flightlines: and
- Geometric errors and misregistration of the image resulted in uncertainty about the accuracy of training points located in small habitat patches.

We had to resolve both of these problems before we could produce an accurate representation of the key habitat classes on the site. Short of saturating the site with training points around the time of the flight, it would have been difficult to avoid the first of these problems, and this is likely to be an issue for any airborne remote sensing exercise. We could have avoided the second problem by locating our training points only within relatively large (>20 x 20 m) and homogeneous stands of each habitat class. This would have been a problem at Kenfig because some of the scarcer habitats occur only within a small-scale mosaic with other habitat classes.

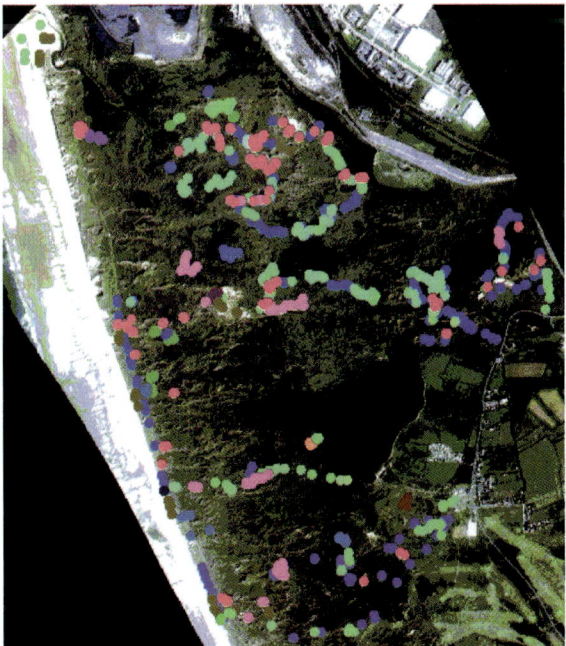

*Fig 32-3.* The ATM image of Kenfig dunes showing the distribution of training points for the key habitats. It is just possible to discern where the two flight strips were joined during mosaicking (parallel to the western edge of the pool). Contrast this with the image shown in Fig. 31-1.

By the time that we received the remote sensing images, it was already too late to take corrective measures before the next field season. In practice, this meant waiting until the following August before we could increase the number of training points in each flightline.

Our decision to fly the site at the end of the growth period for the dune habitats was rewarded, however, as this reduced the risk of the new training points being affected by seasonal periodicity. Nevertheless, the number of habitat classes that we had originally identified for classification was compromised by the delay, as we could not risk compromising the original training data with new points situated in habitats that could have changed class in the time since the flight. As a consequence, we could increase the number of training points only in vegetation that was still 'successionally-young' (as this would also have been 'successionally-young' 12 months earlier), or in vegetation had clearly been in a mature phase of development for several years. Vegetation in the intermediate phases of dune development, such as species-rich grassland and slacks, had to remain poorly represented by training points.

The following August we revisited Kenfig and increased the number of ground truth points in the appropriate habitat classes, and used GPS points taken from the terrestrial monitoring exercise in 1997 to increase the number of ground truth points for successionally-young dune slack habitat, which occurs as part of a mosaic with other habitat classes. We then returned to the computers and used the new training points to identify new training areas for the classification.

After re-running the classification, the signs were encouraging, with the distribution of several habitat classes beginning to reflect what we believed to be their true status. Unfortunately, the distribution of other habitat classes was still inaccurate, so we began the process of combining them to increase the overall accuracy of the image. From the outset, we had expected some of these habitat classes to be problematic, particularly those that were structurally similar and differentiated only by species diversity, e.g. closed rich grassland and closed poor grassland. Others were more surprising. For example, we did not expect to have problems isolating stands of Bracken *Pteridium aquilinum*, which could not be isolated from shaded areas of closed grassland.

Most disappointing, however, was the apparent failure to isolate the young slack habitat, which is the conservation priority at Kenfig. Initially, we assumed that this habitat class had been lumped with the young dune grassland class, compromising the mapped distribution of two of the most important habitats. It was only during the subsequent quality assurance exercise that we realised that the image was correct. Young slack vegetation had virtually disappeared at Kenfig, with only scattered small fragments (typically <1 x 1 m in area) now meeting the criteria that we had given for the habitat class. These fragments were too small to be isolated from the species-rich grassland and slack classes: two more habitat classes that could not be isolated on the image.

We ran several classifications during this phase of the project in an attempt to isolate some of the lumped habitat classes. This involved focusing solely on the larger training areas and deleting training areas based on small habitat patches of <5 x 5 m in area. We also revisited the site one more time to increase the number of training points; on this occasion to search for any large stands of habitat (>25 x 25 m in area) in the classes that had proved difficult to isolate. On completing these tasks, there was nothing more that we

Since the energy gain due to the displacement of a nucleus into the neighboring pinning valley grows with the volume of the nucleus, while the energy cost essentially scales with its surface, for small electric fields ($E \to 0$) a large nucleus can be expected. Depending on the wall dimensionality $d$, balancing the two energy terms $V \sim L^d$ and $A \sim L^{d-1}$, where $L$ is the spatial extension of the nucleus, leads to different behavior. For a one-dimensional domain wall (string, $d = 1$), the nucleus consists of two point-like kinks, whose activation energy therefore always remains finite, and the system exhibits a *linear* response to small driving forces. For a two-dimensional domain wall on the other hand, minimizing (4) gives $L^* \sim 1/E$, showing that the size of the nucleus grows as the electric field decreases. In this scenario the energy barriers to domain-wall motion, and thus nucleus growth, grow as $\Delta(E) \sim 1/E$, using (4), giving a nonlinear response with $v \propto \exp{-1/E}$. The stochastic nucleation proposed by Miller and Weinreich can thus explain the observed nonlinear response *only if* the domain wall itself is a two-dimensional surface embedded in a three-dimensional crystal. This means that the dimensions of the nucleus, at a given field $E$, have to be smaller than the thickness of the system. Otherwise, the energy of the nucleus saturates, resulting in the linear response of the one-dimensional scenario. It is also important to note that if the creep consists of motion in a commensurate potential the dynamical exponent is constrained to be $\mu = 1$. As already mentioned, this particular scenario is microscopically related to the intrinsic periodic pinning of the domain wall by the ferroelectric crystal lattice potential. The strength of this potential was calculated in ab-initio studies of $180°$ domain walls in $PbTiO_3$, showing that the wall energy varies from $132\,mJ/m^2$ to $169\,mJ/m^2$ depending on whether the domain wall is centered on a $Pb-O$ or $Ti-O_2$ plane in the crystal [10].

To test whether the observed creep behavior is indeed due to the stochastic nucleation process, we calculated the size of the critical nucleus, following the formulation derived by Miller and Weinreich for the energetically most favorable dagger-shaped nucleus of horizontal extension $a$, height $l$ and thickness $c$ forming at an existing $180°$ domain wall, as shown in Fig. 6 [33], where $P_s$ is the polarization, $b$ the inplane lattice constant, $\epsilon$ the dielectric constant of PZT at ambient conditions, and $E$ the applied electric field. The depolarization energy can be written as $U_{\text{depolarization}} = \frac{2\sigma_p b a^2}{l}$, with $\sigma_p = [4P_s^2 b \ln{(0.7358a/b)}]/\epsilon$ [33]. By minimizing the free-energy change due to nucleation with respect to the dimensions of the nucleus $a$ and $l$, with $c$ taken as equivalent to the lattice constant $b$ (the distance between two minima in the commensurate potential), the size of the critical nucleus $a^\star$ and $l^\star$, as well as the activation energy $\Delta F^\star$, can be calculated as [33]:

$$a^\star = \frac{\sigma_w(\sigma_w + 2\sigma_p)}{P_s E(\sigma_w + 3\sigma_p)},$$

$$l^\star = \frac{\sigma_w^{1/2}(\sigma_w + 2\sigma_p)}{P_s E(\sigma_w + 3\sigma_p)^{1/2}},$$

$$\Delta F^\star = \frac{4b}{P_s E}\sigma_p(\sigma_w + 2\sigma_p)\left(\frac{\sigma_w}{\sigma_w + 3\sigma_p}\right)^{3/2}. \tag{5}$$

To compute the actual values, standard parameters for PZT ($P_s = 0.40\,\mathrm{C/m^2}$, $\epsilon = 100$, $b = 3.96\,\text{Å}$), and the ab-initio value for the domain-wall energy density[5] $\sigma_w = 0.132\,\mathrm{mJ/m^2}$ can be used. In our case, the applied electric field varied from $\sim 2$ to $20\,\mathrm{MV/m}$ (with the factor 10 correction), depending on the thickness of the sample used and the distance from the AFM tip, with the most intense fields for thin films and small domains. Corresponding values of $\sigma_p$ were between 1.6 and 0.9 J/m$^2$. Since $\sigma_p$ is therefore greater than $\sigma_w$, following Miller and Weinreich, the expressions for the critical values can be simplified to:

$$a^\star = \frac{2}{3}\frac{\sigma_w}{P_s E},$$

$$l^\star = \frac{2\sigma_w^{1/2}\sigma_p^{1/2}}{\sqrt{3}P_s E},$$

$$\Delta F^\star = \frac{8b}{3\sqrt{3}P_s E}\sigma_p^{1/2}\sigma_w^{3/2}. \tag{6}$$

For the field range used, these equations would give critical values of $a^\star \sim$ 12.5–125 nm and $l^\star \sim 53$–710 nm. These results imply that for the given electric field range, the vertical size of the critical nucleus would exceed the thickness of the film itself. This suggests that the films are in a 2-dimensional limit, with the domain walls acting as a quasi-one-dimensional manifold, for which the Miller and Weinreich stochastic nucleation model, or alternatively, weakly driven motion through a commensurate potential, could not explain the nonlinear response observed. In addition, the values for the activation energy of $\sim 0.6 \times 10^{-21}$–$0.5 \times 10^{-20}$ would suggest extensive domain-wall motion as a result of thermal activation already at room temperature, a phenomenon not observed in PZT. Finally, the values of the dynamical exponent we observe, generally not equal to one, are also a strong indication that an alternative microscopic mechanism for the observed creep process should be considered.

---

[5] This value is computed for PbTiO$_3$. The presence of Zr in PZT would lead to local variations of this energy density.

# 6 Domain-Wall Creep in a Random Potential

In the alternative scenario of a canonical "glassy" system, an elastic manifold is weakly collectively pinned by the quenched disorder potential present in the medium, with important consequences for both its static and dynamic behavior. Disorder is present in any realistic system: in $PbZr_{0.2}Ti_{0.8}O_3$, vacancies and other defects in the lattice structure are likely sources of disorder. Another possibility is the presence of Zr atoms (the material is essentially a solid solution of 20% $PbZrO_3$ in 80% $PbTiO_3$), although preliminary studies of domain-wall dynamics in pure $PbTiO_3$ show similar static and dynamic exponents. In ferroelectric films the presence of disorder would dominate domain-wall behavior for both 1- and 2-dimensional walls at large scales. However, given the thinness of the domain wall, we note that the commensurate potential of the crystal is also present in the problem, although possibly at length scales below those experimentally accessible with our current system.

In order to analyze the effects of disorder on domain-wall motion we again consider the energy of a segment of ferroelectric domain wall of length $L$ displaced by $u(z)$ from the elastically ideal flat configuration as shown on Fig. 2. The energy scales as[6]

$$U(u, L) = \sigma_w u^2 L^{d-2} - U_{disorder}[u] - 2P_s EL^d u, \qquad (7)$$

where the first term describes the elastic-energy contribution, and is expressed for a local elasticity.[7] A more accurate description of long-range forces, such as dipolar forces, modifies the elasticity and amounts to replacing $d$ by $(3d-1)/2$ in the following formulas (see [42] and References therein). The second term is due to pinning by the disorder potential, and the third is the energy due to the application of an external electric field. $U_{disorder}$ depends on the precise nature of the disorder.

As detailed in Sect. 3, in the absence of an external electric field $E$, an equilibrium roughness configuration of the domain wall would be expected, characterized by a power-law growth of $B(L)$ with different exponents. For $r$ smaller than a characteristic length, the Larkin length $L_c$ [43,44], $B(L)$ grows as $B(L) \sim L^{4-d}$. Below this length there is no metastability and no pinning of the elastic interface. Above the Larkin length, the growth still follows a power law, but with an exponent $2\zeta$ ($B(L) \sim L^{2\zeta}$) dependent on the nature of the disorder. The Larkin length corresponds to the length for which the displacements are of the order of the size of the interface or the correlation

---

[6] There are constants of order one, dependent on the dimension $d$, which have been omitted from each term in the energy. These constants will not affect the creep exponent $\mu$.

[7] Note that in order to take into account the depolarization effects lengths along the vertical axis have to be scaled by a factor $(\sigma_p/\sigma_w)^{1/2}$, as in (6). Here, $L$ denotes lengths perpendicular to the polarization direction.

length of the random potential[8] $B(L_c) = \max(\xi, r_f)$. The Larkin length is thus the smallest length at which the wall can be weakly pinned, and above which it can adjust elastically to optimize its local configuration.[9] Above $R_c$ one can thus write

$$B(L > L_c) = \max(\xi, r_f)^2 \left(\frac{L}{L_c}\right)^{2\zeta}. \tag{8}$$

$L_c$ is also the length scale at which pinning appears in the system in the presence of a driving force. Using[10] (7) for $u \sim \xi$ and $L = L_c$ one can directly obtain[11] the critical field $E_c$

$$E_c \simeq \frac{\sigma_w \xi}{P_s} \left(\frac{1}{L_c}\right)^2. \tag{9}$$

In our case, a rough estimate of the values of $E_c$ may thus be obtained by extrapolating the linear behavior of the velocity, which occurs at high field values. Although we were unable to extend our measurement significantly into this region, we can nonetheless at least place a lower bound on the value of $E_c$ of 180 MV/m, as indicated on Fig. 7 for one of our thinner films, where higher values of the field could be implemented. Taking $\xi$ to be of the order of a unit cell, we can use the field data to extract an approximate value of $L_c \sim 0.2$ nm, below the limit of resolution of our measurement.

In the creep regime, we can rewrite (7). For simplicity we write formulas for the isotropic case. Using the scaling $u \sim \xi(L/L_c)^\zeta$ one obtains

$$E(u, L) = U_c \left(\frac{L}{L_c}\right)^{d-2+2\zeta} - 2P_s E L_c^d \xi \left(\frac{L}{L_c}\right)^{d+\zeta}, \tag{10}$$

where $U_c = \sigma_w \xi^2 L_c^{d-2}$. Minimizing the energy with respect to the external field $E$, we obtain the size of the minimal nucleus as

$$L_{creep}/L_c = (f_c/f)^{1/(2-\zeta)}, \tag{11}$$

with $f = 2P_s E$. The minimal barrier height to be passed by thermal activation thus corresponds to the length $L^*$, leading to a velocity of the form

$$v \propto \exp\left(-\beta U_c (f_c/f)^{\frac{d-2+2\zeta}{2-\zeta}}\right), \tag{12}$$

---

[8] In this simplified description we assume that the temperature is low enough to neglect thermal effects.

[9] Above $L_c$, the domain wall can also remain locally pinned on individual strong pinning sites, but in the present discussion, only weak collective pinning is considered.

[10] We now denote simply by $\xi$ the $\max(\xi, r_f)$.

[11] As before, the length here is the length perpendicular to the polarization direction.

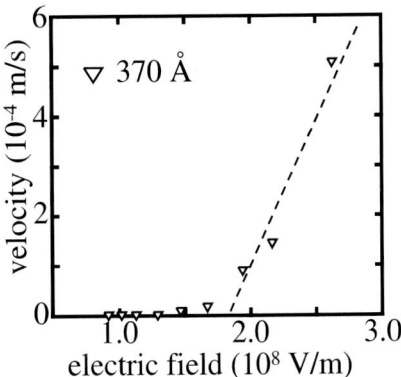

**Fig. 7.** Domain-wall velocity as a function of the applied electric field in a 37-nm $PbZr_{0.2}Ti_{0.8}O_3$ film. Extrapolating the linear behavior at high fields allows the critical field $E_c$ to be estimated as $180\,MV/m$ (figure after [8]).

if one assumes an Arrhenius law in passing the barriers. The very slow (creep) response is due to the fact that for a small force the system would have to rearrange large portions of the interface to be able to find a new metastable state of low enough energy. The barriers a domain wall must pass to make such a rearrangement therefore diverge as the force goes to zero.

The expression (11) gives the critical nucleus size $L_{creep}$ as a function of the applied field $E$ and $L_c$. We note that this expression is independent of the dimensionality of the film, and that the applied and critical fields are present as a ratio, thus removing the uncertainty associated with the correction of the field in the AFM tip–ferroelectric thin-film configuration. As for the case of the periodic potential, these expressions are valid if the size of the nucleus is smaller than the thickness of the sample. Otherwise, one of the dimensions of the nucleus should be replaced by the thickness, transforming a two-dimensional interface into a one-dimensional line. A crucial difference between the periodic and the disordered cases is that creep due to disorder can still exist in the one-dimensional situation, contrary to the periodic case. Note that the question of whether the films should be considered as one- or two-dimensional depends on which mechanism controls the nucleus. A film could thus be in the one-dimensional limit for the periodic potential, thereby invalidating the periodic potential as a possible origin for the creep process, and still be in the two-dimensional limit for the disorder provided that the size of the nucleus due to disorder remains smaller than the thickness of the film. Although creep is still present in the one-dimensional disordered case, the value of the exponent $\mu$ depends on the dimension. Using the values for $E_c$ and $L_c$ we obtained, we can estimate the size of the critical nucleus for the creep process and compare it with that found for the Miller–Weinreich formulation. In our system, the applied field is a function of the distance $r$ away from the tip center. Using the largest possible (random-field) value of $\zeta$

we find $L_{\text{creep}}$ to vary between 0.2 and 1 nm in the thinnest films (29.0–51.0 nm), and 0.2 and 2.5 nm in the thickest films (95.0–130.0 nm). We note that the formalism used in the section on incommensurate pinning of the domain wall was developed in particular to describe linear domain walls, with an applied force, and therefore domain-wall creep, perpendicular to the wall. However, in the case of the circular domains we investigated, $L_{\text{creep}} \sim 0.01r$, where $r$ is the radius of the domain, so the approximation of a linear domain wall seems reasonable.

## 7 Experimental Observation of Domain-Wall Roughness

Although the studies of domain-wall dynamics allowed us to determine that indeed disorder, rather than a commensurate pinning potential, was the mechanism governing the observed radial creep of domain walls in epitaxial thin films, questions about the exact nature of this disorder remained open. In order to ascertain the precise physics of the pinned domain walls and also the possible role of the long-range dipolar interactions that exist in ferroelectric materials, a direct analysis of the static domain-wall configuration was performed, allowing the roughness exponent $\zeta$ and the effective domain-wall dimensionality $d_{\text{eff}}$ to be extracted. To measure domain-wall roughness, we wrote linear domain structures with alternating polarization by applying alternating $\pm$ writing voltages while scanning the AFM tip in contact with the film, then imaged the resulting domain walls with the maximum resolution of our experimental setup ($\sim 2$–5 nm), allowing us to extract the correlation function of relative displacements $B(L)$. As shown in Fig. 8 for the three different films used,[12] we observe the expected power-law growth of $B(L)$ at short length scales,[12] comparable to the $\sim 50$–100 nm film thickness, followed by saturation of $B(L)$ in the 100–1000 nm$^2$ range, indicating the absence of large-scale relaxation of the domain walls at ambient conditions from their initial straight configuration determined by the AFM tip position during writing.

---

[12] We note that that the observed power-law growth nonetheless extends out to length scales above the resolution limit determined by the tip size ($\sim 5$ nm) and its interaction with the ferroelectric film, where inherent noise in the AFM measurement could give rise to artefacts such as spurious correlations at short lenght scales ($\sim 10$–20 nm). We specifically chose a minimal pixel size of 5–10 nm (depending on the two image sizes used) to try and minimize the contribution of such artefacts, with most of our data points for the correlation thus being taken over greater scales. For the correlation function of relative displacements at ambient conditions, we observe the power-law growth behavior from which we derive the 2.5 effective dimensionality out to length scales of the order of 100 nm (6–16 pixel scales), well into the limit where the effects of small length scale noise can be safely disregarded. Moreover, we clearly demonstrate the reproducibility of the imaging of particular nanoscale features in multiple sequential length scans of the same domain wall.

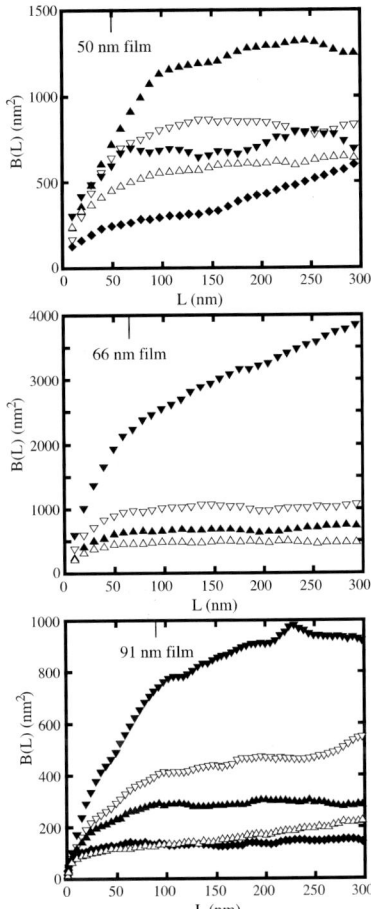

**Fig. 8.** Average displacement correlation function $B(L)$ for different sets of ferroelectric domain walls in 50-, 66- and 91-nm thick $PbZr_{0.2}Ti_{0.8}O_3$ films, shown out to $L = 300$ nm. Power-law growth of $B(L)$ is observed at short length scales, followed by saturation, suggesting a nonequilibrium configuration at large $L$ ((**a**) and (**c**) after [9])

At ambient conditions, no relaxation from this flat as-written configuration at large $L$ was apparent over an observation period of one month [9], indicating that room-temperature thermal activation alone is not sufficient to equilibrate the domain walls over their entire length. These results are in agreement with our previous studies [3,8] in which both linear and nanoscopic circular domains remained completely stable over 1–5 month observation periods. Such high stability is inherent to the physics of an elastic disordered system, where energy barriers between different metastable states diverge as the electric field driving domain-wall motion goes to zero, which makes relax-

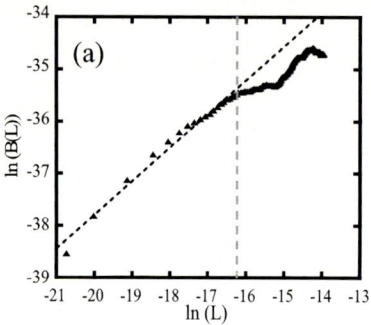

**Fig. 9.** Typical ln–ln plot of $B(L)$. Fitting the linear part of the curve (left of the vertical line) gives $2\zeta$. Average values of the characteristic roughness exponent $\zeta$ extracted from the equilibrium portion of the $B(L)$ data are 0.26, 0.29 and 0.22 in the 50-, 66- and 91-nm thick $PbZr_{0.2}Ti_{0.8}O_3$ samples, respectively (figure after [9]).

ation exceedingly slow. In fact, we believe that even the relaxation leading to the observed power-law growth of $B(L)$ at smaller length scales is not purely thermal, but occurs due to subcritical stray fields during the writing process itself. When the direction of the applied electric field is reversed to form the alternating domain structure, the neighboring region already written with the opposite polarity nonetheless experiences the resulting electric field, allowing the domain wall to locally reach an equilibrium configuration. To ensure that domain-wall relaxation was not hindered by the pinning planes of the lattice potential in the ferroelectric films [10, 45], we also wrote sets of domain walls at different orientations with respect to the crystalline axes. We found no correlation between the roughness of domain walls and their orientation in the crystal. This result is in agreement with the analysis of the previous section, pointing out the negligible role of the commensurate potential compared to the effects of disorder.

From the short length scale power-law growth of $B(L)$, we extract a value for the roughness exponent $\zeta$, characterizing the roughness of the domain wall in the random-manifold regime where an interface individually optimizes its energy with respect to the disorder potential landscape. As shown in Fig. 9, a linear fit of the lower part of the $\ln(B(L))$ vs. $\ln(L)$ curve allows $2\zeta$ to be determined. Average values of $\zeta \sim 0.26$, 0.29 and 0.22 were obtained for the 50, 66 and 91 nm thick films, respectively.

In addition to the investigations of static domain-wall roughness described above, we independently measured domain-wall dynamics in each film, obtaining values of 0.59, 0.58 and 0.51 for the dynamical exponent $\mu$ in the 50-, 66- and 91-nm thick films, respectively.[13]

---

[13] These $\mu$ values are lower than the three values measured in [7], but consistent with all the subsequent measurements performed on nine other films, all grown under similar conditions.

# 8 Domain Walls in the Presence of Random-Bond Disorder and Dipolar Interactions

In this section we show how these data, analyzed in the theoretical framework of a disordered elastic system, provide information on the microscopic mechanism governing domain-wall behavior. The direct measurement of domain-wall roughness clearly rules out the lattice potential as a dominant source of pinning. In that case, the walls would have been flat with $B(L) \sim a^2$, where the lattice spacing $a \sim 4\,\text{Å}$ is the period of the pinning potential [10]. Given the stability and reproducibility of the wall position over time, the effect of thermal relaxation on the observed increase of $B(L)$ can also be ruled out. The measured roughness must thus be attributed to disorder. As discussed in Sect. 3, for random-bond disorder, the roughness exponent $\zeta_{\mathrm{RB}} = 2/3$ in $d_{\mathrm{eff}} = 1$ and $\zeta_{\mathrm{RB}} \sim 0.2084(4 - d_{\mathrm{eff}})$ for other dimensions, while for random-field disorder $\zeta_{\mathrm{RF}} = (4 - d_{\mathrm{eff}})/3$. Should the wall be described by standard (short-range) elasticity, $d_{\mathrm{eff}}$ in the above formulas is simply the dimension $d$ of the domain wall ($d = 1$ for a line, $d = 2$ for a sheet). However, in ferroelectrics the stiffness of the domain walls and thus their elasticity under deformations in the direction of polarization is different from that for deformations perpendicular to the direction of polarization because of long-range dipolar interactions. The elastic energy (expressed in reciprocal space) thus contains not only a short-range term $H = \frac{1}{2}\sum_q C_{\mathrm{el}}(q)u^*(q)u(q)$ with $C_{\mathrm{el}} = \sigma_w q^2$ but also a correction term due to the dipolar interaction $C_{\mathrm{dp}} = \frac{2P_s^2}{\epsilon_0 \epsilon}\frac{q_y^2}{q} + \frac{P_s^2 \zeta}{\epsilon_0 \epsilon}\left(\frac{-3}{4}q_x^2 + \frac{1}{8}q^2\right)$ where $y$ is the direction of the polarization, $P$ is the ferroelectric polarization and $\epsilon$ and $\epsilon_0$ are the relative and vacuum dielectric constants. Because $q_y$ now scales as $q_y \sim q_x^{3/2}$, the effective dimension $d_{\mathrm{eff}}$ to use in the above formulas is $d_{\mathrm{eff}} = (3d - 1)/2$ [42, 46]. Using the above expressions for the roughness exponent we see that the measured $\zeta \sim 0.26$ value would give $d_{\mathrm{eff}} \geq 3$ for random-field disorder, ruling out this scenario. On the other hand, random-bond disorder would give $d_{\mathrm{eff}} \sim 2.5$–$2.9$, a much more satisfactory value, which is compatible with a scenario of two-dimensional walls (sheets) in random-bond disorder with long-range dipolar interactions.

This conclusion can be independently verified by the dynamic measurements, since the creep exponent $\mu$ is related to the static roughness exponent $\zeta$ via $\mu = \frac{d_{\mathrm{eff}} - 2 + 2\zeta}{2 - \zeta}$. The values of these two exponents from the independent static and dynamic measurements can therefore be used to calculate $d_{\mathrm{eff}}$. For the 50-, 66- and 91-nm thick films we find $d_{\mathrm{eff}} = 2.4$, 2.5 and 2.5, respectively, in good agreement with the expected theoretical value for a two-dimensional elastic interface in the presence of disorder and dipolar interactions. Taken together, these two independent analyses provide strong evidence that the pinning in thin ferroelectric films is indeed due to disorder in the random-bond universality class. Note that for the short-range domain-wall relaxation observed, the walls are in the two-dimensional limit. However, if equilibrium domain-wall roughness could be measured for larger $L$, a crossover to one-dimensional behavior would be expected, with a roughness exponent $\zeta = 2/3$.

# 9 Recent Studies
# of Ferroelectric Domain-Wall Dynamics

Until recently, detailed nanoscopic experimental studies of ferroelectric domain walls have been relatively rare. However, the available, and continually expanding technological tools permitting direct access to individual ferroelectric domains at the necessary scales have resulted in significant interest in the subject. In bulk single crystals of lithium niobate, recent work has focused on the ultrahigh-voltage regime leading to "domain breakdown", the formation of filamentous equilibrium length scale domains by rapid forward growth [47]. In the framework of an elastic disordered system, as mentioned in Sect. 3 subcritical domain-wall motion was probed in single-crystal films of triglycine sulfate [34], where thermally activated domain growth and dynamic scaling in agreement with predictions for random-bond disorder were independently observed, concurrently with our studies of the same phenomena in $PbZr_{0.2}Ti_{0.8}O_3$. In triglycine sulfate films, where the disorder potential appears to be relatively weak, the effects of line tension due to domain curvature and thermal evolution even at ambient temperature can be readily accessed [35], in contrast to the epitaxial $PbZr_{0.2}Ti_{0.8}O_3$ films, which show much higher domain stability [48]. Subsequent nanoscopic AFM studies of ferroelectric domains in lithium niobate [40, 49] also showed a similar linear dependence of domain size on the magnitude of the applied voltage, and an exponential dependence of domain-wall velocity on the applied electric field, using the charged-sphere model for the tip, as described in [7]. Domain-wall creep was also accessed by susceptibility measurements in single crystals of periodically poled potassium titanate phosphate [50] and potassium hydride phosphate [51], eliminating many of the possible effects of grain boundaries and other macroscopic defects present in sol-gel films. In these studies, the authors indirectly probe domain-wall behavior in both subcritical and sliding regimes, and find critical exponents, in agreement with quenched random-field models for an elastic interface. More experimental observations in both single crystals and epitaxial thin films, focusing especially on the microscopic nature of the disorder and its interaction with domain walls, would obviously be very useful. In this respect, promising new techniques, such as time-resolved X-ray microdiffraction, allowing ferroelectric domain-wall motion to be accessed in real time,[14] could yield interesting results.

---

[14] private communication with A. Grigoriev (2005)

# 10 Conclusions

Using the unprecedented control and precision provided by AFM, we were able to study the growth of individual nanoscale ferroelectric domains in epitaxial thin films, investigating the static and dynamic behavior of domain walls. Our studies demonstrate that domain-wall motion in ferroelectric thin films is a creep process in which $v \propto \exp(-\beta U_c (E_c/E)^\mu)$, with a dynamical exponent $\mu \sim 0.6$. This process controls the lateral growth of domains in low electric fields applied by an AFM tip. A detailed analysis of the possible microscopic origins of the observed domain-wall creep suggests that it is the result of competition between elastic behavior and pinning in a disorder potential. The reduced dimensionality of our thin films compared to the size of the critical nucleus precludes pinning in the commensurate potential of the crystal itself as the mechanism for the nonlinear field dependence of the velocity. All the domains show high stability (up to 4 months for the longest-duration experiments), inherently explained by the physics of a system in which elasticity and pinning by a disorder potential compete, leading to glassy behavior in the presence of low electric fields. In addition, we were able to extract the power-law growth of the correlation function of relative displacement $B(L) \propto \left(\frac{L}{L_c}\right)^{2\zeta}$ from the short length scale roughness configuration of domain walls, with a static roughness exponent $\zeta \sim 0.2$. Combining these two independent results, a value of 2.5 was obtained for the effective dimensionality $d_{\mathrm{eff}}$, in very good agreement with theoretical predictions for 2-dimensional elastic interfaces in the presence of random-bond disorder and dipolar interactions. However, many intriguing questions about these low-dimensionality systems remain open: the possibility of 1-dimensional behavior at higher length scales, a greater role of the commensurate lattice potential in films where the disorder potential is weaker, and the thermal response of the system are all potential research avenues. The precise control of the crystalline quality and thickness possible with current oxide growth techniques, as well as the nanoscale resolution provided by atomic force microscopy, make epitaxial ferroelectric perovskite thin films a useful and readily accessible model system for the study of elastic interfaces in disordered media, and we hope will be the focus of many future studies.

## Acknowledgements

The authors would like to thank X. Hong and C. H. Ahn for many useful discussions. This work was supported by the Swiss National Science Foundation through the National Center of Competence in Research "Materials with Novel Electronic Properties-MaNEP" and Division II. Further support was provided by the New Energy and Industrial Technology Development Organization (NEDO) and the European Science Foundation (THIOX).

# References

[1] J. F. Scott, C. A. P. de Araujo: Ferroelectric memories, Science **246**, 1400 (1989)

[2] R. Waser, A. Rüdiger: Ferroelectrics: Pushing towards the digital storage limit, Nature Mater. **3**, 81 (2004)

[3] A. K. S. Kumar, P. Paruch, J. M. Triscone, W. Daniau, S. Ballandras, L. Pellegrino, D. Marré, T. Tybell: High-frequency surface acoustic wave device based on thin-film piezoelectric interdigital transducers, Appl. Phys. Lett. **85**, 1757 (2004)

[4] C. Caliendo, I. Fratoddi, M. V. Russo: Sensitivity of a platinum-polyyne-based sensor to low relative humidity and chemical vapors, Appl. Phys. Lett. **80**, 4849 (2002)

[5] T. Giamarchi, A. B. Kolton, A. Rosso: Dynamics of disordered elastic systems, in M. C. Miguel, J. M. Rubi (Eds.): *Jamming, Yielding and Irreversible Deformation in Condensed Matter* (Springer, Berlin 2006) p. 91, cond-mat/0503437

[6] G. Blatter, M. V. Feigel'man, V. B. Geshkenbein, A. I. Larkin, V. M. Vinokur: Vortices in high-temperature superconductors, Rev. Mod. Phys. **66**, 1125 (1994)

[7] T. Tybell, P. Paruch, T. Giamarchi, J.-M. Triscone: Domain wall creep in epitaxial ferroelectric $Pb(Zr_{0.2}Ti_{0.8})O_3$ thin films, Phys. Rev. Lett. **89**, 097601 (2002)

[8] P. Paruch, T. Giamarchi, T. Tybell, J.-M. Triscone: Nanoscale studies of domain wall motion in ferroelectric thin films, J. Appl. Phys. 100, 051608 (2006)

[9] P. Paruch, T. Giamarchi, J.-M. Triscone: Domain wall roughness in epitaxial ferroelectric $Pb(Zr_{0.2}Ti_{0.8})O_3$ thin films, Phys. Rev. Lett. **94**, 197601 (2005)

[10] B. Meyer, D. Vanderbilt: Ab initio study of ferroelectronic domain walls in $PbTiO_3$, Phys. Rev. B **65**, 104111 (2002)

[11] T. Nattermann, S. Scheidl: Vortex glass phases in type-II superconductors, Adv. Phys. **49**, 607 (2000)

[12] T. Giamarchi, S. Bhattacharya: Vortex phases, in C. Berthier, et al. (Eds.): *High Magnetic Fields: Applications in Condensed Matter Physics and Spectroscopy* (Springer, Berlin 2002) p. 314, cond-mat/0111052

[13] G. Grüner: The dynamics of charge density waves, Rev. Mod. Phys. **60**, 1129 (1988)

[14] T. Nattermann, S. Brazovskii: Pinning and sliding of driven elastic systems: From domain walls to charge density waves, Adv. Phys. **53**, 177 (2004)

[15] E. Y. Andrei, G. Deville, D. C. Glattli, F. I. B. Williams, E. Paris, B. Etienne: Observation of a magnetically induced wigner solid, Phys. Rev. Lett. **60**, 2765 (1988)

[16] T. Giamarchi: Electronic glasses, in S. I. di Fisica (Ed.): *Quantum Phenomena in Mesoscopic System* (IOS, Amsterdam 2003) cond-mat/0403531

[17] M. Kardar: Dynamic scaling phenomena in growth processes, Physica B **221**, 60 (1996)

[18] D. Wilkinson, J. F. Willemsen: Invasion percolation: A new form of percolation theory, J. Phys. A **16**, 3365 (1983)

[19] S. Lemerle, J. Ferré, C. Chappert, V. Mathet, T. Giamarchi, P. Le Doussal: Domain wall creep in an ising ultrathin magnetic film, Phys. Rev. Lett. **80**, 849 (1998)

[20] V. Repain, M. Bauer, J. P. Jamet, J. Ferré, A. Mougin, C. Chappert, H. Bernas: Creep motion of a magnetic wall: Avalanche size divergence, Europhys. Lett. **68**, 460 (2004)

[21] D. A. Huse, C. L. Henley: Pinning and roughening of domain walls in ising systems due to random impurities, Phys. Rev. Lett. **54**, 2708 (1985)

[22] M. Kardar, D. R. Nelson: Commensurate-incommensurate transitions with quenched random impurities, Phys. Rev. Lett. **55**, 1157 (1985)

[23] D. A. Huse, C. L. Henley, D. S. Fisher: Huse, Henley and Fisher respond., Phys. Rev. Lett. **55**, 2924 (1985)

[24] D. E. Wolf, J. Kertész: Surface width exponents for three- and four-dimensional eden growth, Europhys. Lett. **4**, 651 (1987)

[25] B. M. Forrest, L. H. Tang: Surface roughening in a hypercube-stacking model, Phys. Rev. Lett. **64**, 1405 (1990)

[26] P. W. Anderson, Y. B. Kim: Hard superconductivity: Theory of the motion of Abrikosov flux lines, Rev. Mod. Phys. **36**, 39 (1964)

[27] L. B. Ioffe, V. M. Vinokur: Dynamics of interfaces and dislocations in disordered media, J. Phys. C **20**, 6149 (1987)

[28] T. Nattermann: Interface roughening in systems with quenched random impurities, Europhys. Lett. **4**, 1241 (1987)

[29] P. Chauve, T. Giamarchi, P. Le Doussal: Creep and depinning in disordered media, Phys. Rev. B **62**, 6241 (2000)

[30] D. T. Fuchs, E. Zeldov, T. Tamegai, S. Ooi, M. Rappaport, H. Shtrikman: Possible new vortex matter phases in $Bi_2Sr_2CaCu_2O_8$, Phys. Rev. Lett. **80**, 4971 (1998)

[31] W. J. Merz: Domain formation and domain wall motions in ferroelectric $BaTiO_3$ single crystals, Phys. Rev. **95**, 690 (1954)

[32] F. Fatuzzo, W. J. Merz: Switching mechanism in triglycine sulfate and other ferroelectrics, Phys. Rev. **116**, 61 (1959)

[33] R. C. Miller, G. Weinreich: Mechanism for the sidewise motion of 180° domain walls in barium titanate, Phys. Rev. **117**, 1460 (1960)

[34] V. Likodimos, M. Labardi, M. Allegrini: Kinetics of ferroelectric domains investigated by scanning force microscopy, Phys. Rev. B **61**, 14440 (2000)

[35] V. Likodimos, M. Labardi, M. Allegrini: Domain pattern formation and kinetics on ferroelectric surfaces under thermal cycling using scanning force microscopy, Phys. Rev. B **66**, 024104 (2002)

[36] D. Damjanovic: Logarithmic frequency dependence of the piezoelectric effect due to pinning of ferroelectric-ferroelastic domain walls, Phys. Rev. B **55**, R649 (1997)

[37] D. V. Taylor, D. Damjanovic: Domain wall pinning contribution to the nonlinear dielectric permittivity in $Pb(Zr,Ti)O_3$ thin films, Appl. Phys. Lett. **73**, 2045 (1998)

[38] V. Mueller, Y. Shchur, H. Beige, S. Mattauch, J. Glinnemann, G. Heger: Dielectric dispersion due to weak domain wall pinning in $RbH_2PO_4$, Phys. Rev. B **65**, 134102 (2002)

[39] P. Paruch, T. Tybell, J.-M. Triscone: Nanoscale control of ferroelectric polarization and domain size in epitaxial $PbZr_{0.2}Ti_{0.8}O_3$ thin films, Appl. Phys. Lett. **79**, 530 (2001)

[40] K. Terabe, M. Nakamura, S. Takekawa, K. Kitamura, S. Higuchi, Y. Gotoh, Y. Cho: Microscale to nanoscale ferroelectric domain and surface engineering of a near-stoichiometric LiNbO$_3$ crystal, Appl. Phys. Lett. **82**, 433 (2003)

[41] S. V. Kalinin, D. A. Bonnell: Local potential and polarization screening on ferroelectric surfaces, Phys. Rev. B **63**, 125411 (2001)

[42] T. Emig, T. Nattermann: Disorder driven roughening transitions of elastic manifolds and periodic elastic media, Eur. Phys. J. B **8**, 525 (1999)

[43] A. I. Larkin: Effect of inhomogeneities on structure of mixed state of superconductors, Sov. Phys. JETP **31**, 784 (1970)

[44] A. I. Larkin, Y. N. Ovchinnikov: Pinning in type-II superconductors, J. Low Temp. Phys **34**, 409 (1979)

[45] S. Pöykkö, D. J. Chadi: Ab initio study of 180° domain wall energy and structure in PbTiO$_3$, Appl. Phys. Lett. **75**, 2830 (1999)

[46] T. Nattermann: The incommensurate-commensurate transition in random-field model, J. Phys. C **16**, 4125 (1983)

[47] M. Molotskii, A. Agronin, P. Urenski, M. Shvebelman, G. Rosenman, Y. Rosenwaks: Ferroelectric domain breakdown, Phys. Rev. Lett. **90**, 107601 (2003)

[48] P. Paruch, J.-M. Triscone: High-temperature ferroelectric domain stability in epitaxial PbZr$_{0.2}$Ti) · 8O$_3$ thin films, Appl. Phys. Lett. **88**, 162907 (2006)

[49] R. J. Rodriguez, A. J. Nemanich, A. Kingon, A. Gruverman, S. V. Kalinin, K. Terabe, X. Y. Liu, K. Kitamura: Domain growth kinetics in lithium niobate single crystals studied by piezoresponse force microscopy, Appl. Phys. Lett. **012906** (2005)

[50] T. Braun, W. Kleeman, J. Dec, P. A. Thomas: Creep and relaxation dynamics of domain walls in periodically poled KTiOPO$_4$, Phys. Rev. Lett. **94**, 117601 (2005)

[51] D. R. Taylor, J. T. Love, G. J. Topping, J. G. A. Dane: Crossover from pure to random-field critical susceptibility in KH$_2$As$_x$P$_{1-x}$O$_4$, Phys. Rev. B **72**, 052109 (2005)

# Index

# APPENDIX A –
# Landau Free-Energy Coefficients

Long-Qing Chen

Department of Materials Science and Engineering,
The Pennsylvania State University, University Park, Pennsylvania 16802 USA

The thermodynamics of ferroelectrics is usually described by the phenomenological Landau–Devonshire theory. Using the free energy for the unpolarized and unstrained crystal as the reference, the free energy of a ferroelectric crystal as a function of strain and polarization can be written as (see, e.g., [1])

$$F(\varepsilon, P) = \frac{1}{2}\alpha_{ij}P_iP_j + \frac{1}{3}\beta_{ijk}P_iP_jP_k + \frac{1}{4}\gamma_{ijkl}P_iP_jP_kP_l$$
$$+ \frac{1}{5}\delta_{ijklm}P_iP_jP_kP_lP_m + \frac{1}{6}\omega_{ijklmn}P_iP_jP_kP_lP_mP_n$$
$$+ \frac{1}{2}c_{ijkl}\varepsilon_{ij}\varepsilon_{kl} - a_{ijk}\varepsilon_{ij}P_k - \frac{1}{2}q_{ijkl}\varepsilon_{ij}P_kP_l + \cdots , \tag{1}$$

where $\alpha_{ij}$, $\beta_{ijk}$, $\gamma_{ijkl}$, $\delta_{ijklm}$, and $\omega_{ijklmn}$ are the phenomenological Landau–Devonshire coefficients, and $c_{ijkl}$, $a_{ijk}$, and $q_{ijkl}$ are the elastic, piezoelectric, and electrostrictive constant tensors, respectively. If the parent phase is centrosymmetrical, all odd terms are absent:

$$F(\varepsilon, P) = \frac{1}{2}\alpha_{ij}P_iP_j + \frac{1}{4}\gamma_{ijkl}P_iP_jP_kP_l + \frac{1}{6}\omega_{ijklmn}P_iP_jP_kP_lP_mP_n$$
$$+ \frac{1}{2}c_{ijkl}\varepsilon_{ij}\epsilon_{kl} - \frac{1}{2}q_{ijkl}\varepsilon_{ij}P_kP_l + \cdots . \tag{2}$$

In (2), the set of coefficients, $\alpha$, $\gamma$ and $\omega$, in the Helmholtz free energy correspond to those measured under a clamped boundary condition.

Under the stress-free boundary condition, the macroscopic shape change of a crystal due to the ferroelectric phase transition is described by the spontaneous strain that can be obtained through the derivative of the Helmholtz free energy (2) with respect to strain, i.e.,

$$\sigma_{ij} = c_{ijkl}\varepsilon_{kl}^0 - \frac{1}{2}q_{ijkl}P_kP_l = 0 . \tag{3}$$

Solving (3) for strain, we have

$$\varepsilon_{kl}^0 = \frac{1}{2}s_{ijkl}q_{klmn}P_mP_n = Q_{ijmn}P_mP_n , \tag{4}$$

where $s_{sijkl}$ is the elastic compliance tensor and

$$Q_{ijmn} = \frac{1}{2}s_{ijkl}q_{klmn} . \tag{5}$$

K. Rabe, C. H. Ahn, J.-M. Triscone (Eds.): Physics of Ferroelectrics
Topics Appl. Physics **105**, 363–372 (2007)
© Springer-Verlag Berlin Heidelberg 2007

Substituting the spontaneous strain (4) back into the free-energy expression (2), we have

$$G(P) = \frac{1}{2}\alpha_{ij}P_iP_j + \frac{1}{4}\left(\gamma_{ijkl} - \frac{1}{2}s_{mnor}q_{mnij}q_{orkl}\right)P_iP_jP_kP_l$$
$$+\frac{1}{6}\omega_{ijklmn}P_iP_jP_kP_lP_mP_n + \cdots, \tag{6}$$

or

$$G(P) = \frac{1}{2}\alpha_{ij}P_iP_j + \frac{1}{4}\left(\gamma_{ijkl} - 2c_{mnor}Q_{mnij}Q_{orkl}\right)P_iP_jP_kP_l$$
$$+\frac{1}{6}\omega_{ijklmn}P_iP_jP_kP_lP_mP_n + \cdots. \tag{7}$$

The fourth-order coefficients are different for the clamped (2) and stress-free (7) boundary conditions, and they are related by

$$\gamma'_{ijkl} = \gamma_{ijkl} - \frac{1}{2}s_{mnor}q_{mnij}q_{orkl} = \gamma_{ijkl} - 2c_{mnor}Q_{mnij}Q_{orkl}, \tag{8}$$

where $\gamma'_{ijkl}$ is the fourth-order coefficient for the stress-free boundary condition. In general, experimentally determined coefficients correspond to $\gamma'$ since it is usually easier to do measurements under stress-free boundary conditions.

In the following, the Landau–Devonshire coefficients are presented for a number of oxides, including the well-studied systems BaTiO$_3$, SrTiO$_3$ and PZT, collected from the open literature. All the data were provided for the stress-free boundary conditions unless noted otherwise. They are all in SI units with the temperature in K.

# 1 BaTiO$_3$

For BaTiO$_3$, a Landau–Devonshire potential up to eighth order has been employed,

$$G(P_x, P_y, P_z) = \alpha_1\left(P_x^2 + P_y^2 + P_z^2\right) + \alpha_{11}\left(P_x^4 + P_y^4 + P_z^4\right)$$
$$+\alpha_{12}\left(P_x^2P_y^2 + P_y^2P_z^2 + P_x^2P_z^2\right) + \alpha_{111}\left(P_x^6 + P_y^6 + P_z^6\right)$$
$$+\alpha_{112}\left[P_x^2\left(P_y^4 + P_z^4\right) + P_y^2\left(P_x^4 + P_z^4\right) + P_z^2\left(P_x^4 + P_y^4\right)\right]$$
$$+\alpha_{123}P_x^2P_y^2P_z^2 + \alpha_{1111}\left(P_x^8 + P_y^8 + P_z^8\right)$$
$$+\alpha_{1112}\left[P_x^6\left(P_y^2 + P_z^2\right) + P_y^6\left(P_x^2 + P_z^2\right) + P_z^6\left(P_x^2 + P_y^2\right)\right]$$
$$+\alpha_{1122}\left(P_x^4P_y^4 + P_y^4P_z^4 + P_z^4P_x^4\right)$$
$$+\alpha_{1123}\left(P_x^4P_y^2P_z^2 + P_y^4P_z^2P_x^2 + P_z^4P_x^2P_y^2\right). \tag{9}$$

Two sets of coefficients for (9) are given in Table 1. The elastic and electrostrictive coefficients are listed separately in Table 2. The free energy under

# 5. HOW TO RECOGNISE THE COASTAL LAGOONS HABITAT IN KVARKEN

## 5.1 Definitions

In the Gulf of Bothnia, the coastal lagoon habitat is synonymous with the Natura 2000 sub-habitat 'flads and gloes, and lagoon- like bays'. As a consequence of the ongoing land upheaval process, the water basin (coastal lagoon) passes through different morphological and ecological developmental stages (Munsterhjelm, 1997). The 'flad, gloe, and lagoon-like bay' sub-habitat includes the developmental stages of coastal lagoons, i.e. 'juvenile flad', 'flad', 'gloe-flad' and 'gloe', as well as the streams that connect gloes and flads to each other and to the sea. Gloe-lakes and other coastal lakes are excluded from the Annex I habitat.

Systems with several water basins in different developmental stages, connected to each other and to the sea, are regarded as one site. The following characteristics are typical of flads, gloes, and lagoon-like bays:

- Shallow ($\leq$ 4-m) water basins with some sea contact;
- Seawater exchange is restricted by bottom sills, straits, or other structures (e.g. underwater moraine ridges, sandbanks, vegetation);
- Flads and gloes always have some kind of threshold formation that restricts sea water exchange;
- Lagoon-like bays have no restrictive threshold formation; the seawater exchange is restricted by, for example, straits. Lagoon-like bays never develop into gloe-flads or gloes.

## 5.2 Delimitation

The habitat starts at the threshold formations that restrict seawater exchange: these are visible above or below the seawater surface or straits. The habitat ends where the water basin is no longer in contact with the seawater, which is when the altitude of the water basin, or its discharge, is the same as the maximum seawater level (m.a.s.l.). In the Kvarken region this altitude is 1.4 m.a.s.l.

## 5.3 Characteristic species

The flads, gloes and lagoon-like bays sub-habitat often has well developed submerged vegetation, characterised by Brackish Water-crowfoot *Ranunculus baudotii*, green algae *Vaucheria* and *Cladophora* spp., *Pondweeds* Potamogeton spp., Water-milfoils *Myriophyllum* spp., Autumnal Water-starwort *Callitriche hermaphroditica*, Duckweeds *Lemna* spp. and Stoneworts *Chara* spp.

The sub-habitat is used by several fish species for spawning, migration, and reproduction, e.g. Perch *Perca fluviatilis* and Pike *Esox lucius*, and is important for amphibians, e.g. Moor Frog *Rana arvalis* and Common Toad *Bufo bufo*.

*Table 34-1.* Key attributes for assessing the conservation status of the coastal lagoons habitat and the proposed monitoring methods. The relevant authorities should agree site-specific targets for each attribute before monitoring takes place.

| | Key attributes for the Coastal Lagoons Annex I habitat | Monitoring method/s |
|---|---|---|
| **Area** | Area of habitat (see definition) within N2000 areas | Aerial photography |
| | The presence of sufficient unexploited juvenile flads to secure the future recruitment of the habitat. | To be assessed – perhaps orthorectified photographs |
| | The area of habitat with unfavourable conservation status | GPS+ Mapping |
| | The area of coastal lagoons habitat outside N2000 areas | Multibeam scanning |
| **Structures and functions** | **Physical infringement** | |
| | Dredging operations (particularly at the lagoon entrance), dumping, landing stages and other physical infringements of the water environment that causes permanent alteration to the substrate and shoreline. | EIA - Monitoring |
| | Presence of drainage ditches, road banks, soil/peat extraction, forestry, agriculture and buildings in the catchment area | EIA, Satellite-change analysis |
| | Occurrence of dredging, dumping, landing stages and other exploitation at the lagoon entrance in Habitat 1150 outside N2000 areas | EIA, Monitoring of dredging applications + aerial photography |
| | **Functions and maintenance** | |
| | Occurrence of spawning migration or reproduction of >2 of the species Pike, Perch, Roach, Ide or Burbot and presence of Bleak. | Fish sampling |
| | Number of grazed shores | Agricultural documents |
| | **Pollution** | |
| | Maintenance of natural vegetation | Diversity index |
| | Cover estimates for mats of the algae *Vaucheria dichotoma*, and *Cladophora fracta* at the surface and / or reed and other nitrogen favouring species. | Line-transect Estimates of algal mat cover |
| **Characteristic species** | **Exploitation of species** | |
| | Maintenance of commercial fish species populations | Fish sampling |
| | Depth dispersal and cover estimates for characteristic species (to detect regional eutrophication) | Transect |
| | Site-specific list of characteristic species (Table 34-2) | Field survey |

*Table 34-2.* Characteristic species that act as indicators of change in coastal lagoons.

| Latin name | Comments | Pre-flad | Flad | Gloe-flad |
|---|---|:---:|:---:|:---:|
| *Myriophyllum spicatum* | Saltwater species. Increasing. Occurs locally | X | | |
| *Myriophyllum sibiricum* | Indicator of eutrophication. Also occurs naturally. | | X | X |
| *Ranunculus baudotii* | Saltwater species. Increasing. Occurs locally | X | | |
| *Potamogeton pectinatus* | | X | X | X |
| *Potamogeton perfoliatus* | Indicator of eutrophication. Also occurs naturally on clay substrates (not mud) | (X) | | |
| *Chara aspera* | | X | X | X |
| *Chara baltica* | | X | | |
| *Chara tomentosa* | Increasing. Occurs locally | | (X) | |
| *Najas marina* | Increasing, Occurs locally | | (X) | |
| *Lemna trisulca* | Indicator of eutrophication and very low salinity. Also occurs naturally. | | | X |
| *Callitriche hermaphroditica* | Indicator of eutrophication. Also occurs naturally. | X | X | X |
| *Subularia aquatica* | Saltwater indicator, not found on muddy substrates | | X | X |
| *Eleocharis acicularis* | | | X | X |
| *Cladophora* and *Vaucheria spp.* | Indicator of eutrophication. | X | X | X |
| *Leuciscus idus* | | | | |
| *Abramis brama* | | | | |
| *Lota lota* | | | | |
| *Perca fluviatilis* | | | | |
| *Esox lucius* | | | | |
| *Rutilus rutilus* | | | | |
| *Alburnus alburnus* | | | | |
| *Bufo bufo* | | | | |
| *Rana arvalis* | | | | |
| *Rana temporaria* | | | | |
| *Triturus vulgaris* | | | | |

# 6.    A PILOT STUDY TO DEMONSTRATE METHODS FOR MONITORING COASTAL LAGOONS

During summer 2003, we carried out a small pilot study on the "Coastal Lagoons" (sub-category flads, gloes and lagoon-like bays) in the Kvarken region. The aim of the study was to test the boundary criteria, to trial recording a number of parameters in preparation for monitoring, and to use infrared photographs for the interpretation of above-water vegetation.    As a change to the catchment area of flad-sites affects the quality of the flad itself, we included the catchment in the study.

## 6.1    The study area

In Sweden, the pilot study was carried out in two Natura 2000 areas where the habitat is well represented (Kronören and the Holmö archipelago). In Finland, the study was carried out in the Replot-Björkö archipelago, part of which is located within the Kvarken archipelago Natura 2000 area.

## 6.2    Methods

### 6.2.1    Definitions

*Table 34-3.* Definition of a site with the 'Coastal Lagoons, sub-category flads, gloes and lagoon-like bays' Natura 2000 habitat (Habitat 1150)

| Definition of a site with the coastal lagoons habitat | Site includes water, shoreline and water up to the threshold formation that limits seawater exchange. If several such lagoons at varying stages of development (juvenile-flad, flad, gloe-flad, gloe) lie interconnected with each other and the sea, they are regarded as a single site. The catchment of the site extends 100 m outside its boundary with the sea. If the site is small, the catchment is denoted by a zone of 100 m around the shoreline. |
|---|---|

### 6.2.2    Identification and delimitation

Flad sites and their discharges were identified by aerial photograph interpretation. The digital tools that we used to interpret the photographs are listed below in Table 34-4.

*Table 34-4.* The digital tools used to interpret the aerial photos in Sweden and Finland.

| Sweden | Finland |
|---|---|
| ArcView | ArcView |
| Monochrome orthophotographs (1998), 4600 m | Monochrome orthophotographs (1997), 9600 m |
| Height data, 0.5 m equidistance | Height data, 2.5 m equidistance |
| Linear data from property maps | Linear data from property maps |
| Vegetation maps | Colour aerial photographs, 2001, 8415 m |

# 6.3 Sweden

Coastal lagoon sites were identified at a scale of 1:10 000 and digitised at a scale of 1:6 000 (except for large sites for which the scale was adjusted accordingly). The site boundary was digitised around the shoreline of the incorporated water basins and along the threshold, strait or islands that limit seawater exchange (Figure 34-3). Each site was numbered, and its area estimated in ArcView GIS.

The potential landward extension of the sites was delimited using the 1.5 m contour line. Gloes that lay below this contour line, or with a discharge below this contour, were digitised as coastal lagoon sites. Where several water basins formed a continuous system of flads and gloes connected to the sea, these were identified as a single site. The boundary was roughly digitised along the shoreline, along the banks of interconnected streams, and around formations limiting seawater exchange.

Initially, we used contour lines to locate potential watersheds; this was an easy way of identifying the catchment areas. If there was any doubt in our minds after going through this process, we used linear data (water courses and their direction of flow) to identify the flow direction of small lake outlets. We also used vegetation maps to identify mires and other wetlands: again, these helped us to delimit the catchment areas.

The catchment boundary was digitised at the watersheds and 100 m outside the lagoon's seaward entrance. Interconnected water basins identified as single sites were denoted as belonging to a single catchment area. It was occasionally difficult to differentiate the watersheds of smaller sites. In these cases, we used a 100 m zone surrounding the site to represent its catchment area.

# 6.4 Finland

Coastal lagoon sites were identified at a scale of 1:10 000. Located sites were delimited on digital maps at a scale of 1:6 000. The site boundary was digitised around the shoreline of each water basin, and at the threshold strait or reef that limited seawater exchange (Fig. 34-4). The shoreline was usually clearly visible on the digital images, and in less clear cases the Finnish National Land Survey's shoreline data at a scale of 1:20 000 was used as a guide. Seawater levels at the time of photographing were also taken into account when delimiting the site.

We set a vertical limit of 1 m.a.s.l. for the habitat. In Finland, there is currently no contour database in digital form with a resolution of 1.5 m. Therefore, we used contour lines of 2.5 m, together with indications of wetlands, to estimate which water basins were connected to the sea (i.e. below 1.5 m.a.s.l.). These height differences were not distinguishable from aerial photographs.

The catchment area of flads was delimited using aerial photographs, contour data, and ditch-lines from digital maps at a scale of between 1:10 000 and 1:50 000 (Fig. 34-4). Arrows on maps denoted the direction of water flow, and the position of wetlands proved useful when estimating the outer boundary of sites. The seaward boundary of the catchment area was set at 100 m outside the flad's seaward outlet.

*Photograph by Juha Katajisto*

*Figure 34-3.* The inner part of the flad at Rydskärdsfjärden in the Kvarken area of Finland.

Flads and gloes forming a network were denoted as belonging to a single catchment area. Small, separate flads with indistinct catchments were regarded as having a theoretical catchment area consisting of a land area within a radius of 100 m from the flad's shoreline.

## 6.5    The parameters used for pilot monitoring

In Sweden, the monitoring methods were piloted at 23 flad sites in Kronören Natura 2000 area. In Finland, the monitoring methods were piloted at one flad site, Rydskärsfjärden-Finnvikarna, which forms part of the Kvarken archipelago Natura 2000 area. The monitoring trial was augmented by an experiment using infrared images to interpret the above-water aquatic macro vegetation. Twelve non-biological parameters were used (Table 34-5), and the interpretations and measurements were done digitally (Table 34-6).

*Table 34-5.* Parameters used for the pilot monitoring in Sweden and Finland.

| Within the site and within 100 m of the shoreline | Within the catchment area |
|---|---|
| Area of the site | Roads (total length). |
| Occurrences of dredging beside landing stages (number). | Road culverts/road banks (number). |
| Other occurrences of dredging (number, total length). | Forestry areas (area). |
| Landing stages/piers (number). | Agricultural areas (area). |
| Buildings (number). | Ditches (total length). |
| Occurrences of dumping (number). | |
| Other physical disturbance. | |

*Table 34-6.* Digital tools used for measuring the monitoring parameters (in italics).

| Sweden | Finland |
|---|---|
| ArcView | ArcView |
| Monochrome ortho-photographs (1998), 4600 m (roads, ditches, agriculture and forestry, dredging, quarries, dumping, road banks) | Monochrome ortho-photographs (1997), 9000 m (roads, ditches, agriculture and forestry, dredging, quarries) |
| Linear and point data from property maps (roads, ditches, agriculture and forestry) | Linear and point data from property maps (roads, ditches, fields, buildings, bridges, power lines) |
| Vegetation map (forestry) | Colour photographs, 2001, 8415 m height (mires, roads, fields, agriculture and forestry, aquatic macrophytes) |
| * Västerbotten County Administration's GIS layer marking piers/landing stages and associated dredging | |
| * Västerbotten County Administration's GIS layer marking buildings near the shoreline | |

* Dredging, piers/landing stages and buildings near the shoreline (< 100 m from the shoreline) were digitised in spring 2003 by Västerbotten County Administration using digital ortho-photographs (4600 m, 1997-1998) and property maps (2001).

# 7.     RESULTS

## 7.1     Sweden

Twenty-three flad sites were identified within the Kronören Natura 2000 area, making up a total area of 258 ha. The County Administration's estimation of the total area of habitat in this area is 233 ha. In the Holmö Archipelago Natura 2000 area 157 flad sites were identified, with a total area of 549 ha. A previous estimation of the total 1150 habitat in this area was 750 ha.

Note that sites identified as coastal lagoons in the pilot study should be considered as 'potential' 1150 sites, since biological criteria (i.e. the characteristic species) were not used. Future biological surveys may result in a number of sites being deleted if they do not meet the criteria for the habitat.

*Table 34-7.* A summary of the monitoring results for the 23 flad sites at Kronören, Sweden.

| Monitoring parameters Within sites: | Total number | Total length (m) | Total area (ha) | Disturbed sites (%) | (no.) |
|---|---|---|---|---|---|
| Dredging near landing stages | 14 | | | 21.7 | 5 |
| Other dredging | 3 | 335 | | 8.7 | 2 |
| Landing stages | 109 | | | 34.8 | 8 |
| Buildings (near shore) | 229 | | | 47.8 | 11 |
| Other physical disturbance | 2 | | | 8.7 | 2 |
| Monitoring parameters Within catchment area: | Total number | Total length (m) | Total area (ha) | Disturbed sites (%) | (no.) |
| Ditches | | 10 264 | | 17.4 | 4 |
| Roads | | 33 138 | | 43.5 | 10 |
| Road culverts/road banks | 14 | | | 21.7 | 5 |
| Forestry | | | 316.8 | 43.5 | 10 |
| Agriculture | | | 12.6 | 21.7 | 5 |

## 7.2    Finland

The total area of all 1150 sites in Rydskärsfjärden-Finnvikarna was 1.7 $km^2$, and the catchment area 17.3 $km^2$. The whole site, but only 3.3 $km^2$ of the catchment, was located within the Natura 2000 area, together with wetlands and primary succession-stage forest.

*Table 34-8.* Summarised monitoring results for Rydskärfjärden-Finnvikarna, Finland (one site).

| Monitoring parameters Within sites: | Total number | Total length (m) | Total area (ha) |
|---|---|---|---|
| Dredging near landing stages | 0 | | |
| Other dredging | 4 | 944 | |
| Landing stages | 0 | | |
| Buildings (near shore) | 32 | | |
| Other physical disturbance | 0 | | |
| Monitoring parameters Within catchment area: | Total number | Total length (m) | Total area (ha) |
| Ditches | | 109 560 | |
| Roads | | 23 026 | |
| Road culverts/road banks | 2 | | |
| Forestry | | | 350 |
| Agriculture | | | 30 |

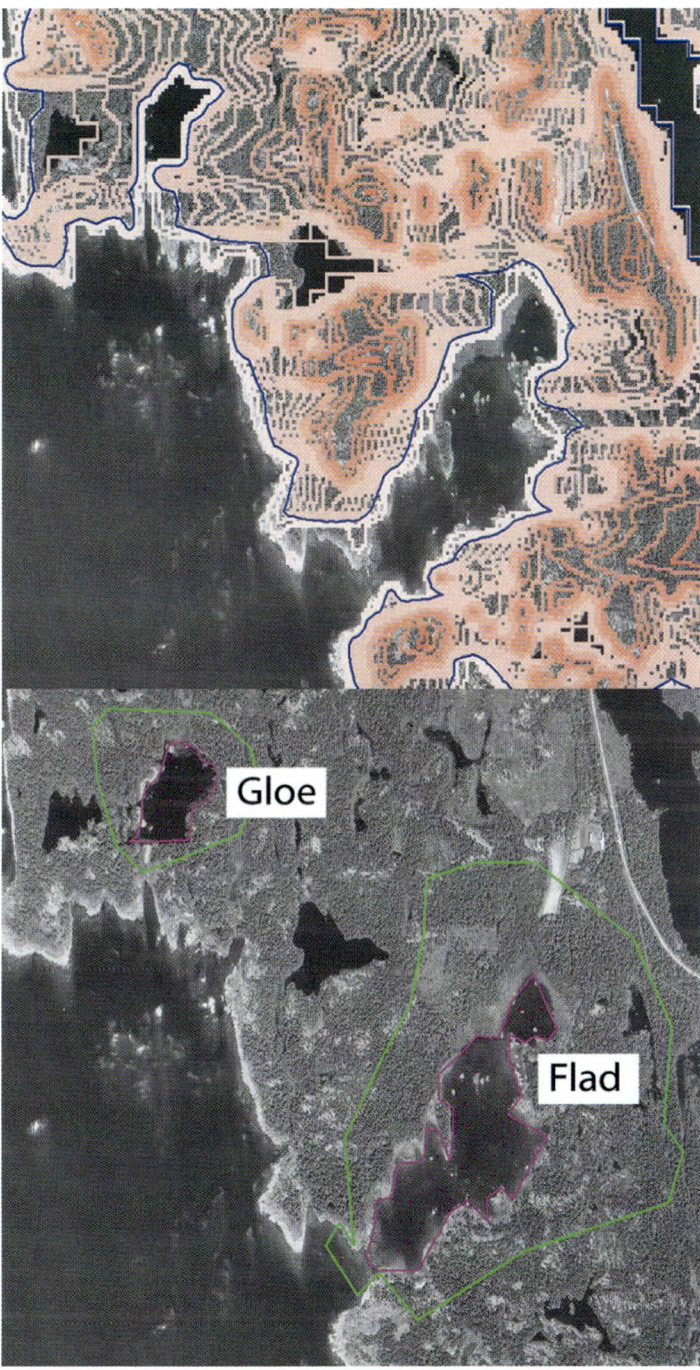

*Figure 34-4.* An example of a delimited 1150 site within the Kronören Natura 2000 area, Sweden. Above: contours (pink); the contour line at 1.5 m.a.s.l. marked in blue. Below: the delimited 1150 site (purple) and its catchment (green).

The influence of human activity is very apparent within the catchment area (Fig.34-6). Forestry clear-cuts, ditches, pine plantations, roads and culverts are all present. The site itself is also influenced by building and dredging. However, since the thresholds at the flad outlet are still intact, dredging has had little effect on the seawater exchange.

## 7.3    Using infrared images to identify macro vegetation

This experiment was initiated with a field survey, in which we mapped the vegetation along five transects within the site (Fig. 34-5). Along these transects, we recorded species presence, and mapped, approximately, the distribution of the dominant species. Subsequently, we transferred this information onto a map, and used this as a template for interpreting the infrared images. By projecting the infrared images onto the map template we were able to provide an approximate delimitation of the distribution of dominant species within the entire Rydskär area.

The shoreline within both Rydskärsfjärden and Finnvikarna was dominated by Common Reed *Phragmites australis*, which could be identified on the infrared images by its light shade of colour. At locations where Bulrush *Typha latifolia* or Spike-rushes *Eleocharis* spp. were prominent, the image showed a redder colour. Reeds greatly dominated, and the variation of species in different areas was low. As water appears black on the infrared images, we could not see the submerged vegetation.

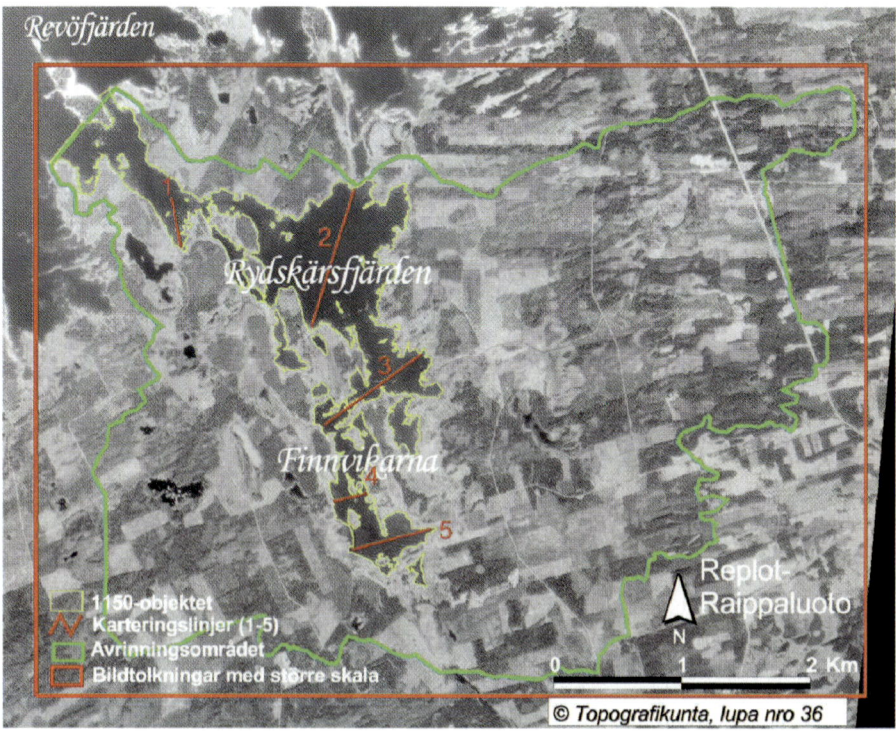

*Figure 34-5.* An example of a delimited coastal lagoons site (yellow) and its catchment area (green) in Finland. The transects used for vegetation mapping are shown in red.

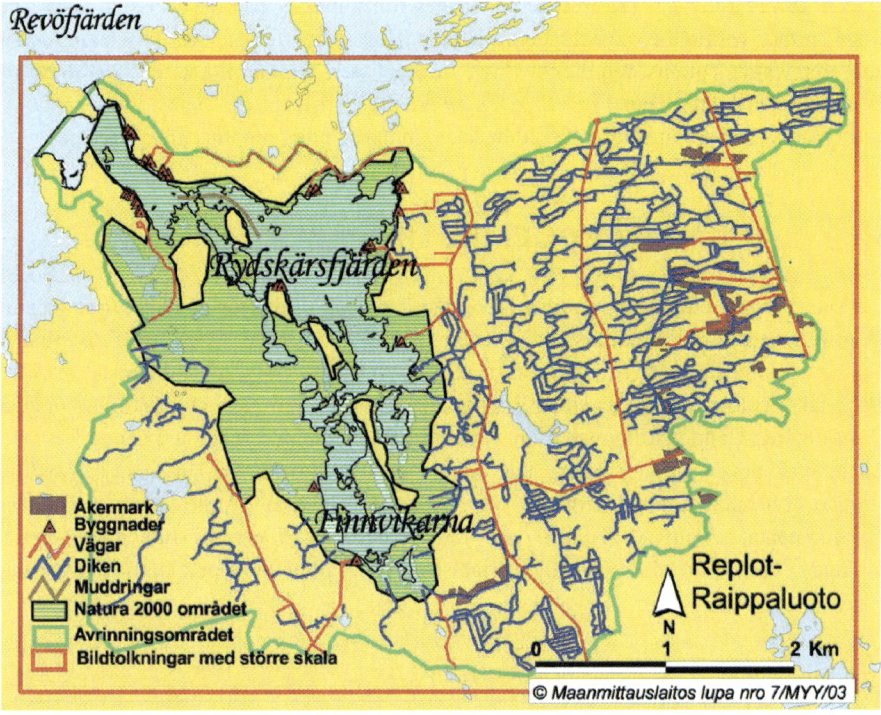

*Figure 34-6.* Physical impact at Rydskärsfjärden-Finnvikarna and their catchments showing: Arable land (purple areas); Buildings (purple triangles); Roads (red lines); Ditches (blue lines); Dredging (brown line); Natura 2000-area (stripes); Catchment area (green).

## 8. CONCLUSIONS

Delimiting habitat type 1150 landwards, and locating flads and gloes, was a straightforward procedure when height data was used. The precision of the height database is low (approx. ± 2.5 m), which adds a degree of uncertainty, particularly when studying gloes. However, the advantages outweigh these disadvantages, since the method is quick and easy to use. Juvenile flads are more difficult to delimit since we do not know the penetration depth of the orthophotography, and a criterion for the habitat type is the presence of visible thresholds that limit seawater exchange. Juvenile flads, i.e. flads in the developmental stages, are probably under-represented.

Dredging, roads, ditches, road banks and forestry activity were clearly visible in the monochrome orthophotographs taken from 4600 and 9600 m. Buildings near the shore and landing stages were clearly visible from 4600 m, but from 9600 m were difficult to distinguish from the large boulders that are common in the Finnish study area. Mires, arable land and other open land were also difficult to differentiate from 9600 m. Quarries and dumps were not distinguishable in orthophotographs.

We can recommend infrared aerial images for mapping macro vegetation above the water surface within the coastal lagoons habitat. However, a field survey will be needed to interpret which colours represent which macrophytes. Furthermore, the usefulness of infrared images will depend on when they were taken. Images for vegetation mapping in coastal areas and archipelagos should be taken in the summer months, after mid-June.

# 9.    ACKNOWLEDGEMENTS

Many people outside the project group participated in this work and contributed with their knowledge and enthusiasm. Particular thanks go to Kajsa Berggren (Västerbotten County Administration, Sweden), Lars Ericson (Umeå University, Sweden), Anders Haglund (Ecology group, Swedish Environmental Protection Agency), Gunilla Forsgren (Västerbotten County Administration, Sweden), Richard Hudd (Fish and Game Research Institute, Finland), Lena Kautsky and Hans Kautsky (Stockholm University, Sweden), Riggert Munsterhjelm (University of Helsinki, Finland), Anne Raunio (Finland Environmental Centre, Helsinki), and Ralf Wistbacka (Larsmo Municipal Council, Finland). Thanks are also due to Tim Hipkiss for translating this report from Swedish into English.

# 10.    REFERENCES

Ericson, L. & Wallentinus, H-G. (1979). Sea shore vegetation around the Gulf of Bothnia. Guide for the International Society for Vegetation Science, July-August 1977. *Wahlenbergia* **5**: 1-142. ISBN 91-85410-03-9

Munsterhjelm, R. (1997). The aquatic macrophyte vegetation of flads and gloes, S coast of Finland. *Acta Botanica Fennica* **157**: 1-68. ISBN 951-9469-51-6

# PART VII

## LOOKING TO THE FUTURE

# CHAPTER 35

# THE CHALLENGES AND OPPORTUNITIES THAT LIE AHEAD

## CLIVE HURFORD

*Countryside Council for Wales, Plas Penrhos, Ffordd Penrhos, Bangor, Gwynedd, LL57 2BQ*
*clive.hurford@serapias.net*

This book has focused almost exclusively on protected sites: this was a conscious decision. Away from the more remote and mountainous areas of Europe, most of the rare and threatened habitats and species persist primarily within protected areas, which are refuges for species that can no longer survive in the wider countryside. Therefore, in the short-term at least, our goal must be to ensure that these protected areas are managed sympathetically for the species that we would like to see repopulate the wider countryside.

A logical conservation strategy would be to secure the management of the protected sites, and then target the land adjoining them for appropriate management agreements. We could then gradually expand out into the wider countryside. The role of agri-environment and Forest Stewardship Council schemes is to ensure that the wider countryside is suitable for repopulation. However, unless we initially secure the habitats and species in the protected areas, attempts to restore the wider countryside will meet with only limited success.

The EC Habitats Directive, a powerful and carefully constructed piece of legislation, presents us with an unprecedented opportunity to protect the fauna and flora of Europe, primarily through the Natura 2000 network. Furthermore, the Directive recognises the importance of monitoring, and was the catalyst for much of the developmental work described here: without financial support from the Life-Nature fund, we would not have had this opportunity to explore the possibilities for monitoring.

*C. Hurford & M. Schneider (eds.), Monitoring Nature Conservation in Cultural Habitats, 379–384.*
© 2007 *Springer.*

As a general rule of thumb, committing 10% of the financial resources available for conservation management to ensure that the management has achieved its aims would be a good investment. Under-committing resources for monitoring is a false economy as without a reliable check to assess whether the management is delivering effective conservation, we risk:

- Wasting the resources committed in the first management cycle;
- Repeating the mistake when the management is reviewed;
- Losing credibility with funders and policy regulators; and ultimately
- Degradation or loss to the conservation value of our sites.

The process outlined in Part III of this book will facilitate a carefully considered and transparent conservation strategy in a way that is conducive to efficient and reliable monitoring. It will also ensure that the vital components of an operational site management plan are in place. These are, that:

- The most important habitats or species are clearly prioritised;
- There are maps showing both the current distribution and the desired distribution of the priority habitat or species; and
- There are condition indicators for the priority habitat or species that include an unambiguous definition of optimal condition.

In addition, the plan should contain details of the management intended to deliver optimal condition. Few recent conservation management plans contain this information. With the exception of the condition indicator table, which adds a critical level of detail to the information in the maps, these recommendations have been in the literature for many years. For example, Usher (1973) suggested providing guidance on the prioritisation of resources, while the Nature Conservancy Council (1989) recommended using maps to illustrate management aims. All early management planning guidance advocated including details of the intended management programme.

With respect to the challenges facing conservation, the monitoring process has drawn attention to four key areas that must be addressed before we can expect to achieve effective conservation of our fauna and flora. These are the tendency towards:

- *Ad hoc* prioritisation;
- Management indecision;
- Inexperience in land management;
- A diminishing base of field expertise.

In the absence of a clear conservation strategy and monitoring of that strategy, resources will continue to be allocated for management on an *ad hoc* basis. This will result in individual interests dictating the state of the conservation resource and lead to further management discontinuity on our most valuable sites. Conservation losses originating from this source can be attributed directly to conservation managers.

*Photograph by Clive Hurford*

*Figure 35-1.* The meadows and pastures, and indeed the way of life, in the more remote mountain areas of Europe are under threat as the local farmers grow older and their children move to the cities seeking a more affluent lifestyle.

*Photograph by Clive Hurford*

*Figure 35-2.* Orchid-rich lowland meadows and pastures, like this with abundant Heart-flowered Orchids *Serapias cordigera* near the coast in northern Spain, are under threat from agricultural improvements.

Formalised prioritisation is a fundamental requirement of effective conservation. We can no more manage for everything on every site than we can monitor everything. To attempt to do this would guarantee only that we discriminated against every habitat and species equally. Prioritisation should be based on a dispassionate overview of the international and national conservation resource (Chapter 7).

Management indecision is another significant barrier to effective conservation. The vast majority of sites that come under the influence of conservation managers have a history of cultural management. On taking management control of a site we have only three options available: to carry on with the existing management, to change the existing management, or to remove the management altogether. The same level of responsibility is attached to each of these options.

If we know that the conservation value of a site is degrading and we decide not to change the management, that decision will inevitably lead to further degradation and loss. Similarly, if the conservation value of a site is strongly linked to a habitat with a history of cultural management and we remove that management, our decision discriminates against the species associated with that habitat. In cultural habitats, the option most likely to achieve conservation gains is to change the existing management and ensure that it is compatible with the conservation aims for the site.

If we accept that all three of these management options carry the same level of responsibility, and they clearly do, we have everything to gain from taking positive management action.

With the exception of conservation managers who do not believe in conservation management, most management indecision stems from a lack of confidence. The reasons underlying this are deep-rooted and need to be addressed on several levels. In the UK, many conservation managers join conservation bodies straight from university and are handed the responsibility for managing several sites on their first day in post. Nothing in their education will have prepared them for this level of responsibility. The education system has been phasing out ecologically based disciplines from the curriculum since the early 1980s. Few universities now teach botany, other than as a module on a broad-based Environmental Biology course, and field trips are often seen as expensive luxuries. I understand from colleagues elsewhere in Europe, that this problem is not confined to the UK. Therefore, conservation bodies cannot wait for the education system to recognise and address this problem; it has to be dealt with 'in house'. This will require a change of culture, however, so that possessing field expertise is recognised as a fundamental requirement of the job and not an optional extra. New technologies, such as remote sensing, have the potential to make a new level of information available to conservationists, but without field expertise we will not have the ability to classify, interpret or ground-truth the images. Fortunately, the more perceptive conservation bodies have recognised this and there are signs that the necessary change of culture is starting to happen.

Practical land management skills are in similarly short supply. Few conservation managers now have farming or forestry backgrounds, so they have little practical experience to call on when setting up management agreements. This problem is compounded by the fact that only the older farmers will have practical experience of management that is sympathetic to nature conservation.

While there is no shortage of guidance on habitat management in the literature, we should also take the opportunity to gather information on management techniques from areas where traditional farming methods are still in operation and maintaining habitats of conservation value. When we visit these areas, e.g. the more remote parts of northern Spain, Romania, Bulgaria and Slovenia, it is clear that the management is very different to that carried out elsewhere in Europe (Fig. 35-3). If we want to maintain the conservation value of these areas, the relevant international conservation organisations must find ways of ensuring that current management practices can persist. There are already signs that traditionally managed hay meadows are being abandoned in the mountain areas of Europe, primarily as a result of:

- An aging farming population and an exodus of their offspring to seek a more profitable existence in cities; and
- The pressures of the Common Agricultural Policy as accession countries join the European Community.

This should be a concern, and perhaps a priority, for conservationists throughout Europe.

*Photograph by Clive Hurford*

*Figure 35-3.* Many of the more spectacular hay meadows in Europe are still being managed by traditional labour-intensive methods. The challenge is to ensure that these survive intact without discriminating against the farmers that manage them.

# IN CONCLUSION

The approach recommended in this book makes full use of our existing knowledge to increase the efficiency of a monitoring project. It is difficult to see how we could collect less information without compromising the monitoring result. Some might argue that we are already collecting less information than we should, but that is open to debate.

For the purposes of conservation management, we simply need to know whether our management is achieving its aims. This is not a statistical question. Farmers and foresters do not use complex statistics to assess whether the harvest has met their expectations; they simply weigh it. This book has focused on describing, and illustrating, a similarly reliable and efficient approach to monitoring nature conservation.

# REFERENCES

Nature Conservancy Council. (1989). Site management plans for nature conservation: a working guide. NCC report. Peterborough.

Usher, M.B. (1973) Biological Management and Conservation: theory application and planning. Chapman and Hall. London.

# APPENDICES

# APPENDIX I

# GLOSSARY AND ABBREVIATIONS

**Accuracy:** how close a single measurement, or group of measurements, is to the true value of a parameter.

**Area of study:** the area we actually look at in the field, including every point which might be included as part of a sample.

**ATM:** Airborne Thematic Mapper. A sensor used for Earth Observation.

**Attribute**: measurable characteristics of a habitat or species.

**AVHRR:** Advanced Very High Resolution Radiometer.

**Base-rich:** referring to non-acidic soils or waters: typically having a pH approaching or exceeding 7.

**BACI:** Before-After-Control-Impact designs. A design for field studies analogous to a controlled experiment. The effect of an impact such as an oil spill is investigated by comparing measurements before and afterwards at a control site and impacted site.

**Bi-directional Reflectance Distribution Function.** The function describing the reflectance of a surface at a given viewing and illumination geometry.

**Calcareous:** soils or water rich in calcium carbonate, usually derived from limestone or chalk.

**CASI**: Compact Airborne Spectrographic Imager. A type of sensor used for Earth Observation.

**Code phase data:** GPS data based on the coarser pseudorandom code in contrast to that based on carrier phase data.

**Competitors:** species that exploit conditions of low stress and low disturbance.

**Composite attributes:** a small suite of site-specific attributes used to determine the condition of a key habitat at each monitoring point. The co-occurrence of these attributes defines good quality habitat.

**Conservation manager:** the person responsible for making conservation management decisions for a habitat or species on a site.

**Coppice:** as a system of woodland management, coppice involves repeated cutting, on a rotation basis, in which multiple stems are allowed to grow up from the cut stumps (stools) of a felled broad-leaved tree. The rotation depends on the end product.

**Cover targets:** used to assess whether the cover of a species, or related group of species, e.g. grasses or ericoids, is above or below a critical threshold that will impact on the conservation value of a key habitat or species.

**C-S-R model:** in this model, each species is classified into one of seven categories according to its lifestyle. Three primary categories and four intermediates are identified. The primary categories are a) competitors, b) stress-tolerators, and c) ruderals.

**Cultural habitats:** for the purposes of this book cultural habitats are defined as those habitats which are derived from human management activities, modified by human activities or impacted on by human activities.

**DBH:** diameter at breast height, a measure used for assessing the girth of a tree.

**Digital elevation model (DEM):** a regularly sampled grid of elevation values above a given level, usually a map datum, can be used to produce 3D images of terrain in GIS.

**Earth Observation:** the process of measuring reflected electromagnetic radiation from the surface of the Earth using an imaging sensor.

**EGNOS:** European Geostationary Navigation Overlay Service.

**EOS:** Earth Observing System.

**Epixylic:** living on dead wood.

**Eutrophic:** describes ecosystems that are nutrient-rich and highly productive.

**Favourable condition:** this is used to describe a habitat or species (of secondary importance on a given site) which is in a desired or acceptable state. This desired state is defined in the condition indicator table.

**Feature (interest feature):** a general term sometimes used in this book to mean a habitat or species of conservation importance on a given site.

**Fixed-point photography:** Precisely repeatable habitat photographs, taken from a location that can be re-found. Used to support habitat monitoring or surveillance projects.

**GIS (Geographical Information Systems):** Sophisticated computer systems able to manipulate, analyse and display spatial information.

**Geostatistical data:** Data made up of measurements of a variable at a series of known points and only at these points. For example, the distribution of birds nests (spatial location) with the number of eggs (variable). Compare point data (location of a feature, no variable) and lattice data (information for every point, not just some points).

**GLONASS:** Global Orbiting Navigation Satellite System. A GPS satellite system.

**GPS (Global Positioning System):** A highly accurate satellite navigation system. The receiver calculates a position on the ground by timing signals that are broadcast from a series of special GPS satellites.

**Hibernacula:** Sites where bats congregate to hibernate during the winter months.

**HyMap**: Hyperspectral Mapper. A sensor used for Earth Observation.

**Image registration:** The process of calculating the numerical relationship between image pixels and a given map projection or another image data set.

**Indicator assemblage:** A suite of co-existing species that will respond negatively to factors expected to impact on the condition of a habitat.

**Indicator species:** A species known to respond, either positively or negatively, to factors expected to impact on the condition of a habitat – or that of associated species.

**Interpolation:** The estimation of values of an attribute at unsampled points from measurements made at surrounding sites.

**Land manager:** The person responsible for carrying out the conservation management, most often this will be a farmer, a forester or a nature reserve warden.

**Lattice data:** A data type stored as an array of values covering the whole area of interest. In a geographical information system this can include both vector and raster data, since both include data values for every point. Contrast point data and geostatistical data.

**Masking:** the production of a set of image pixels that are either processed together, or omitted from a given processing step.

**Mesotrophic:** the prefix "meso" means mid-range. The mesotrophic state is defined as having a moderate supply of nutrients and therefore moderate biological productivity.

**Metapopulation:** a group of intermittently linked subpopulations belonging to the same species, each isolated in a patch of habitat. The long-term survival of the metapopulation depends on the balance between local extinctions and re-colonisations in the patchwork of a fragmented landscape.

**MODIS:** Moderate Resolution Imaging Spectrometer. Used for Earth Observation.

**Monitoring:** For the purposes of this book monitoring is defined as assessing the condition of a habitat or species against a predetermined standard.

**Mosaicking:** the process of combining multiple image data sets together based on their spatial coverage. Spatial coverage is usually expressed in geographical coordinates.

**Nemoral:** pertaining to, or living in, a forest or wood.

**NIR:** near infrared.

**NNR:** National Nature Reserve. A site owned or managed by the UK conservation agencies.

**NOAA:** National Oceanic and Atmospheric Administration.

**NVC:** National Vegetation Classification (NVC). This is a comprehensive floristic classification of terrestrial and freshwater vegetation types in the UK. It recognises roughly 400 separate plant communities, many further divided into sub-communities.

**Offset:** recording a GPS location some distance from a target, and then attaching the distance and direction to the target to this remote location. This is a way of recording the location of point that cannot obtain a reliable position fix because of difficult circumstances, such as being situated at the base of a tree or cliff.

**Oligotrophic:** describes ecosystems that are nutrient-poor and have a low productivity. Often referring to rivers and lakes that have clear water and low biological productivity (oligo = little; trophic = nutrition).

**Optimal condition:** used to describe when a habitat or species of conservation priority on a site is in the best condition that we could hope to achieve, as defined in the condition indicator table. The term 'favourable condition' suggests something less than this, and is often interpreted as an 'acceptable' state.

**Post-processed differential correction:** the process of correcting the position of a GPS field receiver relative to a base station, undertaken after fieldwork is completed.

**Photomonitoring:** taking a repeatable series of photographs to provide a 360$^{\circ}$ panoramic record of a habitat from sites that can be precisely re-located.

**Pin-frame:** a device used to record accurate and unbiased measures of cover in quadrats.

**Pixels:** Short for picture elements. A familiar example is the small squares which are shown in a digital photograph. In a remote sensing image, pixels represent regularly sampled values of a given property, usually values of reflectance or radiance.

**Point data:** Data made up primarily of locations of a feature, such as the distribution of birds nests or of ancient trees.

**Position fix:** a GPS field kit uses satellite signals to calculate a single position by averaging a series of these individual position fixes.

**PPP sampling:** Probability Proportional to Prediction sampling. An advanced sampling method that weighs the effort and advantages of collecting particular samples in designing a sampling scheme. Also known as 'list sampling'.

**Precision:** how close repeated measurements of the same attribute are to one another, though not necessarily to the true value of the attribute: we may not know the true value.

**Quadrat:** a sampling unit for vegetation, usually a square frame with an area of 50 x 50 cm, 1 x 1 m, or 2 x 2m. Typically used for habitat research and surveillance.

**Radiance:** a physical measure of the electromagnetic energy emanating from a surface.

**Raster data grid:** A type of lattice data made up of a rectangular array of values, rather like the image in a digital photograph.

**Red-listed species:** in Sweden, species assigned to the following categories by the Swedish Threatened Species Unit: RE=Regionally Extinct, CR=Critically Endangered, EN=Endangered, VU=Vulnerable, NT=Near Threatened and DD=Data Deficient.

**Reflectance:** a measure of the ratio of incoming electromagnetic energy to outgoing electromagnetic energy.

**Remote sensing:** broadly used to describe any measurement taken of an object at some distance, rather than by direct contact.

**RGB:** red, green, blue.

**Ruderals:** species that exploit conditions of low stress and high disturbance.

**SAC:** Special Area of Conservation. A site protected under Natura 2000 legislation.

**Sample layout:** The spatial design and distribution of sample points e.g. random, stratified random, regular grid.

**Sampling frame:** the set of all points available to be chosen as part of a statistical sample.

**Semi-improved:** this term is usually used to describe agricultural habitats which have been modified to increase their productivity, but which retain some of the species characteristic of their more natural precursors.

**Semi-natural:** this term describes habitats or ecosystems, which are composed largely of native species, but where the structure and floristic and faunal composition has been profoundly influenced by the actions of people.

**Semivariance:** variance taking into account spatial scale in a grid of geostatistical data. Calculated for a particular point spacing as half the variance of the differences between values at all possible points spaced a constant distance apart.

**Seral stage:** a recognisable stage in the development, or succession, of a habitat.

**Site:** used in this book to describe the area of conservation interest under discussion. In general, it has been used to refer to any area protected for nature conservation.

**Site manager:** a person, or organisation, responsible for carrying out conservation management on a site, often this is the landowner, but it could also be a reserve warden.

**SLR:** Single Lens Reflex, referring to a type of camera.

**Snag:** any dead or dying standing tree. Snags should be at least 8 cm in diameter at breast height (dbh) and at least 2 m tall. At Biskopstorp, most beech snags refer to broken dead trees (caused by infection by the fungus *Fomes fomentarius* and subsequent windthrow).

**SPOT:** Système pour l'observation de la Terre. A sensor used for Earth Observation.

**SSSI:** Site of Special Scientific Interest. A site of national conservation importance protected by legislation in the UK.

**Staking out:** navigating to pre-determined points in the field. A term used in surveying with a geographical positioning system (GPS).

**Stands:** discrete blocks of habitat, for example, a wood or a meadow.

**Stress-tolerators:** species that exploit conditions of high stress and low disturbance.

**Surveillance:** typically, a series of repeat surveys used to track trends of habitats or species. Differs from monitoring by not measuring against a predetermined standard.

**Survey:** a set of standard observations, usually obtained with a standard method and within a restricted time period, typically a one-off exercise, e.g. mapping a habitat or compiling a species list.

**Swath width:** the width of image data acquired by an imaging sensor during an operational pass over a site, given in degrees or as a physical distance on the ground.

**Thematic map:** A map dividing the ground into classes rather than showing variables at points, perhaps showing land cover or land use types.

**TIR:** thermal infrared. A type of sensor used for Earth Observation.

**Training areas:** usually used in image classification to mean the areas of a site where the land cover type is known, either through being visually recognisable, e.g. forests, or through the collection of vegetation data with high accuracy GPS. When the vegetation in part of an image has been identified, e.g. as species-rich grassland, by either of these methods, a 'training area' is then set up to identify other areas of species-rich grassland in the image.

**Windthrow:** trees uprooted by strong winds. Rot in the roots, or a shallow rooting pattern, is often associated with windthrow.

# Index